ECOLOGY AND UTILIZATI
OF DESERT SHRUB RANGEL
IN IRAQ

ECOLOGY AND UTILIZATION OF DESERT SHRUB RANGELANDS IN IRAQ

by

D.C.P. THALEN

Dr. W. JUNK B.V. – PUBLISHERS – THE HAGUE
1979

ISBN 90 6193 593 8

Cover design Max Velthuijs

ACKNOWLEDGEMENTS

On the completion of this study I would like to thank all those who have in one way or another taught, coached, guided, encouraged and, most important, corrected me in some 30 years of education and learning. Each of them has contributed in some way to make the publication of this book possible. There are so many people to whom I owe a debt of gratitude that I cannot mention them all by name. Space only permits a modest sample but I hope here to have included at least those who have most directly helped me with the work that is reported on the following pages.

On the completion of my primary school and grammar school education, my studies at Groningen University equipped me professionally for my work in Iraq. In this respect I wish to acknowledge Dr. E. van der Maarel and Prof. Dr. J.P. Kruijt for teaching me the principles of vegetation science and ethology respectively.

I am grateful that the name of the three persons who have probably influenced me most in my professional thinking are officially connected with this study. Prof. Dr. D. Bakker introduced me in plant ecology and supported and encouraged me throughout my university studies. I am thankful for his help and guidance received before, during and after the work in Iraq, for his interest in this somewhat unusual study for the Laboratory of Plant Ecology of Groningen University, for his patience in waiting for drafts and for the personal way in which the final manuscript could be discussed.

To Mr. R.N. Kaul I am much indebted for the way in which he introduced me in the subject of rangelands, deserts, and man as an ecological factor in arid lands, and for being an example for me of a dedicated worker under conditions which were not always ideal. His personal friendship and his readiness to let me share his wide knowledge of arid lands have had a most profound influence on my work in Iraq, an influence for which I cannot be grateful enough.

The fact that Dr. Ir. I.S. Zonneveld acts as a co-referent for the present study is an understatement dictated by the law. He had a continuous interest in the study, based amongst other things on a personal acquaintance with the local situation, and was willing to discuss the problems which arose at any time. I wish to express my appreciation for the freedom I was given in the Department of Vegetation Survey at ITC to complete the manuscript as well as the stimulating criticism I received on the subsequent drafts.

Of the Iraqi colleagues and assistants who have contributed to the field data collection that served as a basis for his study, I wish to thank the following: Dr. Tariq A. Al-Ani for organizing from his side the many fieldtrips; Mr. Muhaned M. Al-Mufti, Mr. Mohammed G. Ahmed and Mr. Samir H.A. Al-Thanun for their part in the organization of the work in the field and taking part in the actual data collection; and Mr. Khaleel Al-Nawadi for measuring thousands of shrubs. Without their help and the work of the drivers and labourers, often under harsh conditions, this study could not have been accomplished. The hospitality of the Bedouins, villagers and police-officers is something I will never forget. I am also grateful to Dr. Mahmood A. Clor, Mrs. Fawzia M.R. Al-Cherchafchi and Mrs. Maeda Al-Jawad for their assistance in the Plant Ecology Laboratory of the Institute for Applied Research on Natural Resources. The help received from the staff of the National Herbarium in identification of specimens and of the D.G. of Range Management in making their documents available, and facilitating the fieldwork in some cases, is also acknowledged with thanks.

After returning to the Netherlands and during work carried out abroad I discussed ideas and parts of the manuscript with many collegues. I cannot thank them all here. I consider it a privilege to have had twice the opportunity to discuss at Kew a variety of subjects with Mr. Evan Guest who has devoted probably more of his time to the vegetation of Iraq than anyone else. With Dr. H.N. le Houérou and Prof. Dr. K.H. Batanouny I have had many hours of fruitful discussions. I wish to thank Dr. A.D.Q. Agnew for his continuous interest in the study, his comments and the corrections he made in a large part of the final draft. The photographs of the Khedr-el-Mai exclosure of 1960 in chapter X were made by him. I thank Mrs. M.D. Rigby for correcting most of the text of the first six chapters.

My thanks are due to several people at ITC who have helped me: Drs. A.D.J. Schaap for commenting on an early draft of chapter XII; Mr. P.H. van der Hulst and Mr. H.L. Wilbrink for the skilful work they did on some of the photographs; Drs. D. Kovacs and Mr. R.X. Portier for assisting in statistical matters and numerical processing; Mr. E.A. van Leeuwen for drawing early versions of some of the figures; Mrs. Drs. N.J. Wentink for making some of the Russian literature accessible; Mrs. A.R. Morgan for correcting the 'summary'; Ir. H.L. Rogge for critical reading and improving the 'samenvatting'; the N-secretariaat, especially Mrs. H. Hofmans-Assink and Mrs. A.M. van der Linde, for typing parts of the manuscript; and Ing. J.H. de Vos tot Nederveen Cappel and Dr. H.G.J. Huizing for commenting on a draft of the land evaluation section.

Without the contribution received from the Netherlands Organisation for the Advancement of Pure Research (Z.W.O.) during the first nine months after returning from Iraq the present study would probably never have been written up.

I thank Mr. R. de Jonge and Mr. J.P.R. van den Worm for their skilful and speedy drawing of the figures. Gustav Fischer Verlag in Stuttgart kindly granted the permission for reproduction of Fig. VI–1, Fig. VI–3 and Fig. VI–5.

This study could have never been reproduced and published in its present form without the efforts of Mr. W.R. Peters and, at an earlier stage, of Mr. S.P. Bakker of Dr. W. Junk b.v. Publishers. I am much indebted to Mr. P. Zoutendijk for the work he did in transforming into the final product a draft which was submitted in parts, and was not always correctly and neatly formulated, and for his patience in waiting for the last 'few' pages.

CONTENTS*

* A detailed table of contents precedes each chapter.

'En résumé, le phytoécologiste travaillant dans les pays du Proche- et du Moyen-Orient, doit être un explorateur intelligent beaucoup plus qu'un statisticien méthodique. Il doit avoir une assez bonne résistance physique pour supporter de longs voyages dans des conditions peu confortables . . . . . . A mesure que progressent ses recherches sur le terrain, et qu'il visite de nouveaux territoires, il doit pouvoir faire constamment l'autocritique des interprétations et des conclusions qu'il a pu faire antérieurement en ce qui concerne les types de végétation et l'écologie des especes importantes, modifiant ainsi progressivement ses points de vue en fonction des observations nouvelles, afin d'arriver au maximum d'objectivité possible. Enfin, il doit se garder de tout académisme, et ne jamais oublier que, sans se départir d'une certaine rigueur scientifique, il doit pouvoir fournir aussi tôt que possible des suggestions pratiques aux forestiers, range-managers et agronomes chargés des travaux de remise en valeur et d'aménagement.'

H. PABOT (1962)

Author's address:

Dr. D.C.P. Thalen
International Institute
for Aerial Survey and Earth
Sciences (ITC)
350 Boulevard 1945 – P.O. Box 6
Enschede – Netherlands

PART ONE

Introduction

CHAPTER I

INTRODUCTION

I. INTRODUCTION

I.A. Scope of the study

The degradation of renewable natural resources in the arid areas of South West Asia has become a matter of great concern. Locally the effects of careless utilization of the land and its resources had been felt long ago. It is, however, only relatively recently that the gravity of the situation is being generally and fully realised. It is now well understood that action is required to save what is left and possibly to restore what was once there. Such action requires organization and coordination, but above all knowledge of the present state of the resources and the impact of utilization processes.

Part of this knowledge is already available. The problems are not confined to this part of the world's arid lands, but exist equally well elsewhere. Especially over the last decades an ever increasing number of studies have been published dealing with aspects of arid zone research. Problem analyses were followed by compilations of knowledge in particular fields (White 1956; Hills 1966; Mc Ginnies *et al.* 1967, 1969, 1971; Kaul 1970; Clawson *et al.* 1971; UNESCO Arid Zone Research Series, e.g. UNESCO 1955, 1960a). For the United Nations Conference on Desertification in 1977 the state of knowledge on certain aspects was updated (e.g. Warren & Maizels 1977).

Efforts are now being made towards improved management and sustained optimal use of natural resources in arid lands. Vegetation holds a key position where grazing is the accepted land use. In the arid areas of Iraq, not within reach of irrigation water, this is the situation. The present study deals with these areas, concentrating on a number of closely linked aspects of the ecology and utilization of its vegetation, with improved management as the final aim. They may be formulated as follows:

- What is the present situation of the vegetation cover of these areas in relation to the environmental conditions, particularly when the vegetation is considered from the resource point of view?
- How is the vegetation being utilized at present, how was it utilized in the past, and what are the effects of the utilization?
- What management practices could be adopted for improvement and conservation of the range resources in order to ensure optimum return in the future?
- What are the inventory and evaluation methods, especially with respect to the vegetation, on which such management practices could be based?

Although data have been collected and are presented which provide information on all these points, the study has no pretension of exhaustively covering this array of problems and possible solutions in all details; such a treatise is not possible for the arid zone vegetation of Iraq. It will be shown that data from previous studies, necessary to allow comparison with the present situation, are too scarce and incidental. Experimental evidence, apart from the work reported in this study, is virtually lacking. In the following chapters therefore an attempt is made:

(i) to describe the vegetation and its environmental setting and clarify how the present state of this resource of the desert areas of Iraq can be explained as the result of continuous exploitation by man and his livestock, especially during the twentieth century, and

(ii) to indicate how, based on knowledge of the situation, the processes taking place, in combination with new inventory and evaluation techniques, improvement might be achieved through proper management.

The information presented bears primarily on Iraq. Desert shrub rangelands, however, occupy a vast area of the world's land surface and most of the patterns and processes studied are not limited to Iraq; thus part of the findings presented in this account may well apply to other areas too.

I.B. Background

The field data presented and discussed were collected over the period April 1970 – June 1973 within the framework of a research programme on the desert rangelands of Iraq, developed at the Division of Ecology and Micrometeorology of the Institute for Applied Research on Natural Resources, established at Abu-Ghraib near Baghdad*. This Institute obtained expert support through the United Nations Technical Assistance Programme over the period 1963–1966, amongst others in the field of plant ecology late 1965 (Knetsch & Gounot 1963). Following a request of the Government of Iraq to establish a National Institute, in 1967 a United Nations Development Programme – Special Fund Project with the United Nations Educational, Scientific and Cultural Organization (UNESCO) as executing agency became operational for a period of three years to strenghthen the research programme of the Institute. The Project was later extended into a second phase till the end of 1974.

General objective was to develop at this National Institute a research programme for collecting basic data on the renewable natural resources – climate, soil, water, and vegetation – in order to provide the country's planners with the necessary data about the resources available in each particular region and to advise on their most promising utilization. One of the main activities outlined in the Plan of Operation was to improve natural pastures used by nomadic tribes. The research programme in plant ecology at the Institute, formulated and initiated in October 1965, remained in principle the same when in 1967 the Project started, but more emphasis was placed on the practical and applied aspects of research concerning the improvement and utilization of the desert range resources. Data collection in the initial phase was hampered by a number of factors, e.g. lack of trained staff and equipment, and poor facilities for fieldwork in the desert regions.

When the author joined the Project as Associate Expert in Plant Ecology in April 1970 the general points for a research programme in range ecology had been

---

* At present: Institute for Applied Research on Natural Resources, Jadiriyah, Baghdad.

listed and a number of detailed outlines given for investigating the requirements for plant establishment in arid rangelands (Winkworth 1969; Kaul 1969a, b). A number of fieldtrips had been undertaken for a reconnaissance survey of the desert rangelands and progress had been made with a programme of plant introduction (Al-Ani *et al.* 1970 b; Al-Mufti *et al.* 1970).

Against this background and based on observations made during fieldtrips as well as on information collected from the literature, a well defined research programme was gradually developed on the vegetation of the desert rangelands as a resource. It can be summarized in three headings: (i) identification of the resource, (ii) quantification of the resource, (iii) improvement of the resource.

Details of the planning can be found in Kaul & Thalen (1971a,b) and Thalen (1973). Time schedule and methods are given in chapter III. During the execution of the research programme the 'natural pastures', mentioned in the original Plan of Operation, turned out to consist almost completely of low shrub vegetation types. Perennial grasses and forbs were by and large either completely lacking or scarcely present. The study of the natural pastures became, therefore, in fact a study of desert shrub rangelands.

I.C. Desert shrub rangelands

I.C.1. *Definition*

The words *range* and *rangeland* are usually applied to indicate areas where grazing and browsing of the semi-natural or natural vegetation is the accepted form of land use. The term may originate from ranging in the sense of wandering or roaming in search for food, as given in definitions quoted in Dudley Stamp (1966):

> 'An area of uncultivated ground or wild country over which domestic or wild animals range for food, now especially a cattle range'
> and
> 'An area, space, or stretch of ground over which ranging (*i.e.* wandering or roaming of men or animals) takes places or is possible'.

Other definitions and descriptions can be found, given from different points of view (Lewis 1969; Vallentine 1971; Winkworth 1971), but more precise definitions tend to be either inconveniently long or to cover only a particular type of rangeland.

Shrubs can make up the major part of the plant cover on rangelands. They are perennial, woody plants which branch from the base and do not reach a very large size. In an Australian classification shrubs have been defined as woody plants eight metre or less in height, multistemmed, or, if single stemmed, branched in the lower half (Moore & Perry 1970). In dry environments the height is usually much less. Most of the shrubs of the desert rangelands of Iraq are of this low type, normally less than one metre in height. For such low forms the term dwarfshrub is often applied. Shrubs are found all over the world and certain species show a remarkable ability to survive under harsh environmental conditions. Vegetation types in which shrubs are dominant are particularly widespread in the arid areas. These are areas where precipitation is low, irregular and local. Relative humidity may drop below ten per cent and evaporation is high. Water is often a limiting factor for production, not only for the vegetation but also for the human and animal populations who usually depend on both, water and plant production. The vegetation is sparse and often poor in species. Most soils are shallow and poor in organic matter and salinization is common. Watercourses are widely spaced, vary highly in their water content over the seasons and often do not drain to the sea but into lakes, depressions and other inland basins. Although these general characteristics are easily given, an exact delineation of the arid zone is virtually impossible and varies with the criteria applied.

A term often used synonymously with arid areas and even more difficult to define is *desert*. In the context of the present study the word is used in its ecological sense, as defined by Blakelock *et al.* in Guest (1966):

> 'A region unable to support vegetation throughout its surface, *i.e.* where plants cannot form a close cover but are widely-spaced with bare patches between them, or where plant life may be largely dependent on topography, growing mainly in depressions and other favourable habitats'.

In Fig. I–1 an example is shown of a desert area where the shrub vegetation exhibits both these patterns.

The highest aridity is found in the absolute (extreme) desert, a desert which supports little or no, certainly not a perennial, vegetation. At the other side are the semi-arid transitional formations to woodland and grassland. They may be grouped under the terms semi-desert or sub-desert.

With the above descriptions desert shrub rangelands are broadly defined. A strict definition would be artificial and of little or no practical value. A fairly precise picture, however, of the vegetation formations dealt with in the present study is obtained from a combination of the description of two formations distinguished in UNESCO (1973) for an international classification of vegetation, *viz.* extremely xeromorphic (subdesert) shrubland and extremely xeromorphic dwarfshrubland:

> 'More or less open or very open stands of shrubs and dwarfshrubs with various adaptations to survive or to avoid a long dry season such as strongly reduced leaves, green branches without leaves, succulent stems, etc., mostly found in sub-desert areas'.

I.C.2. *World distribution of desert shrub vegetation*

A delineation of areas where desert shrub rangelands are found is not easily given. Desert shrub vegetation types are found throughout the arid zones of the world

Figure I-1. Typical patch of desert shrub vegetation in Iraq, a stand of *Rhanterium epapposum* in the Southern Desert near Samah.

and a delineation of areas with desert shrub rangelands, therefore, coincides more or less with a delineation of arid areas (Fig. I–2).

Such a delineation has been attempted by many different approaches and the results depend on the criteria applied. Criteria have been derived from the fields of climate, soils and interior drainage, and also vegetation.

Dzerdzeevski (1958) makes a mention of 19 indexes of aridity designed to provide a climatic classification (cf. Reitan & Green 1967). Most classifications use combinations of temperature and precipitation, expressed in numerical values. The systems of de Martonne (1926), Köppen (1931), Thornthwaite (1931, 1948) and Emberger (1955) have been widely used and are discussed amongst others in Meigs (1964), Wallen (1966), Reitan & Green (1967) and Walton (1969). Most widely accepted as satisfactory is the aridity classification of Meigs

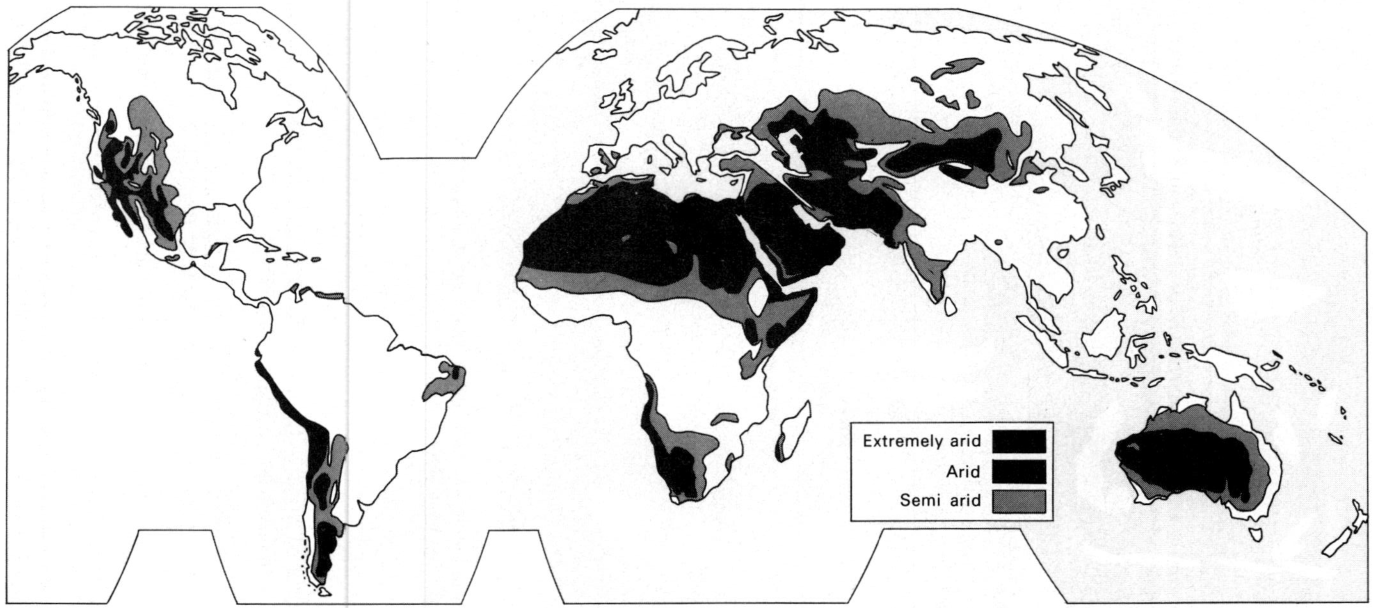

Figure I-2. World distribution of arid areas (after Meigs 1953).

(1953), who constructed maps to delineate the arid zone at the onset of UNESCO's arid zone programme. He computed values according to the index of Thornthwaite (1948).

This index expresses the adequacy of precipitation in relation to the need of plants. He distinguished a.o. dry subhumid, semi-arid and arid climatic types. A subdivision was made in the arid areas by introducing extremely arid areas. In such areas there should not be a regular seasonal rhythm of rainfall and at least 12 consecutive months without rainfall must have been recorded. Details of this procedure of mapping arid homoclimates have been given in Meigs (1953). The areas mapped on this basis as semi-arid, arid and extremely arid are shown in Fig. I–2.

Delineation of the arid zone based on interior basin drainage has been published by de Martonne & Larfrere (1927). Based on soils, Shantz (1956) makes an estimate of 43 per cent of the land area of the earth being of Pedocals, defined as soils with dry subsoils, but such soils extend beyond the arid zone.

Vegetation reflects environmental conditions with their spatial and temporal variations. Vegetation can, therefore, be used as an indicator for arid conditions. However, not only climatic and edaphic conditions are reflected in the vegetation cover, but also the interference of man and animal in the climate – plant – soil complex. In some arid areas many centuries, and especially the last few centuries, of vegetation utilization by man and his livestock have changed the original cover considerably, even to an extent that makes detection of this original cover virtually impossible.

This situation and the fact that the vegetation of large areas under arid conditions has not yet been adequately studied mean that, at the present state of knowledge, a delineation of arid areas on a world scale, based on vegetation alone, is as questionable as one based on any of the other criteria mentioned.

The world distribution of arid zone shrub vegetation is also not easily seen from existing vegetation maps because of the fact that shrub formations are most of the time included in, and depicted as part of, broader defined legend units, usually more than one. In one of the first world vegetation maps, the one by Schimper (1898), arid zone shrub formations are most likely partly included in three of the 15 more or less ecological legend units *viz.* in (i) thorn forest and scrub, (ii) steppes and transitions between steppes and deserts (semi-deserts), and (iii) dry deserts. Similar problems arise when referring to later maps of this kind. There is great lack of consistency between the various authors in dealing with arid zone shrub vegetation types, perhaps partly because until recently little was known of large areas. Le Houérou (1972a) classified the shrublands of the Mediterranean arid zone (between the isohyets of 100 and 400 mm) as depleted forest, chamaephytic steppes, nanophanerophytic steppes, and halophytic crassulescent steppes.

Soviet geobotanists use the term 'shrubdeserts' for sandy, rocky and saline deserts where annual precipitation exceeds 100 mm and shrubs determine the aspect of plant growth (Petrov 1972).

On a regional basis, however, for parts of the world the distribution of arid zone shrub vegetation types has been well outlined or mapped, e.g. North America (McGinnies 1972), South America (Soriano 1972), Australia (Moore *et al.* 1970; Everist 1972) Central Asia (Petrov 1970, 1972), Indo-Pakistan (Dakshini 1972) and South West Asia and North Africa (UNESCO 1969; le Houérou 1972a). From these sources the picture emerges that shrub vegetation types, in a variety of forms, are found all over the arid and semi-arid areas of the world. They may be found as extensive stands, but are often confined to the more favourable habitats in areas where only drought evading species can grow and show up after occasional rains. The distribution, therefore, of arid and semi-arid areas, as shown in Fig. I–2 also reflects the geographical world distribution of arid zone shrub vegetation in a generalized way.

Estimates of areas in the world under a particular vegetation formation have been made, especially during the last few years, for purposes of production estimates on a worldwide basis. Figures pertaining to the present study, derived from various authors are given in table I–1. In interpreting these figures the same problems are being faced as outlined above for mapping. Desert shrub vegetation is never included as one clearly defined formation.

Shantz (1956) made a class 'desert grass-desert shrub', covering 27.4 million km^2. Bazilevich *et al.* (1970) are more specific and put 3.7 million km^2 under shrub-steppe formations and another 8.8 million km^2 under grass and shrub savanna. Lieth & Whittaker (1975) finally present a figure of 18 million km^2 being under desert scrub. Here the difficulty of the use of scrub and shrub comes up. The word scrub is generally used for non succulent plants of stunted growth, but other definitions can be found e.g. Blakelock *et al.* in Guest (1966): a plant community dominated by shrubs.

From the above it may be concluded that probably roughly one third of the total arid zone or about 10% to 15% of the total land surface of the earth is covered by plant communities in which shrubs are dominant. This percentage is considerably higher than that of the total area under cultivation.

I.C.3. *Resource value of desert shrubs*

Until recently relatively little research had been carried out on the possible utilization and development of shrub vegetation types in arid rangelands. However, their resource values and potential are being increasingly realized, as reflected in the proceedings of three symposia; on the Management and Development of Salt-desert Shrub Ranges, on the Biology and Utilization of Wildland Shrubs, and on the Arid Shrublands in Australia and the USA (US Dept. of the Interior 1966; Mc Kell *et al.* 1972; Hyder 1973).

The following values of desert shrub vegetation may be mentioned:

– Shrubs provide *forage* for wildlife and livestock.

– Under severe moisture limitation, the perennial shrubs with their extensive root system can use moisture from a greater volume of soil than grasses or forbs. Shrubs, therefore, offer good possibilities for *forage production under severe arid conditions.*

– Certain shrub species are palatable and may provide forage the whole year round. Other species reach their highest production in specific seasons just when other plant groups are dormant or dried up. Such desirable forage shrubs are especially valuable, during seasons when no grazing is available in the form of grasses or forbs, *in tiding over the periods of limited forage supply.*

Table I-1. Estimates of areas under major plant formations in the world's arid areas, according to various authors. Formations including desert shrub vegetation are printed in italics.

Author(s)	(Bio-) climatic region, biome or plant formation	Area (10^6 km^2)	Part of total land area* (%)
Meigs (1953)	Semi-arid	21.1	15.9
	Arid	21.8	16.3
	Extreme arid	5.8	4.3
	Total arid zone	48.8	36.5
Shantz (1956)	Semi-arid	7.0	5.2
	Sclerophyll brushland	3.0	2.2
	Thorn forest	0.9	0.7
	Short grass	3.1	2.3
	Arid	33.4	25.0
	Desert grass savanna	6.0	4.5
	Desert grass-desert shrub	*27.4*	*20.5*
	Extreme arid	6.3	4.7
	Desert	6.3	4.7
	Total arid zone	46.7	34.9
Bazilevich *et al.* (1970)	Semi-arid	32.4	24.2
	Subboreal belt	8.1	6.0
	Subtropical belt	8.3	6.2
	Shrub-steppe formations	*3.7*	*2.8*
	Other formations (total)	4.6	3.4
	Tropical belt	16.0	12.0
	Grass and shrub savanna	*8.8*	*6.6*
	Other formations (total)	7.2	5.4
	Arid	29.5	22.1
	Subboreal belt	7.0	5.2
	Subtropical	9.7	7.3
	Tropical belt	12.8	9.6
	Total arid zone	61.9	46.3
Lieth & Whittaker (1975)	Forest and woodland	57.0	42.6
	Dwarf and open scrub	26.0	19.5
	Tundra	8.0	6.0
	Desert scrub	*18.0*	*13.5*
	Grassland	24.0	18.0
	Tropical grassland	15.0	11.2
	Temperate grassland	9.0	6.7
	Desert (extreme)	24.0	18.0
	Dry desert	8.5	6.4
	(Ice desert)	(15.5)	11.6
	Cultivated land	14.0	10.5
	Fresh water	4.0	3.0

* Estimates of the earth's total land area vary considerably, dependent on whether glaciated areas (about 13.9 million km^2, see data Lieth & Whittaker (1975) and fresh water are included, see discussion in Bazilevich *et al.* (1970).
For uniformity in the present tables, the most common figure of 133.7 million km^2 has been used which excludes the glaciers.

– Many shrub species have a *high nutritive value* (Chapline & White 1947; Dietz 1972; Cook 1972). Some parts are especially highly nutritious (Short *et al.* 1972) and are collected for supplementary feeding and for feeding animals suffering from diseases.

Apart from their forage value, mentioned above, shrubs provide the necessary *cover*, especially for wildlife, but also for livestock. They give *protection against predators and the elements* (Robinette 1972).

– Shrubs play an important role in *soil protection and stabilization.* They assist in the protection of soil against movement by wind or water. Often shrubs are planted for this purpose as windbreaks or shelterbelts, but the natural shrub cover serves the same function (Kaul 1970; Van Dersal 1972).

– Many shrub species have *medicinal values* for man or have a *potential industrial value*. The desired parts may be collected in the wild or harvested from plantations. In many regions medicinal use of shrubs is well known to local people but has not yet been applied at larger scale (UNESCO 1960b; Krochmal 1972; Jones & Barclay 1972).

– Ove· vast stretches in the arid zone shrubs are the only woody material available. In such areas they may be applied as *small timber for construction*, but serve in particular as a *source of fuel* when no other material for heating and cooking (oil, gas, dung) is available.

– Although difficult to prove or to measure, it can be stated that shrubs have a high *aesthetic value* in arid areas. They bring diversity and colour to the often vast, barren, monotonous, arid landscape. Many species are cultivated for their ornamental value.

The resource aspects listed show the scope for utilization under proper management. However, in many arid areas of the world the shrub vegetation has been used, mainly as a source of forage and fuel, without any further management for many centuries. This has resulted in depletion of the resource, often beyond capacity of regeneration. It happened and still continues over vast areas in the arid zones of South West Asia and North Africa. The shrub vegetation of the Iraqi desert rangelands is only an example.

I.D. Iraq, a brief introduction to the country*

I.D.1. *Environment*

The physiographic situation of Iraq in South West Asia and the position of its territory in relation to the surrounding countries is shown in Fig. I–3. The coun-

* For local names as far as not included in Fig. III–1 reference can be made to the gazetteer in Guest (1966).

Figure I-3. Position of Iraq in South West Asia and location of the study area.

try covers an area of 438 466 km^2 between the longitudes 38°42' and 48°23' E and between latitudes 29°27' and 37°23' N, including 924 km^2 of territorial water and 3 522 km^2 for half of the Neutral Zone in the south. In the north, Iraq borders on Turkey over 305 km in the Kurdistan part of the Armenian-Anatolian mountain region. In the north-west with Syria, in the west with Jordan, and in the southwest and south with Saudi Arabia and Kuwait, the borders have been fixed across the Syrian Desert and the northern part of the deserts of the Arabian Plateau, over 603, 147, 895 and 254 km, respectively (Central Statistical Organization 1970). The diversity in environmental conditions is indicated by the situation of these boundaries being placed in mountainous and lowland areas, desert and sea.

Accounts of the physiography of the country, especially in relation to land use and agricultural development can be found in Fisher (1950), Buringh (1960),

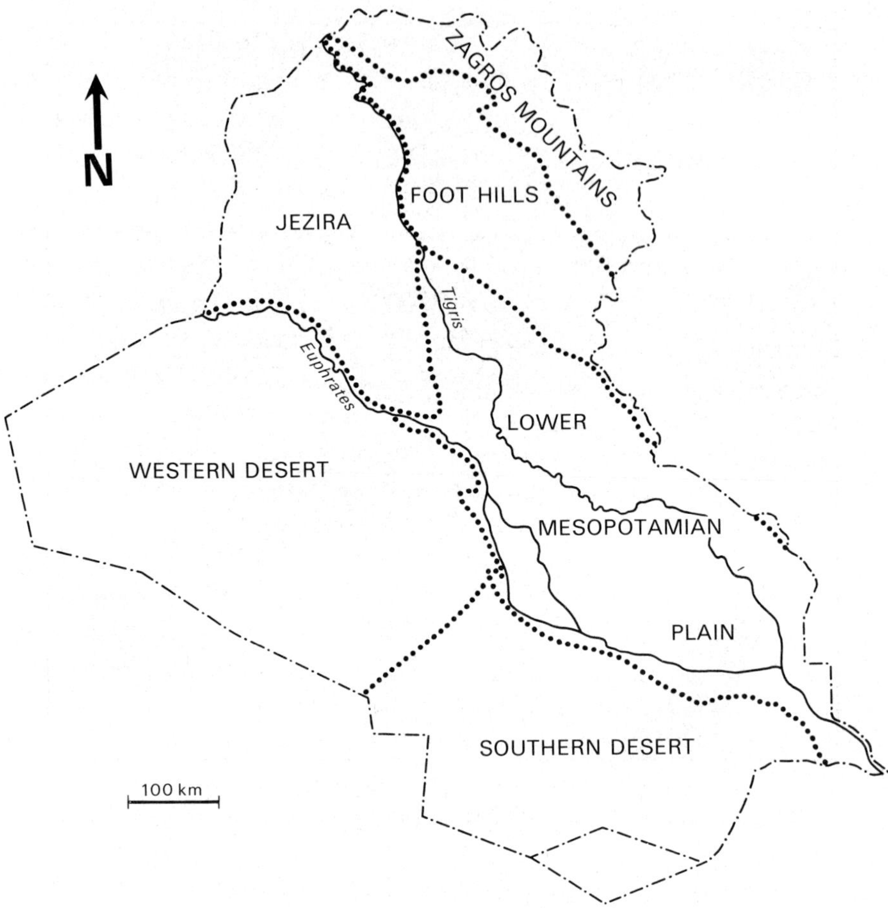

Figure I–4. Physiograpic regions of Iraq (after Buringh 1960).

Wirth (1962), Brice (1966), Clawson *et al.* (1971) and Beaumont *et al.* (1976). Physiographically Iraq can be divided in five regions, roughly situated in northwest-southeastern belts (Fig. I–4),*viz.*:

Zagros mountains area in the north and northeast

Foothills zone, situated as a belt in front of the mountains,

Jezira, Iraqi part of an area between the Tigris and Euphrates Rivers (in Arabic Jezira = island)

Western and Southern Deserts, occupying in the southwest and south about one third to one half of the total area of the country.

Lower Mesopotamian Plain.

The Zagros mountains are the result of folding of the earth's crust mainly during the Miocene. Sediments deposited in the Tethys Sea over millions of years were shaped into parallel mountain ranges and valleys when compressed against the stable continental block of the Arabian foreland. The so formed mountainous landscape was later considerably modified by the continuous process of erosion.

The erosion material partly filled up the valleys but a large amount of debris, gravel, sand, etc. also accumulated and formed deposits at the front of the mountains area. These layers have undergone some slight folding later, resulting now in the Foothills, a landscape of low hill ridges dissected by small streams and rivers which originate from the mountainous area.

By rise of the land and processes of sedimentation in the Tethys sea, smaller inland seas formed, where gypsum, salt and limestone layers accumulated. The Iraqi Jezira forms the remains of such an area. It has a flat to gently rolling topography.

The Western and Southern Deserts were formed on the Arabian continental block. On the nearly horizontal surface a succession of mainly limestone layers was deposited up to eight kilometres thick (Buringh 1960). They reflect the small rises and falls of sea level during the geological eras and are locally overlain by aeolian deposits. Deep wadis have been cut out in these relatively flat areas, especially in the western part, during periods of excessive rainfall.

The Lower Mesopotamian Plain is in principle a partly in filled extensive geosyncline. Sediments of the Tigris and Euphrates Rivers with their tributaries have formed a flat river plain, It has long been assumed that a delta has gradually been extending southward into the Arabian Gulf. However, evidence has been presented that the coastline of Iraq has not changed much, but that sedimentation kept pace with the sinking of the bedrock, resulting in the relatively constant height of the area above sea level (Lees & Falcon 1952).

Iraq has a sub-tropical, continental, arid climate. Rainfall is mainly in winter and early spring (November–April) and ranges from below 100 mm in the south to over 1300 mm in the north, annually. The summers are dry and hot. Temperatures of 45° C are common in July and August, while in December and January frost is regularly recorded, even in the lowland area. Evaporation in Iraq is in the range of 2000–3400 mm annually. Climate of the northern mountainous area is markedly different from the rest of the country, amongst others through higher rainfall and lower temperature. In the southeast the Arabian Gulf has a buffering effect with respect to temperature extremes.

I.D.2. *Population*

The Mesopotamian Plain and the Foothills area form one of the oldest permanently inhabited areas of the world. Some of the earliest successful attempts of plant and animal domestication took place in this part of the world. Agriculture and livestock raising were developed here many thousands of years ago. Mesopotamia is best known from its past great civilizations (Beek 1960).

From the thirteenth century the area of the present Iraq was under permanent occupation of which the Turkish one from 1534 to 1917 was the last. After the First World War the State of Iraq was formed out of three vilayets of the Ottoman Empire: Basra, Baghdad and Mosul. It became a Sovereign State in 1932 and a Republic in 1958.

As the result of constant movement of peoples into and within the area, the population of the country is a conglomeration of ethnic, linguistic and religious groups (Lawless 1972). This population is increasing rapidly, growing over the last thirty years at a rate of over three per cent per year. According to the available census figures the population was 4.8 million in 1947, 6.5 million in 1957 and 8.2 million in 1965. Later figures suggest a population of 10.4 million in 1973, over 13 million in 1978 and an annual population increase of approximately 3.4%. Migration from rural areas to the cities takes place at a large scale. It has been estimated that around 1800 in the present Iraq about 15% to 20% of the total population lived in towns of over 10 000 inhabitants, and that in 1970 this figure was about 47%(Beaumont *et al.* 1976).

By 1980 the population of the capital may reach over four million, more than one third of the population of Iraq (Jones 1969). The above figures indicate an increasing demand for agricultural and livestock products by a relatively decreasing rural population. Discussions on population distribution and trends have been given by Lebon (1953), Hasan (1958) and Lawless (1972).

I.D.3. *Land use*

The use of the land in Iraq is, apart from climatic conditions, mainly governed by availability of water, soil conditions and vegetation cover. Rainfed agriculture can only be practised in the mountainous and foothills area in the northeast. Mean annual rainfall varies here from about 300 to over 1 300 mm yearly. The good and fertile soils are restricted to the valleys and terraces. Outside the rainfed area irrigation water is required. The combination of a certain amount of dissolved minerals in the irrigation water, thousands of years of irrigation, high evaporation from the soils and transpiration of the plants, and certain soil conditions, has caused salinization all over the Mesopotamian Plain. It determines the present land use pattern in this region. (Buringh 1960; Clawson *et al.* 1971; van Aart 1974). Cultivation in the remaining part of Iraq, the Jezira and Western and Southern Deserts is restricted to the Upper Jezira (rainfall-dependent with some minor local supplementary irrigation) and to irregular cultivation at lower situated places which receive runoff water from the surrounding areas.

The total area of Iraq under cultivation in 1970 is given in table I-2. An account of land use patterns and crop distribution can be found in Wirth (1962). Buringh (1960) presented a general classification of soil suitability for *present* agri-

Table I–2. Cultivated land in Iraq in 1970 (calculated from Central Statistical Organization 1970).

Type of cultivation	Area (km²)	Part of total area of the country (%)
Cereal crops (wheat, barley, rice, maize, giant millet, millet)	32 122	7.34
Industrial crops (cotton, tobacco, linseed, sesame, beet, sunflowers, peanuts).	1 198	0.27
Other crops	725	0.17
Vegetables	1 890	0.43
Palmtree orchards	1 455	0.33
Other fruit orchards	505	0.12
Total area cropped	37 895	8.64
Total area fallow	37 470	8.55
Total area cultivated	75 365	17.19

cultural use in Iraq and another one of the *potential* suitability for agricultural use. The classifications are relative and based on the information collected for the explanatory soil map, combined with a general knowledge of agriculture in Iraq. The first classification is, in a summarized form, given in Fig. I–5. It can be seen that 74.2% of Iraq was classified under sub-marginal soils. In the potential soil suitability classification (an evaluation for future agricultural use, assuming considerably improved soil conditions and farm management) the corresponding figure is little higher, 74.6%. The area indicated coincides almost completely with class 4 of Fig. I–5.

From these sources it can be seen that about three quarters of Iraq, comprising mainly the Western Desert, Southern Desert and Jezira (extensions of the Arabian and Syrian Deserts), consist of soils without agricultural potentialities situated in an area with very low rainfall. In this part of the country the remaining renewable natural resources to be utilized are the (semi-)natural vegetation and the animal life that depends on it.

I.D.4. *Vegetation*

In the vegetation of Iraq the great diversity in environmental conditions is reflected. The plant cover ranges from forests on the hardly touched mountain slopes to halophytic communities in areas where activities of man have been going on for millennia, from aquatic communities to the xerophytic and extremely sparse cover of desert areas. The following can, therefore, only be a brief and general description of the vegetation of the country as a whole. It is presented in order to place the vegetation cover of the desert rangelands in perspective to the vegetation

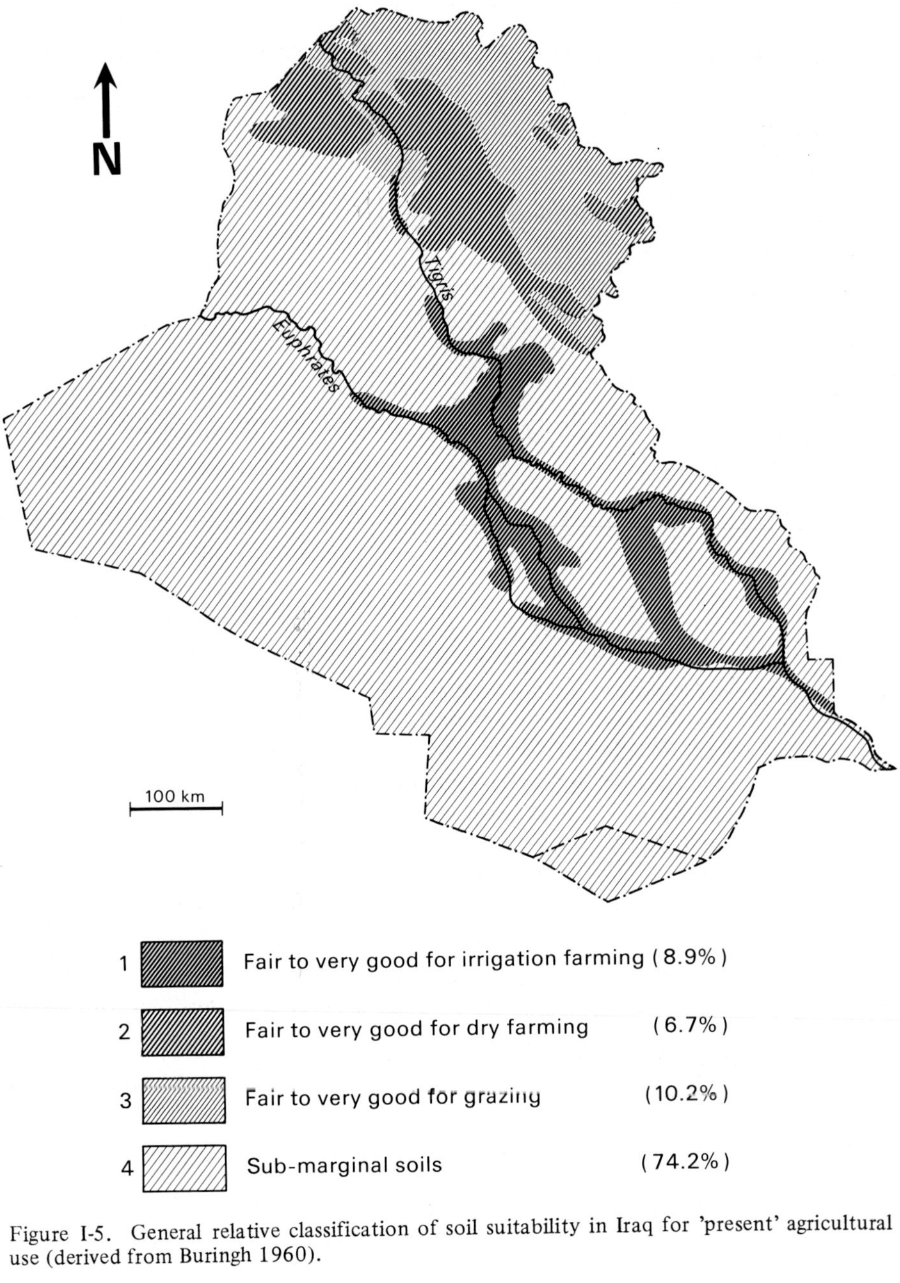

Figure I-5. General relative classification of soil suitability in Iraq for 'present' agricultural use (derived from Buringh 1960).

of the rest of the country. Some emphasis is given to the use of the different types of vegetation.

For the present purpose the following categories of plant cover, modified from Guest (1966), may be distinguished:

i. Vegetation above the timber-line
ii. Forest
  a – Mountain forest
  b – Riverine forest
iii. Hygrophilous vegetation
iv. Halophytic vegetation of lower Iraq
v. Segetal and ruderal vegetation
vi. Desert and steppe vegetation

ad i. Knowledge on the vegetation of the *highest areas* of Iraq, at an elevation between 2750 m and 3730 m, is scanty. Some collectors have brought specimens but no descriptions of the plant cover are available. Guest (1966) lists some herbaceous species which are predominant in summer on rocky mountain summits. They mainly belong to the families Compositae, Labiatae, Cruciferae and Caryophyllaceae.

One of the most striking formations is the so called *thorn-cushion* vegetation, found in the zone of about 1750–3000 m, where low growing spiny herbs and dwarf shrubs form compact cushions, measuring over one metre across. A large number of *Astragalus* spp. has been collected here, with *Acantholimon* and *Onobrychis* as common other genera. In between these shrubs, herbs and grasses are found.

In meadows a rich mixture of species may be found including valuable forage species, e.g. of the genera *Trifolium* and *Medicago*. For extensive species lists see Handel–Mazzetti (1914b), Guest (1953a,1966), Springfield (1954) and Zohary (1973).

ad ii. Generally available information on the *forests* of Iraq was limited to notes included in more general publications (Handel–Mazzetti 1914b; Guest 1932, 1933, 1953a; Springfield 1954; Mooney 1956), until the report by Chapman (1957) was published which proved invaluable for later reference on the subject. The same author published a map of forest areas at scale 1 : 750 000, later included in a simplified version in Guest (1966), where forest deterioration is also discussed.

ad iia. *Mountain forest*, in northern Iraq is generally found at elevations between 500 m and 1800 m under an annual rainfall of 700–1400 mm. It is a *Quercus* forest with the species *Q. aegilops*, *Q. infectoria* and *Q. libani*. Tree density can range from closed forest in relatively undisturbed areas to scattered stunted oak shrubs, as remnants of the original forest, near villages. The total area under mountain forest has been estimated at 20 000 km^2 half of which is relatively unexploited. The remaining part is in various stages of degradation and actually represents potential forest areas (Chapman 1957).

Deterioration of the Iraqi forests may have been going on for many centuries but the oldes traced records are from travellers in the nineteenth century. Guest (1966) quotes them in detail and makes a remark that extensive tracts of land which are now completely denuded of trees and even of the shrubby remnants (*e.g.* in the Sulaimaniya region) should actually be added to this potential forest area. The ground vegetation of the mountain forest is rich and diverse. Springfield (1954) and Guest (1966) give lists of species. Most of these provide excellent grazing and some plants are cut, dried and used as forage in times of scarcity.

Brichambaut(1956a) wrote a note on the perennial grasses, while Kittani (1956) discussed some aspects of grazing in the forest area. Further information on the mountain forest can be found in Handel–Mazzetti (1914b), Guest (1932, 1933, 1953a, 1966), Chapman (1957), Agnew (1962b), Hadac & Agnew (1962, 1963), Oedekoven (1970) and Zohary (1973).

ad iib. *Riverine forest* is found in the mountains as well as in the plains. Along the streams in the north, narrow belts of trees can be found of *Salix* spp., *Platanus orientalis* and *Populus euphratica.* Shrubs are also common here, e.g. *Nerium oleander, Vitis vinifera, Rosa canina* (Guest 1966). In the plains about 200 km^2 has been estimated to be under 'ahrash', lowland riverine forest. Is is found irregularly distributed along the streams and on islands. Most common species are *Populus euphratica, Tamarix* spp. and *Salix* spp. Part of the year these forest areas may be flooded and locally they are exploited as fuel. Two characteristic cultivated tree species, locally dominating the landscape should be mentioned here. In the north *Populus nigra* is widely grown in dense stands for timber. In the southern part of the country orchards of date palm (*Phoenix dactylifera*) are found which make Iraq to the most important exporter of dates in the world.

ad iii. As *hygrophilous* could be classified the vegetation of lakes, rivers, canals and of areas bordering these waters and regularly submerged for long periods. The most extensive vegetation types of this category are found in the southeast, in the area roughly between Amara and Basra, known as 'the Marshes'. In this area thousands of square kilometres are covered by extensive shallow lakes, partly changing into mudflats during dry periods. *Phragmites communis** and *Typha angustifolia* are the most common species and grow in dense stands over vast areas. Young shoots of these and other species are grazed by waterbuffaloes, while large stands of full

* = *P. australis*

grown plants are harvested for several uses *e.g.* mat making, house building and the paper industry. Smaller patches of hydrophytic vegetation are found all over Iraq, especially in and near the regularly irrigated (or flooded) areas in the Lower Mesopotamian Plain. A sharp distinction with the two categories briefly discussed under (iv) and (v) below cannot be made. The most common trees found in the marshlands and near water courses have been mentioned under (iib). There seems to be little variation in the phanerophytes of the aquatic environment, but a good botanical exploration of the region has yet to be made.

ad iv. Salinization has, locally, taken place all over Lowland Iraq, partly as the result of thousands of years of (man-made) irrigation, partly as a natural phenomenon in depressions. *Halophytic vegetation types* are widespread and common, especially in the Mesopotamian Plain. A vegetation of *Halocnemum strobilaceum, Seidlitzia rosmarinus, Suaeda baccata* and *S. vermicularis*, of the higher shrub *Nitraria retusa*, the grasses *Aeluropus littoralis* and *A. lagopoides* and the low *Cressa cretica*, can be found in particular species combinations as well as in virtually monospecific stands. Studies on the use of some of these and other natural plant species as indicators of soil type and the degree and kind of salinization have been initiated (Harris 1960; Al-Ani *et al.* 1970a; Habib *et al.* 1970; Chaudri *et al.* 1971). Well known extensive saline depressions are found near Najaf, Shitatha, Kerbala (Abu-Dibbis) and in the Lower Jezira (Boara depression and others). Vegetation descriptions of selected sites have been given by Chapman (1960) and Abul-Fatih (1975). Detailed studies, combining data collected over large areas and enabling a first classification of halophytic vegetation types, are still awaited. Some saltbush vegetation types are grazed by camels (and to a minor extent by sheep and goats) during part of the year. Dried up shrubby species are being cut and collected as fuel.

ad v. Under *segetal and ruderal vegetation* the weeds of the cultivated fields and waste land may be grouped. Large species collections have been made and lists published (Chapman 1954; Guest 1966; Zohary 1973). A monograph on the weeds of Iraq was prepared (Hassawy *et al.* 1968). (Semi-) natural species as weeds in the cultivated fields have a function in the 'stubble grazing'. Their value as additional fodder is not easily assessed but should not be underestimated. Common genera include *Trifolium, Trigonella, Medicago, Vicia, Matricaria, Malva* and the grasses *Avena, Alopecurus* and *Hordeum* (Bor 1968). The deeprooting *Lagonychium farctum* is one of the most common and noxious weeds, especially in the irrigated area. Under favourable conditions and without cutting, dense impenetrable stands of a few metres high may develop. Outside the irrigated area low forms of this species are found on heavy soils, sometimes saline, even in the desert areas (Harris 1960). Another very common species of irrigated land, often found together with *L. farctum* is the camelthorn: *Alhagi maurorum*. Even these species, although often difficult to collect and handle, are locally used as fuel. Two species commonly occurring after local cultivation in the rainfed areas of the desert (and as such useful as indicators) are *Peganum harmala* and *Artemisia scoparia*. They cover vast stretches in the northern part of the Jezira and are a transition to the last category of vegetation types, mentioned below.

ad vi. *Desert and steppe vegetation types* in Iraq are by far more extensive than the other categories, mentioned above, together. They are found in the Western and Southern Deserts and Jezira, but also over less extensive areas in the Mesopotamian Plain which have never, or at least for a long time not, been irrigated or flooded. Part of the Falluja Desert between Baghdad and Falluja and the Ghurfa Desert between Tigris and Adhaim Rivers on one side and Diyala River on the other are examples. This vegetation and its utilization are the subject of the present study. The area studied in the field does not include the Ghurfa Desert. For a vegetation description of this part of Iraq reference can be made to Batanouny & Hilli (1973).

As pointed out by Guest (1966), a gradual transition exists between the typical desert and the steppe vegetation. In the Southern Desert, Western Desert and Lower Jezira an open cover of low shrubs and a few perennial grasses and forbs form the typical vegetation, often confined to areas which receive additional water through runoff. After the rains the open space between these plants can be filled up temporarily by an ephemeral vegetation composed of species that survive the long hot summer as seeds. The area is relatively poor in species, some 250–300 at the most according to Guest (1966), of which some 60% are annuals. Trees are absent with the exception of some isolated locations in wadibeds and planted ones in and near permanent settlements. Chenopodiaceae and Compositae are well represented in the shrub cover. *Haloxylon salicornicum, H. articulatum, Artemisia herba-alba* and *A. scoparia, Achillea* spp. and *Rhanterium epapposum* are amongst the most common species, in almost monospecific stands or as dominants in a mixed cover. Perennial grasses are rare with few exceptions, especially *Stipagrostis* spp. of which *S. plumosa* is the most common. The vegetation is discussed in detail in chapter V and following chapters.

In the Central and Upper Jezira, with a rainfall gradient from about 150 mm to 300 mm, the semi-desert shrub vegetation changes into what might be called a steppe vegetation. A somewhat denser cover is found here, including a thin sward of *Poa bulbosa, Carex stenophylla* and perennial forbs. Well developed vegetation is only found as relicts because most of this area is regularly ploughed.

In its diversity the natural and semi-natural vegetation of Iraq serves a number of purposes for the local population. It provides amongst others fodder for live-stock, fuel for heating and cooking, and materials for house construction. The presence of the plant material has always been taken for granted, but large areas of Iraq already reflect the human impact on the vegetation resource. For the (potential) forest areas the deterioration is obvious, for the (sub) desert areas the present study may contribute to an understanding of processes, resulting patterns and possible remedies.

CHAPTER II

LITERATURE REVIEW

II. LITERATURE REVIEW

II. A. Introduction

Original contributions to the knowledge of the desert areas of Iraq are scattered over a few books, journals and printed reports. Information on the ecology and utilization of the vegetation in particular is by and large limited to mimeographed, typed, or handwritten reports. Retrieval of this information is difficult. To facilitate future reference and as an introduction to the present study, a literature review is given. In this review emphasis has been laid on ecological work, in particular that of an applied nature.

Documented research on the natural plant cover of Iraq, including that of the desert areas, started about 400 years ago with taxonomical studies on plant collections brought together by explorers. In the 19th century, gradually more attention was paid to aspects of geographical distribution of the species. These first steps in phytogeography led to more ecologically orientated studies which have been undertaken since the beginning of this century. The first publications dealing with the vegetation of the desert areas from the resource point of view date back to only 35 years ago. In the observations of the various authors no strict seperation in the contribution to these fields of study can be made. The relevance of including in this review a survey of the many taxonomical and descriptive vegetation studies is debatable. They have been included in order to present as complete a picture as possible of the development of botanical studies in Iraq.

II. B. Sources

Guest & Blakelock (1954) compiled a first detailed 'Bibliography' of 'Iraq', which was later extended by the same authors in Guest (1966). They listed 'references to works dealing with or touching on aspects of the botanical exploration of Iraq . . .'. Brief supplements to this bibliography are being published in the later volumes of the 'Flora of Iraq'. The brief list of Kreeb (1961) and its annotated extension by Hadac (1965) give together 34 references to phytosociological studies in Iraq. A long list of references on geobotanical aspects of the Middle East is included in Zohary (1973). McGinnies attached to his compilation on the state of knowledge of the vegetation of the world's deserts an annotated bibliography, which proved extremely useful (McGinnies *et al.* 1967). A similar study on South West Asia dealing mainly with plant ecology was published by Delbès (in UNESCO 1955), while for vegetation descriptions and maps the UNESCO/FAO study 'Vegetation Map of the Mediterranean Zone' is a starting point (UNESCO 1969). The first bibliography on natural resources of Iraq was compiled by Srivastava (1969). References to work on the desert range resources of countries in the Middle East, listing many unpublished reports, were given by Ibrahim (1967). This source,

Winkworth (1969), and Kaul & Thalen (1971a) provide the most extensive lists of studies dealing with the range resources of Iraq, while the reference lists of the geographical studies by Wirth (1962) and Al-Kasab (1966) also give information on this point. For references on environmental conditions of the desert areas, Buringh (1960) on soils, and Parsons (1955 a,b,c,d) and van der Sluijs (1969) on hydrology and geology should be mentioned.

II. C. Review of available data

II. C.1. *Species collection and identification*

The present Republic of Iraq, in earlier centuries in its entirety or partly included in regions referred to as Northern Arabia, Mesopotamia, Babylonia, Assyria, etc., was covered by journeys of early explorers and plant collectors. Rauwolff has been mentioned as the first naturalist who visited the region and collected plants. The specimens he collected during his travel in 1574-1575 from Aleppo in Syria to Ana, Baghdad, Kirkuk and Mosul in Iraq are kept at Leyden and were published by Gronovius (1755) in his 'Flora Orientalis' (Guest 1966; Zohary 1973). Roux (1964) remarks that Rauwolff's travel was an 'isolated episode', and the next traveller referred to in Guest (1966), is Dapper (1680), more than a century later. A few collectors followed late in the 18th century and an increasing number in the 19th and 20th centuries. A list of over 150 names of plant collectors in Iraq with details on their itineraries, covering a period from the 16th century to 1965, has been compiled by Blakelock and Guest (in Guest 1966). More than half of the 150 persons listed, however, contributed only 'a few specimens' and another quarter 'some specimens'. The four names given of the collectors in the 18th century are J.G. Bruguière and G.A. Olivier in 1795, l'Abbe Sestini in 1781-1782, and A. Michaux in 1782, all of them travelling either along the Tigris or along the Euphrates Rivers and, therefore, of little interest for information on the desert regions. Another 18 collectors are listed for the 19th century of which P.M.B. Aucher-Eloy in 1835, F.R. Chesney in 1835-1837, C. Texier in 1835-1840, T. Kotschy in 1841 and J. Bornmüller in 1892-1893, are the most important. For details see Blakelock and Guest in Guest (1966).

All plants collected by the early travellers are in European herbaria, a.o. at Geneva, Paris, Vienna and Kew. The Rustam Herbarium, established by Guest, when resident in Iraq (1929-1933) with the help of others (more than 30 names of collectors are listed for that period), was the first collection of plant specimens to stay in Iraq. It was in 1946 incorporated in the National Herbarium. This herbarium was extended and had already a collection of over 25 000 specimens in 1967 (McGinnies *et al.* 1967). Specimens have been largely identified with the help of the Kew Herbarium, and the Institution has since been the most important source of reference in Iraq for plant identification. Other herbaria, established later in Iraq, are now found at University Institutions, the College of Agriculture and College of Science of Baghdad University, and the Universities of Basra and Mosul.

Apart from Guest, who did invaluable compilation work on the flora and vegetation of Iraq, including the desert areas, other persons should be mentioned who contributed substantially to the knowledge of these areas during the present century. Handel-Mazzetti (1912, 1913, 1914 a) published results from an expedition in 1910 to Mesopotania and Kurdistan. Zohary (1940, 1941, 1950) made extensive tours, partly accompanied by Eig (1935) and Guest. J.B. Gillett, Director of the National Herbarium 1946-1949 and his Iraqi successor Dr. Ali Al-Rawi collected single or jointly all over Iraq. Al-Rawi later continued his collection, often together with botanists from other countries, residing at Baghdad for a few years *e.g.* Guest during a second stay in 1952-1958 and Rechinger in 1956-1957. The first extensive collections from all over the desert areas, especially the Southern and Western Deserts were made by them in 1955-1957. A number of botanists made contributions to the floristic knowledge of Iraq, but relatively little of the desert areas, in the late fifties, early sixties. Most of them had direct relations with Iraqi herbaria, e.g. C. von Regel in 1953-1954, N. Polunin in 1955-1958, A.D.Q. Agnew in 1958-1963, and E. Hadac in 1959-1961. At the same time, however, collections were started in the desert areas in relation with range survey trips by Iraqi (A. Al-Rawi, S.Y. Serkahia, H. Al-Tikriti, M. Al-Khateeb) and foreign specialists (H.W. Springfield in 1953, H.E. Schwan in 1955-1956, G. Long in 1955-1956). Almost all of the collectors, mentioned above for the period since World War II, have published papers on their collections or selected taxonomic groups. When specifically dealing with Iraq, these are listed in Guest (1966). A detailed review of this taxonomic work falls outside the scope of the present study. Over the last decade increasing attention has been paid to taxonomy and a new generation of botanists now in charge of the Iraqi herbaria is regularly collecting in most parts of the country, occasionally including parts of the desert areas.

The specimens that have been deposited as a result of four centuries of travelling and collecting are important for the study of the desert vegetation. They are the only means for studying historical aspects of species distribution and as long as no detailed complete flora for the country has been published, their value for species identification by comparison can hardly be overestimated.

Boissier's 'Flora Orientalis' (1867-1888) has long been the outstanding reference for species identification of the region. In this monumental work most of the early plant collections from Iraq are incorporated. Several collectors afterwards enumerated and elaborated their collected specimens, e.g. Bornmüller and Handel-Mazzetti, but no attempt was made to complete a flora for the country, until rather recently. Blakelock, after having worked on the Rustam Herbarium, started in 1954 a series of papers under the title 'Notes on the Flora of Iraq, with keys', and expressed the hope:

> '. . . that the two series of papers ('The Rustam Herbarium, Iraq' and the present publication) taken together will provide a useful, if provisional flora of Iraq'.

The two series of papers became an important source of reference when work was begun on the 'Flora of Iraq', a joint project of the National Herbarium in Iraq and the Kew Herbarium. Sofar, four volumes have been published of the nine volumes planned (Guest 1966; Townsend & Guest 1966, 1974; Townsend *et al.* 1968). Volume one (Guest 1966): 'Introduction to the flora; an account of the geology, soils, climate and ecology of Iraq with gazetteer, glossary and bibliography', will remain for many years to come the basic reference for most botanical studies in Iraq and has proved invaluable for the present study. Dr. K.H. Rechinger started the preparation of a working flora of Iraq in 1956, when resident at Baghdad. He had to limit the area covered to 'Lowland Iraq', excluding the mountain area in the North, but including Jezira and Western and Southern Deserts (Rechinger 1964). No ecological information is given, but keys and species descriptions are provided and collection sites enumerated. For the time being it is the only recent flora covering the desert areas of Iraq. A.D.Q. Agnew initiated the preparation of the first regional flora, the 'Flora of the Baghdad district', but more than Part I, Monocotyledons, has not been published (Agnew 1962a). Finally, mention should be made of studies carried out on groups of plants, e.g. Medicinal Plants (Al-Rawi & Chakravarti 1964) and Weeds (Hassawy *et al.* 1968).

II. C. 2. *Vegetation descriptions and ecological studies*

A number of scientists listed in the previous section has not only collected and identified specimens, but also given descriptions of vegetation in relation to environment, brief and general (Nabelek 1923) or lengthy and detailed (Handel-Mazzetti 1914b). The last authors account is the first comprehensive vegetation description of the desert areas of Iraq. He distinguished in the plains of Mesopotamia (versus the mountains of Kurdistan) deserts in the south and steppes in the north. Each of them, desert and steppe, is subdivided into edaphic types, characterized as sand desert, gravel desert, gravel steppe, humus steppe, gypsum steppe, etc. Handel-Mazzetti gives species lists for each of the edaphic types, without however, any indication of their relative contribution to the communities.

Information on the vegetation of the northern part of the Jezira and of the foothills region was given by three authors surveying or studying in more detail the environmental conditions of the Moroccan Locust (*Dociostaurus maroccanus*) in its outbreak areas. Uvarov (1933) in presenting a physiographic description notes:

> 'Most of the level country is usually called desert, but this is something of a misnomer, since it represents only a seasonal desert, during the summer drought, while the winter and spring rains transform most of it into more or less acceptable pastures, with a short but fairly dense cover of ephemeral grasses which are soon burnt by the sun'.

He gives *Artemisia herba-alba* as the landscape-plant of the semi-desert in the altitudinal zone next to the Mesopotamian alluvial plain, and *Prosopis stephaniana* (Syn. *Prosopis farcta, Lagonychium farctum*) for deeper, perhaps more fertile soil. The major part of his paper, however, describes conditions in the foothills area. More details are given in Eig (1935). Following a general phytogeographical outline of the area, he discusses vegetation – locust relations centered around the most characteristic species combination, abbreviated as P.- C.- R., standing for *Poa bulbosa, Carex stenophylla, Ranunculus asiaticus*, which he studied in the Mosul liwa. As common associates are listed the bulbous *Gagea reticulata* and *Ornithogalum ulophyllum*, and the annuals *Minuartia picta, Filago spathulata, Koeleria phleoïdes, Adonis dentata, Helianthemum salicifolium, Garrhaddiolus hedypnois* and *Erodium cicutarium*. Bodenheimer (1943) published an account of about the same area, but merely listed plant species collected at each 'reservation' with their Arabic names. A first list of species collected in spring 1933 by Eig and Zohary in the desert of Basra and between Basra and Ur was published by Zohary (1941). It provides some general phytogeographical information and ecological notes on some leading species observed, including *Haloxylon salicornicum, Rhanterium epapposum, Zizyphus nummularia, Lycium barbarum, Tamarix articulata* and *Artemisia herba-alba*. The list of annuals included in this paper is still a good guideline for the most common species encountered in the area at the end of the rainy season. One year earlier, Zohary (1940) published a 50-page geobotanical analysis of the Syrian desert. This thorough study, rather than dealing with general geobotany, is an analysis from the phytogeographical standpoint and as such very valuable. Zohary interpreted the Syrian Desert so widely that virtually the entire desert area of Iraq is included. A later phytogeographical study by this author (Zohary 1950) lists all the plant species collected up to the early forties in Iraq with their distribution in the broad phytogeographical regions and includes a useful description of the Iraqi environment and certain characteristic species combinations. In a more recent, extensive, work (Zohary 1973) the author compiled his geobotanical knowledge of the Middle East. This work includes special sections on Iraq.

Phytogeographical accounts discussing Iraq have been published by von Regel (1953, 1954). Guest (1933) published a bulletin 'Notes on Plants and Plant

Products with their colloquial names in Iraq'. It is the first study of its kind in Iraq, and although in principle an alphabetical species list, it contains invaluable (ecological) notes for botanical workers in the country. Its function will only be overtaken when all notes have been incorporated in the Flora of Iraq, presently being published. In two later papers Guest (1953 a,b) discussed the environmental conditions in Iraq (physiography, climate, geology, soils) followed by an outline of species associations found in the various broad ecological regions, and the (mostly cultivated) plants of the country. Most of this knowledge is in a largely extended form incorporated in the 'Introduction' volume to the Flora of Iraq (Guest 1966). This volume for the first time encompassed the botanical knowledge on Iraq. Before it was published the summarized and general vegetation description of the country by Springfield (1954) proved sometimes useful.

Since the late fifties more detailed ecological work on selected areas has been reported. Haines (1957, 1958) studied autumn flowering of some annuals and perennials especially in the Falluja desert. Together with Agnew (Agnew & Haines 1960) he reported on the influence of perennials on annuals in the Jezira. Species distribution around *Astragalus spinosus* was considered and the positive relationship (likely due to higher organic material content and fertility of the soils, reduced wind velocity and thus lowering of water loss) outlined. In a later paper, Agnew (1961) pointed out the capacity of *Haloxylon salicornicum* to accumulate mounds of wind-borne debris and described the association of other species with these mounds. He showed that cyclic changes may take place leading to successional changes in the vegetation as a whole, ultimately incorporating the low tree *Zizyphus nummularia* and so giving a more complex community structure.

Phytosociological studies, relating plant communities to soils and physiography have been carried out by Batanouny & Sheikh (1972) and Batanouny & Hilli (1973). In the first paper the vegetation of four 'ecosystems' is described, *viz.* Desert Plains, Plateau, Wadis and River Plains, subdivided into a number of habitats identified by 'correlation to the differences in physiography, edaphic and biotic factors'. The floristic composition of one or two stands (in one case four) in each habitat is presented with a value of the combined cover-abundance scale according to Braun-Blanquet (1964) and an indication of soil texture. The second study is of a similar type, outlining 14 major plant communities in the Ghurfa-desert, distinguished on a rather subjective basis and named after the dominant species. A general paper on reconnaissance work in the desert is Monakhov (1972). The author states that:

> 'Even with strip reconnaissance of the vegetation in the Western Desert, it proved to be possible to establish a rather clear association of certain plant communities with lithologic-geological or hydro-geological conditions'*

It proved of little value for detailed reference.

Al-Ani & Haddad (1973) reported on flowering periods and life forms in the Falluja and Iskanderya gypsum desert.

---

* Translated from Russian

Relations between vegetation and salinity have been studied in the Mesopotamian Plain, but also in local saline depressions in the desert (Habib *et al.* 1970; Al-Ani *et al.* 1970a; Chaudri *et al.* 1971; Abul-Fatih 1975). The last study gives valuable diagrams of the excavated root systems of ten species occurring in the Najaf saline depression, where the author distinguished three community types named after the dominant species *Bienertia cycloptera, Seidlitzia rosmarinus* and *Salsola crassa-Zygophyllum coccineum.* Harris (1960) compared the distribution of three annual and six perennial species with climate and soil type. Salinity and degree of aeration proved to be most important, and mechanical penetration and water absorption by roots comparatively unimportant.

Ten years ago, the first M.Sc. studies for the Botany Dept., College of Science, Baghdad, were carried out in selected areas of the desert: Baiji sand dunes (Al-Hilli 1970), Ukhaidir area (Nasserula 1971) and Najaf saline depression (Abul-Fatih 1970). These studies are of a rather detailed descriptive nature and give extremely useful information on plant distribution in relation to habitat conditions. The authors are the frontline of a sofar lacking generation of Iraqi field ecologists studying the vegetation there where it should be studied.

Reports and publications on general descriptive and ecological studies, undertaken as part of the fieldwork of which the present study is another result, may finally be mentioned here: reconnaissance survey of the Southern and Western Deserts (Al-Ani *et al.* 1970b, 1974); data on exclosures (Thalen & Al-Mufti 1973) and on the most important shrub species *Rhanterium epapposum* (Thalen 1974a) *Haloxylon salicornicum* (Thalen 1974b) and *Artemisia herba-alba* (Kaul & Al-Mufti 1974).

The foregoing review may show that, with exceptions (Agnew 1961, Agnew & Haines 1960), the ecological work on the desert areas of Iraq has by and large been of a broad, mainly floristic, descriptive type, until rather recently. Thus the detection of changes in the vegetation cover over the last half century is made difficult because of the lack of comparable data.

II. C. 3. *Range resources and related fields*

It was already mentioned that Guest in his bulletins for the first time wrote about the vegetation cover and the separate species as a natural resource (Guest 1932, 1933). His notes cover species collected from all over the country. Gillett (1948 a, b, c) was the first author to emphasize the value of the vegetation cover of the arid rangelands in three mimeographed reports on (1) the desert vegetation (2) the steppe vegetation and (3) the pastures and wild fodder plants of Iraq. The first two reports are rather general; they pin-point the role played by the natural vegetation in livestock production also dealing with general ecological characteristics of desert and steppe. The last study, very useful, discusses the grazing provided by the different types of land. The author distinguishes three main groups in the plants of importance as fodder of deserts and dry steppes, *viz.*

(i) Perennial herbs whose perennating buds lie below, or at the ground level (geophytes and hemicryptophytes) (ii) Annuals (iii) More or less woody plants which persist from year to year above the ground (nanophanerophytes and chamaephytes).

The most important species in each group are listed. For the last group the following remark is made:

> 'The importance of the third group, the perennial woody or half-woody plants is that they, provide, or used to provide some grazing in the summer and autumn when the herbaceous plants have disappeared. The most important of these plants belong to the Chenopodiaceae and to the Compositae'.

The species referred to in this group are the subject of the present study. The remaining four groups of land which provide grazing, *viz.* irrigated cultivated lands, rainfed agricultural lands, oak forest zone and mountain pastures are briefly discussed with extensive species lists, including scientific as well as colloquial names.

In the fifties, range management specialists from the United States of America studied the rangelands of Iraq and advised on improvement (Springfield 1954, 1957; Bryan & Springfield 1955; Schwan 1955, 1956, 1957; Anderson 1958). Their reports, generally speaking, clearly list some outstanding problems on which the authors agree:

1) The ranges are generally in depleted condition, mainly as a result of destructive grazing (too early, too long, too heavy), harmful cultivation of non-agricultural lands and destruction of valuable woody grazing plants (shrubs) by people for fuel.
2) Livestock production is unsatisfactory and wasteful (too low yields per animal, high death rates, low reproduction rates, parasites, diseases).
3) Animal feed is in short supply, especially in summer and early winter and often inadequate year long.

Although several times mention is made of studies being initiated or undertaken e.g. in a number of fenced areas (see Thalen & Al-Mufti 1973), unfortunately no published results of such work on the *natural* vegetation were found. Advice on measures to be taken for range improvement had, therefore, to be based on experience in similar areas elsewhere. The range study programme carried out between 1950 and 1960 can also be traced from a number of mimeographed reports issued by the Range Management Section of the Ministry of Agriculture, listed in Ibrahim (1967). They contain very little original data on the natural ranges.

During the same period FAO carried out some consultants' missions in the region, resulting in reports. Amongst these, Long (1955, 1956) should be mentioned. As grassland and fodder expert he reported the preliminary results of a mission to the desert rangelands, undertaken to study vegetation development of pilot exclosures, get acquainted with the major vegetation types and map grazing areas. Together with Iraqi personnel he worked in the region of Nasiriyah, Busaiya and Khedr-el-Mai,

Zubair and between As-Salman, Shabicha, Nukhaib and Rutba. Although it is indicated that these areas (or rather itineraries) were 'mapped' at a scale 1:500 000, no such maps have been published. Six vegetation types (associations) were distinguished in the first region, four in the second. Valuable suggestions are made for management, repeated in Long (1957). Of the Iraqi authors who wrote about the natural rangelands Saadun Yousif Serkahia should be mentioned first. Serkahia (1956) covers the whole desert region, divided by the author into three parts, from north to south:

1. the smooth desert with a pebbly or gravelly surface extending to about 250 km southeast of Rutba.
2. the stony desert, covered by limestone and reaching to Busaiya, and
3. the sandy desert, most productive part, between Busaiya and Zubair.

The value of the report lies in brief characteristics given of the most important species that were encountered, viz. *Artemisia herba-alba, A. scoparia, Salsola rigida, S. vermiculata, Achillea fragrantissima, Haloxylon articulatum, H. ammodendron, H. salicornicum, Rhanterium epapposum, Astragalus spinosus, Teucrium polium, T. oliverianum, Zilla spinosa, Zizyphus nummularia, Poa sinaica, Aristida plumosa* and *A. obtusa*.

The untimely death of this author left a lot of well-planned quantitative work uncompleted and unpublished. His textbook on Range Management, published in Arabic after his death, contains a useful introduction to the Iraqi desert rangelands with data on important species (Serkahia 1972).

Al-Rawi, at the time Director of the National Herbarium, covered the desert vegetation and its use in some brief accounts in Arabic (cf. Ibrahim 1967). Mahmood (1967) studied and reported on the carrying capacity in fenced areas in the Western Desert. It is the first study of its kind in Iraq and clearly shows the need for more and better planned work along this line. Emberger (1958) published a generalized account of use and misuse of natural resources with examples from the Mediterranean area and Iraq. Van der Veen (1962) briefly reviewed livestock production in Iraq, based on his four years' experience in the country. In an earlier paper (van der Veen 1960a) he provides in a summarized form a very useful review of the grazing and fodder resources of Iraq, listing 27 references. From the animal production side a series of reports has been published, more often than not only in mimeographed form e.g. Ali (1955), Neale (1948), Al-Hashimi (1972).

A field closely related to use and management of the natural rangelands that has been comparatively well studied in Iraq is that of growing fodder crops. Experimental work was carried out and a good number of performance trials of introduced species and strains have been undertaken and reported (Brichambaut 1956; van der Veen 1957, 1959, 1960a, b; Springfield 1954; Post *et al.* 1968; Al-Amari *et al.* 1968; and others). The use of date-products and by-products has even been studied and proved useful (Fine *et al.*, undated; Sarsam *et al.* 1956).

On grazing as a form of landuse in Iraq, discussed in a broader context, Kellersohn

(1952) and the review by Lebon (1964) may be consulted. Detailed information, in this respect, is found in Wirth (1962) who used the 1952/3 census data as the basis for a book, including many maps. Al-Kasab (1966) covered in a thorough and comprehensive study the Bedouin settlement in the northern Jezira. In his introductory description of the physical geography of the region (in fact the whole of Iraq is treated) vegetation is very generally dealt with in a few paragraphs.

From the above discussion it may be clear that observations and studies on the vegetation of the desert rangelands of Iraq up to 1970 were almost exclusively of a qualitative, descriptive nature. For the present study more quantitative and/or production orientated work carried out in surrounding and nearby countries with comparable environmental conditions was very useful for reference (e.g. Kuwait : Kernick 1966 ; Syria : van der Veen 1965, 1967 ; North Africa : le Houérou 1959, 1969, 1975; le Houérou & Hoste 1977).

CHAPTER III

RESEARCH APPROACH

III. RESEARCH APPROACH

III. A. Organization

III.A.1. *Field data collection*

III. A. 1.a. Inventory of the desert rangeland vegetation

Between April 1970 and June 1973 about 40 fieldtrips ranging from three days to two weeks in duration provided field observations and other inventory data. The desert areas were systematically traversed and broadly surveyed along routes selected from available small scale maps. Repeated visits were often necessary to cover the seasonal aspects. Fifteen one-day trips were especially made for production measurements. A complete list of fieldtrips with brief itineraries can be found in Thalen (1973). Fig. III-1 shows the tracks covered in the period April–June 1973. Routes were travelled by four wheel drive terraincars, usually following the desert tracks or roads as directed by police guides or villagers well acquainted with the area and local conditions. Records of observations on vegetation types with an indication of soil type and use factor were made continuously on standard forms along the routes. Locations were noted down according to the distances read from the odometer of the car. Unknown plant specimens were collected for later identification.

This intensified reconnaissance survey programme was the follow up of observations carried out between May 1967 and December 1968 by Winkworth and Al-Mufti (Winkworth 1969) and in 1969 by Kaul, Al-Mufti, Al-Ani and others in the context of the UNDP – assisted research programme as outlined above in I.B.

III. A. 1.b. Quantification of selected stands and utilization studies

Measurements on stands in the field, initially to develop proper methods, started in January 1971 at sites selected during the reconnaissance survey trips undertaken the year before. Regular measurements on aerial biomass and production of a few key cover types started in spring 1971. This was continued until seasonal aspects had been covered by the end of 1972. At the same time sampling of plant material for laboratory analyses proceeded. Early in 1972 quantitative characterization of plant communities was started, and continued until June 1973. This characterization included soil sampling and a brief description of the environmental conditions including the use factor.

Experimental work in the field (simulated grazing and fuel cutting) was started mid-December 1971. Other field studies aimed at understanding the present and improving the future condition of the desert rangelands included measurements on stands inside and outside fenced areas and observations on phenology and reproduction of selected key species. Programmes of introducing and testing exotic

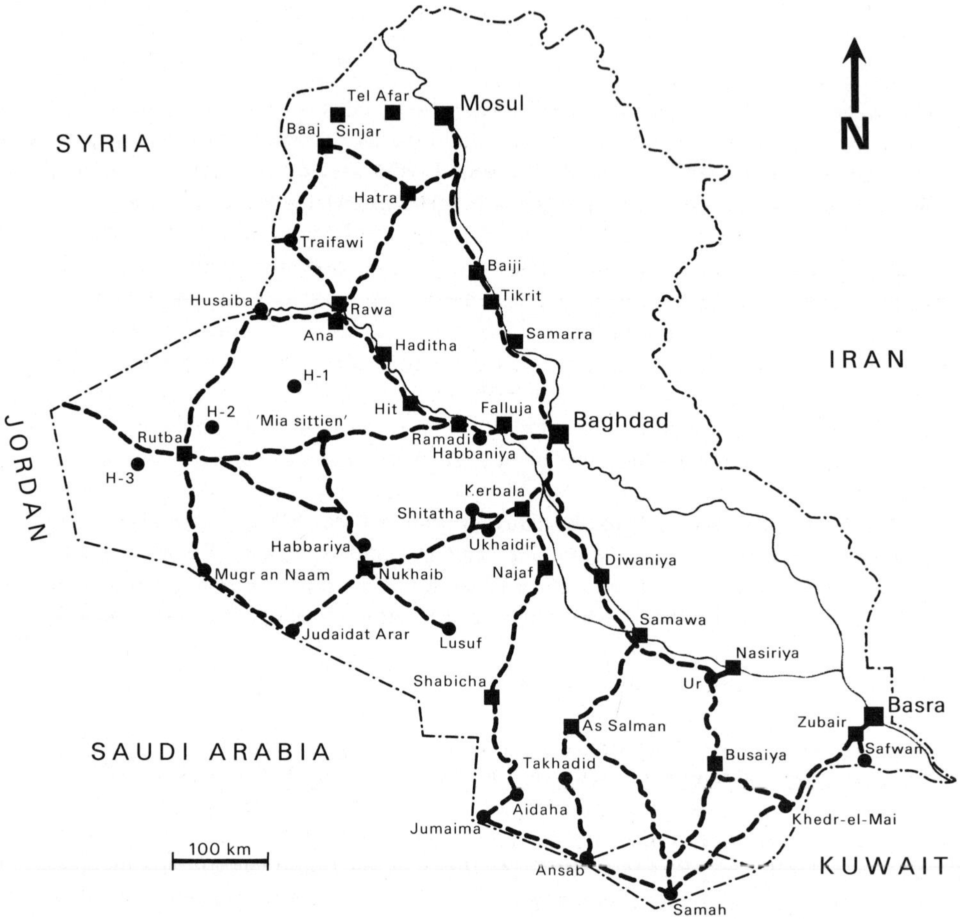

Figure III-1 Routes covered by fieldtrips for the present study between April 1970 and June 1973.

species from isoclimatic regions for range reseeding and a better forage supply, and of laboratory studies on aspects of the autecology of key species were run simultaneously with the programme outlined above. Most of the results of this work have been published elsewhere (Al-Mufti *et al.* 1970, 1971a; Clor *et al.* 1973, 1974 a, b). In the final stage of the fieldwork, May-June 1973, aerial photo interpretation and limited groundtruth collection was carried out in the Nukhaib-Athaman area, which had been selected and preliminary surveyed in February 1972.

III.A.2. *Data analysis, literature study and reporting*

Data analysis and literature study commenced soon after the first fieldtrip had been undertaken. Limited access to modern calculating equipment and library facilities were limiting factors in this first part of the study which was later compensated for.

Some preliminary results of studies on selected topics were reported in technical reports of the Project (Al-Ani *et al.* 1970 a, b, 1971a, b, c; Kaul & Thalen 1971 a, b; Thalen & Al-Mufti 1973). Intensive literature study, extended numerical processing and data analysis were carried out from July 1973 to 1978 in the Netherlands. Additional laboratory work on germination and study of aerial photography was also completed during this period. The literature study, personal communication with colleagues working in similar environments elsewhere, and visits to and fieldwork in (semi-) arid rangelands in Spain, North America, Tanzania, Mali and Botswana in the same years have been of great influence on the analysis, interpretation and final presentation of the data. They also partly account for the time lag between field work and publication date of this study.

III. B. Methods

III. B. 1. *Identification of plant species and nomenclature*

No up-to-date, complete, reference work for the identification of plant species in Iraq was available when the fieldwork was carried out. Main sources of reference were the published volumes of the 'Flora of Iraq' (Guest 1966; Townsend & Guest 1966; Bor 1968), and Rechinger's 'Flora of Lowland Iraq' (Rechinger 1964). Occasionally, Floras of neighbouring countries were consulted (e.g. Post 1932; Rechinger 1963-1970; Täckholm 1956). A limited reference collection of herbarium specimens was brought together and regularly consulted (Al-Mufti *et al.* 1971b). Invaluable assistance was rendered by the National Herbarium at Abu-Ghraib where dried specimens of unknown species were compared and identified with the help of staffmembers and the extensive collection kept there. Unless otherwise mentioned, in the present study nomenclature follows the only fairly complete reference volume for the country, the 'Flora of Lowland Iraq' (Rechinger 1964). Wherever considered useful, synonyms and vernacular names are given.

III.B.2. *Vegetation sampling*

Most plant populations, together constituting a particular vegetation type, are too large to measure the whole population. Estimates of the relevant vegetation attributes can be obtained from representative samples. The values arrived at in this way are subject to error and bias and are indicated as estimates. The accuracy of these estimates depends amongst others on the size, shape and number of sampling units.

The sampling methods adopted for the present study varied considerably, depending on the object of study and study objective. Selected plant units of individual species were sampled as well as stands as a whole. Shape, size, number and distribution of sampling units used for the various special purposes are given under the sections concerned. For most observations preprinted observation sheets were used.

III.B.3. *Measurement of vegetation attributes*

III. B. 3.a. Presence or absence, frequency of occurrence

Records were made on presence or absence of a certain species within an area (sampling unit, quadrat) under examination. A species was recorded present if at least (part of) a plant unit in vertical projection was observed inside such an area. After examining a number of areas, the number where a species was found present could be compared with the total number of areas examined, giving the (shoot-) frequency of the species, a proportion with values ranging from zero to one. It is common usage to express such values as a percentage by multiplying the frequency by 100, giving the per cent frequency. It may be mentioned that frequency values, although easy to obtain and objective, depend on a number of variables, e.g. size of the sampling unit, size of the plant unit, number of plants per unit area and the pattern of distribution of the individual plant units. This has to be considered in an interpretation of the data and restricts its usefulness (Perry 1962; Greig-Smith 1964).

III. B. 3.b. Number of plant units, density

When the limits of an individual or plant unit have been defined, the number of such units per unit area can be determined. Observations on this attribute, the density, were made in most stands where detailed work was carried out. Usually the individual plant units could be clearly delineated. Whenever special criteria were adopted for this purpose, this is mentioned.

III. B. 3.c. Height

Height can be measured as an attribute of single plants as well as vegetation types. In the latter case it is often a value derived from measurements on single plants. For the present study height measurements were taken with a steel tape (scale in cm and mm), reading the perpendicular distance from the outer end of the highest branch or stem to the groundsurface. Hummocks formed by soil particles accumulated around the plants sometimes provided difficulties, but were taken into account, in such cases reading the distance to the hummock surface.

III. B.3.d. Area or cover

Three different interpretations of area occupied by vegetation or cover are most commonly used in vegetation studies, *viz.*

basal area or stem cover – the area occupied by the bases of the plants, in trees often measured as diameter at breast height;

(foliage) cover – the area of the vertical projection of the aerial living parts;

canopy cover or crown cover – the area within the vertical projection of a line just circumscribing the plants.

Basal area finds its first application in forestry studies and is of little use in desert shrub vegetation types. Foliage cover takes into account the open space between leaves and branches. It is a widely used vegetation attribute, which can be estimated visually or measured by using a point quadrat method. Both techniques have their restrictions e.g. difference between observers in estimating and time needed in actual measurement. For a detailed discussion see Brown (1954), Perry (1962) and Mueller-Dombois & Ellenberg (1974). Canopy cover measurements in desert shrub vegetation can be made easily and accurately, because of the usually wide spacing and clearly defined circumference of the individual shrubs (Fig. III-2). Canopy cover was determined extensively for various purposes in the present study. Two techniques were used, line intercept and diameter method.

The line intercept method was applied as described in Brown (1954). In the crown diameter method two diameter measurements on the crown spread of individual shrubs are used (Thalen 1974a). The first measurement gives the maxi-

Figure III-2. Stand of *Rhanterium epapposum,* illustrating the suitability to delineate clearly the circumference and determine crowncover by either line intercept or diameter method.

mum value of the diameter (D_1). The second is taken perpendicular to the first, giving the maximum value in this direction (D_2). Crown cover (cc) is then calculated, using the formula for the area of an ellipse.

$$cc = \frac{1}{4} \pi D_1 D_2$$

III.B.3.e. Volume

Crown volume (cv) was determined from crown cover and height (h) estimates, as shown in Fig. III-3, using the formula:

$$cv = {}^1/_6 \pi D_1 D_2 h$$

The attribute was mainly used in relation to aerial biomass and production estimates (see chapter XIII).

III. B. 3.f. Weight

Weight determinations on plant material were carried out for the present study on either clipped material of individual species or plants, or on mixed vegetation samples collected within sampling units. A simple balance (capacity to 10 kg, read to

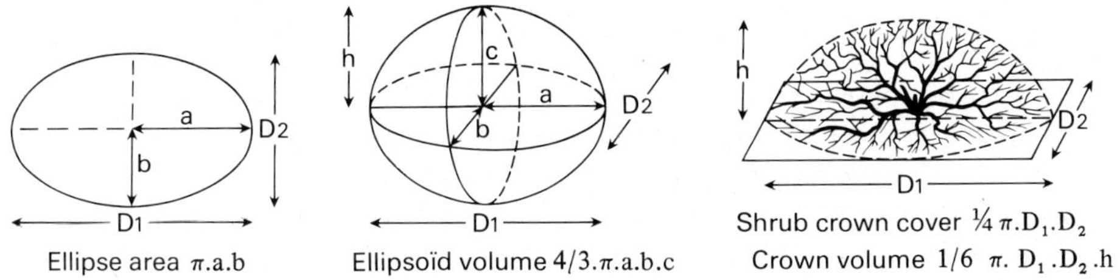

Figure III-3 Measurement of diameters and height in individual shrubs for calculating crown-cover and crownvolume as derived from the area of an ellipse and the volume of an ellipsoïd, respectively.

1 g precise) was used to determine fresh weight in the field immediately after cutting. Samples were taken to the laboratory in paper bags, dried for at least 24 hrs at 70 °C in a forced-draft oven and then the dry weight was determined.

III.B.4. *Chemical analysis of plant material*

In a number of cases plant material was collected in the field, dried as mentioned above, and further treated for chemical analysis. The dried tissues were for this purpose ground to pass a 2 mm sieve. The ground material was then dry-ashed, as described by Johnson & Ulrich (1959). Potassium and Sodium content were measured by flame photometry, Calcium and Magnesium by EDTA (Johnson & Ulrich 1959). Protein was determined by Kjeldahl method (Jackson 1958), and Crude Fibre and Crude Fat by methods described in Horwitz (1960). Analytical results for plant material were expressed on dry matter basis.

III.B.5. *Experimental procedures*

Experimental work was undertaken in the field as well as in the laboratory. The procedures followed vary widely with the objective and are, therefore, discussed under the headings concerned.

III.B.6. *Soil sampling and soil analysis*

The amount of soils data collected was restricted by facilities in the field and in the laboratory, especially with respect to the number of samples which could be collected and processed.

In some cases where detailed or repeated vegetation studies were carried out, a pit was dug, the profile examined, and samples taken from each layer for analysis in the laboratory. On most study sites, however, soil samples were taken (in duplicate) from two depths only: the topsoil, 0-30 cm depth, and the subsoil, from 30 to 60 cm. The samples were collected and transported to the laboratory in nylon bags. Before analysis they were air-dried and passed through a 2 mm sieve. Mechanical and chemical analysis were carried out in the Soil Physics Laboratory and in the Central Chemical Laboratory of the Institute. The mechanical analysis was carried out routinely by sieving and pipette method and the soil textural class expressed according to the USDA system (USDA 1975). The pH and electrical conductivity (EC) were determined in saturated soil extracts and gypsum by precipitation with acetone, following Richards (1954). Other procedures for soil chemical analysis followed Jackson (1958).

III.B.7. *Selected environmental data*

Meteorological data for selected stations were kindly provided by the Iraq Meteorological Department. Occasionally observations on air and soil temperatures were made and once a thermo-hydrograph was used in the field. All other environmental data (geological, hydrological) has been taken from existing reports and other publications as shown by reference.

III.B.8. *Numerical processing*

Processing of numerical data was hampered in the early stages of the study by lack of proper calculation facilities. Paper and pencil and mechanical desk calculators were the tools. This largely improved in later phases of the study and most of the results presented were obtained by using electronic desk calculators. In some cases a PDP 11/45 computer assisted, programmed in FORTRAN or BASIC. These cases are mentioned separately.

PART TWO

Ecology

CHAPTER IV

ENVIRONMENTAL SETTING

IV. ENVIRONMENTAL SETTING

IV.A. Introduction

Compilations of knowledge on climate, geology, hydrology and soils for the Jezira and the Southern and Western Deserts have been given in a number of country-wide studies as well as in local studies on a particular aspect of these areas (Buringh 1960; Guest 1966; UNESCO 1975; Parsons 1955 a, b, c). Descriptions of the biotic factors, particularly animals and man, are not so common.

For an understanding of pattern and process in the desert shrub vegetation, both the abiotic environment and biotic factors are important. They are covered in this chapter. Special emphasis is laid on the ever changing situation of man's impact in the area. The chapter aims at giving the overall picture of the environmental setting of the desert shrub vegetation in Iraq, rather than describing the Jezira, Western and Southern Desert regions in great detail. For such more comprehensive detailed descriptions, especially of the physical environmental conditions in the area, reference can be made to the literature mentioned under the headings concerned.

IV.B. Climate

IV.B.1. *Available data*

Meteorological observations in Iraq were recorded as early as 1888, but only about half a century later, in 1936, with the organisation of the present Iraq Meteorological Department, Directorate General of Civil Aviation, a firm basis for systematic observations in a good number of stations was established. A review of available data from early years can be found in MacFadyen (1938), Dennis (1953) and Guest (1966). Most publications dealing with Iraq's natural resources contain sections on climate, but lack of adequate, longterm and reliable records is constantly pointed out (Buringh 1960; Parsons 1955 a, b, c). A gradual improvement has taken place, reflected in the publications of the Meteorological Department (Meteorological Department 1962), where a reasonable data base is now available with some records covering 30 years or more. Based on these data, Gangopadhyahya has recently given a first account of the agroclimatological features of Iraq (in UNESCO 1975). Well equipped meteorological stations are now maintained at 17 locations and a start has been made with agroclimatological stations. Rainfall observations are available for about 75 points, but some of these are unreliable, of short duration, or both. Gangopadhyahya used data from 55 stations with a more or less acceptable record over 15 years.

The desert areas are extremely poorly represented in the meteorological observation network. In the Southern and Western Deserts long records have been kept only in Rutba (since 1929), while observations for short periods have been recorded in Busaiya, Nukhaib, As Salman and Shabicha, and pumping stations H1, H2, H3. A good number of stations are located at the fringe of the desert areas *e.g.* Sinjar, Tel Afar and Mosul in the Upper Jezira; Ana, Haditha and Habbaniya near the Lower Jezira; Kerbala, Najaf, Nasiriya and Basra near the Southern Desert.

This data base has been used for the account below, although in some cases other stations are mentioned for comparison.

IV.B.2. *Seasons*

For Iraq four seasons can be distinguished as follows:

Winter (December to February)

Winters in the area can be cold and are, compared to the other seasons, wet. About half of the annual rainfall occurs during this time of the year. The lowest temperatures are usually recorded in January; frost is not uncommon. Average minimum temperature for January is 2.5°C for Mosul, 7.0°C for Basra and 1.7°C for Rutba (Fig. IV–1). Weather during this part of the year is highly variable, depending on the flow of air masses. Cool, moist air can be brought east from the Mediterranean. Warm, moist air masses can come from the southeast, from the Arabian Gulf. Cold and dry polar continental air masses can also determine the weather.

Spring (March to May)

About one third of the annual rainfall occurs in this time of the year, often in heavy showers accompanied by thunder. Comparatively large amounts can fall within a short period of time (Fig. IV–5).
There is a gradual increase in temperature in March and April, but in May temperatures rise quickly. In this month mean minimum temperatures are for the stations in or near the desert areas 15.0°C for Mosul, 20.0°C for Baghdad and 23.7°C for Basra; absolute maxima have reached over 45°C for certain stations.

Summer (June to September)

The rapid change in weather during the month of May announces the beginning of the hot, dry summer. Normally no rain falls anywhere in the country in these cloudless months. July and August are the hottest. Temperatures over 50°C are regularly recorded and average mean temperature ranges between 30°C and 35°C. Tropical continental air, originating in Saudi Arabia and Africa determines these weather conditions.

Autumn (October and November)

In September temperatures drop quickly and cloudiness increases. Some 15% of the annual precipitation falls in this period, milder than in spring, not in torrents. Mean temperatures drop some 15–20°C between August and December.

IV.B.3. *Solar radiation and sunshine*

From May to September, Iraq receives amongst the highest incidents of solar radiation of northern hemisphere countries. In the peak of summer it may reach 800 $mWh.day^{-1}.cm^{-2}$, in winter it drops to about 240 $mWh.day^{-1}.cm^{-2}$, and under heavy cloudiness to as little as 40 $mWh.day^{-1}cm^{-2}$. The seasonal total radiation for three north-south selected stations is given in table IV–1.

Table IV–1. Solar radiation and sunshine for some selected stations in Iraq; seasonal and annual values (after UNESCO 1975).

Station		Mean total radiation ($mWh.day^{-1}.cm^{-2}$)					No. of years observation
		Spring	Summer	Autumn	Winter	Annual	
Mosul		563.0	678.9	361.1	299.8	502.2	4
Baghdad		629.1	737.7	405.2	342.3	556.0	7
Nasiriya		687.1	752.1	508.1	448.1	619.2	4
		Sunshine					
Mosul	A*	8.3	11.9	7.5	5.2	8.2	28
	%**	64	88	69	50	69	
Baghdad	A	8.8	11.6	8.1	6.8	9.1	30
	%	68	86	74	65	74	
Basra	A	8.8	10.7	8.5	7.3	9.1	30
	%	68	79	78	70	74	

* A – Actual; mean number of hours per day
** % – Percentage of possible duration

The same table shows the hours of sunshine. The whole desert area falls in a zone of about 70–80% bright sunshine a year of the possible sunshine duration. In the summer months this can reach 90%. Highest cloudiness is from December to April with December as the most cloudy month.

IV.B.4. *Temperature*

IV.B.4.a. Air temperature

Information on air temperatures measured for four stations in or near the desert areas is given in Fig. IV–1. The values range between the absolute extremes of -14.4°C measured at 7 January 1942 at Rutba and +51.1°C measured at 21 July 1937 at Mosul. Normal yearly temperatures in the area range from 19.6°C (Mosul) to 24.0°C (Nasiriya). For normal yearly maximum this is between 26.8°C (Rutba) and 31.6°C (Nasiriya), for normal yearly minimum 11.7°C (Rutba) and 17.8°C

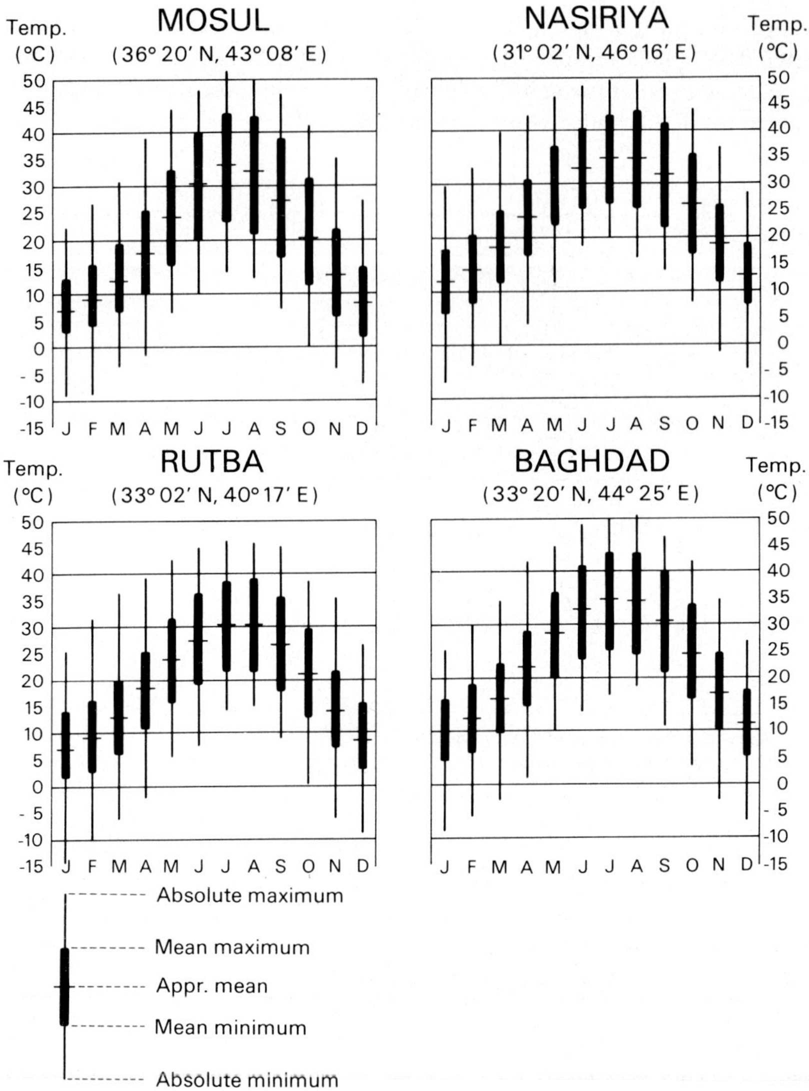

Figure IV–1. Mean and extreme temperatures recorded at Mosul, Nasiriya, Rutba and Baghdad.

(Basra). Although the range in these values is not so much, a great variation is possible, particularly in the transition months in spring and autumn. The range between absolute minimum and absolute maximum, measured for Rutba in March and October is about 40°C. Mean minimum temperatures for the coldest month (January) and mean maximum temperatures for the hottest (July) for the study area are shown in Fig. IV–2.

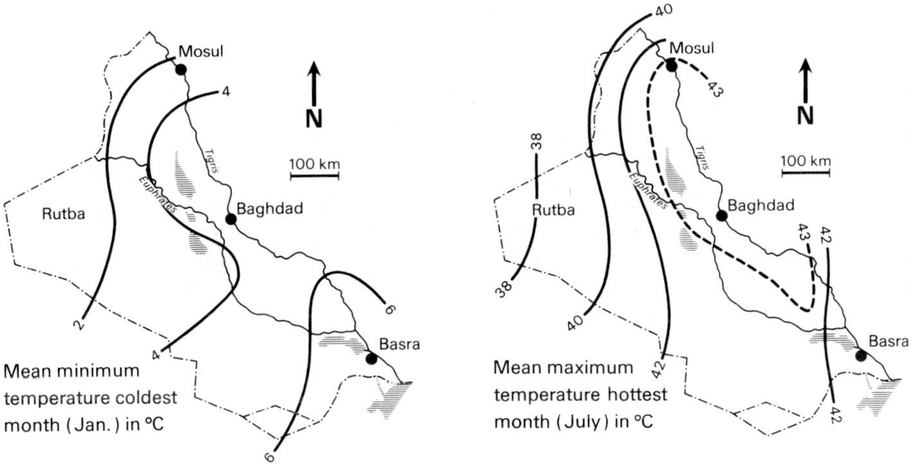

Figure IV–2. Mean minimum temperature of the coldest month (Jan.) and mean maximum temperature of the hottest month (July) in °C in the study area.

IV.B.4.b. Soil temperature

Monthly means of average daily soil temperatures at different depths can be read from table IV–2 for Baghdad and Rutba. It is obvious that temperature ranges

Table IV–2. Monthly means of average daily soil temperature (°C) at different depths (from UNESCO 1975).

Station	Depth. (cm)	Jan.	Febr.	Mar.	Apr.	May.	June	July	Aug.	Sept.	Oct.	Nov.	Dec.
Baghdad	10	10.7	11.2	17.6	25.0	31.0	34.9	36.1	36.1	32.8	27.2	19.9	11.8
(4 years)	20	11.7	11.5	17.4	23.9	29.4	33.6	35.4	35.6	32.8	27.7	20.6	13.1
	30	12.4	13.4	18.7	22.9	29.3	33.6	35.3	35.5	32.7	27.7	20.5	14.5
	50	14.2	13.7	13.0	23.6	28.5	32.8	34.8	35.5	33.5	29.3	23.3	16.8
	100	16.9	17.4	18.7	21.8	26.2	29.5	32.0	33.5	33.1	30.2	25.2	20.2
Rutba	5	6.8	10.6	19.0	20.0	29.2	32.5	34.4	35.5	31.8	25.3	15.0	8.4
(2–3 years)	10	6.1	9.6	17.6	18.7	27.3	31.0	33.0	33.8	30.5	24.5	14.7	8.2
	20	6.9	10.0	16.6	18.7	27.0	30.6	32.7	33.8	30.7	25.0	15.3	9.3
	50	9.1	11.4	17.9	18.6	26.3	29.7	31.9	33.1	30.9	26.3	18.2	18.2
	100	11.7	12.9	17.5	17.9	23.8	27.4	30.2	30.9	29.9	26.6	20.7	15.5

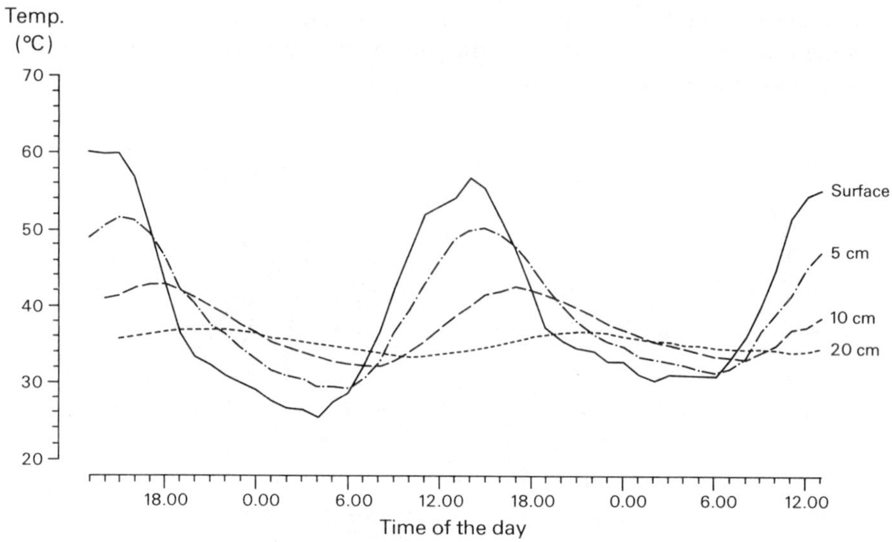

Figure IV–3. Soil temperatures at different depths, recorded over 48 hrs in mid-summer (6 – 8 August 1972) in a sandy soil in the Ukhaidir – Abu Dibbis area.

decrease with increasing depths. This can also be seen from Fig. IV–3. This figure shows the result of 48 hours continuous hourly measurements in a dry sandy soil during the hot summer season. At the surface, temperatures fluctuated between 60.5°C and 25.5°C, at 20 cm depth only between 33.5°C and 37.2°C.

The lower values recorded so far for the monthly means of average daily soil temperature at different depths are in the range of 5.0–10.0°C all in the first 50 cm. The higher values are in the order of 35.0–37.5°C all in the first 30 cm (UNESCO 1975). It should be emphasized that these values are based on only two to four years' observations for ten stations, certainly not fully representative for the desert areas.

IV.B.5. *Precipitation*

IV.B.5.a. Amount

Mean annual rainfall for Iraq as a whole has been calculated to be 260 mm following the Thiessen method and using the values of 66 raingauges (Hassan & Mashkour 1972). The whole desert area falls below this mean value. In the north – south precipitation gradient the northernmost stations in the Zagros mountains have values over 1000 mm, where Penjwin (35°26'N, 45°58'E, at 1311 m) with

1360 mm is an extreme, while for the Southern Desert values of only about 50 mm have been estimated.

Precipitation records over periods of at least ten years for stations in or near the Jezira, Western and Southern Deserts are given in table IV–3 as mean monthly and annual values. Of these 15 stations only Rutba (120.0 mm) and Nukhaib (56.2 mm) are situated in the desert areas, all others at the fringe and near the rivers. Mean values of records kept for a few years only, available for some other stations in the desert, show that in the northern part of the Western Desert rainfall ranges between 100 mm and 150 mm, while in the Southern Desert means of only 50 mm to 100 mm are normal (table IV–4). Isohyets for the mean annual rainfall in the area concerned are given in Fig. IV–4a.

IV.B.5.b. Seasonal distribution and variability

Isohyets for precipitation in autumn, winter and spring are shown in Fig. IV–4 b, c and d. Low predictability of rainfall in arid areas is a well known phenomenon. For Iraq this is illustrated in table IV–5 for seven stations with long observation records. A coefficient of variation of about 0.40 was found for the variability in mean annual rainfall. The stations in the north show less variation than those in the southern part of the area. For the Southern and Western Deserts 60% probability has been calculated for an annual rainfall equal or less than 100 mm. For the Jezira there is a 60% probability for a rainfall between 100 mm and 250 mm or more. The variability is perhaps best shown in Fig. IV–5, where two conse-

Table IV–3. Mean monthly and annual precipitation for 15 stations with records of ten years or longer in or near the Jezira, Southern and Western Deserts.

Station	Location		Elevation (m a.s.l.)	Mean monthly precipitation (mm)								Mean annual prec. (mm)	No. of years obser-vation
	N	E		Oct.	Nov.	Dec.	Jan.	Febr.	Mar.	Apr.	May		
Mosul	36°20'	43°08'	223	10.2	36.5	65.6	67.7	64.6	69.6	50.8	24.9	390.2	30
Sinjar	36°19'	41°50'	476	8.8	31.6	70.8	77.7	60.7	67.0	53.0	33.1	403.2	22
Baiji	34°56'	43°29'	143	1.5	23.5	26.9	34.6	24.2	33.5	25.3	8.5	190.0	13
Haditha	34°07'	42°23'	140	7.8	16.5	16.2	17.9	15.2	16.3	9.2	5.9	122.9	10
Hit	33°39'	42°50'	58	1.1	25.0	16.0	24.0	15.4	20.7	20.3	7.1	119.6	18
Ana	34°28'	41°58'	150	5.3	9.8	22.5	18.6	16.9	20.4	22.2	5.7	121.6	22
Baghdad	33°20'	44°25'	34	3.7	17.2	22.9	25.3	24.4	22.7	22.3	8.1	146.9	30
Habbaniya	33°23'	43°33'	44	2.3	20.2	13.1	27.9	12.0	10.5	19.9	6.2	111.9	30
Rutba	33°02'	40°17'	615	5.4	13.3	16.3	13.6	13.6	15.4	17.6	15.0	120.0	29
Nukhaib	32°02'	42°15'	305	1.9	7.8	5.4	10.1	5.0	1.5	5.7	2.3	56.2	11
Kerbala	32°37'	44°02'	29	0.3	8.9	19.3	16.7	15.9	11.5	18.2	3.9	94.3	10
Diwaniya	31°59'	44°55'	20	3.9	15.5	20.2	21.2	15.0	16.9	17.8	8.4	119.0	30
Nasiriya	31°02'	46°16'	3	2.2	16.8	20.4	19.2	13.4	15.8	16.6	7.1	111.5	30
Samawa	31°18'	45°17'	6	2.6	25.1	25.6	17.5	16.8	7.8	9.7	5.2	130.2	11
Basra	30°33'	47°48'	2	1.0	22.8	30.2	22.8	13.8	20.2	20.4	7.8	139.5	30

Table IV–4. Annual rainfall for six stations in the Western and Southern Deserts with only few years records; mainly derived from Guest (1966).

Station	Location N	Location E	Elevation (m a.s.l.)	Annual rainfall (mm) Max.	Min.	Mean	Source
H–1	33°49'	41°28'	409	233	76	132	Oil Company 1933–44
H–2	33°17'	40°35'	593	196	66	113	Oil Company 1933–44
H–3	32°55'	39°42'	778	249	47	123	Oil Company 1933–44
Shabicha	30°45'	43°40'	287	149	94	114	Police 1935–40
				–	–	51.9	Met. Service 2 years after 1940
As Salman	30°28'	44°43'	202	128	26	88	Police 1935–40
				–	–	71.4	Mentioned in Parsons (1955b)
Busaiya	30°08'	46°09'	144	189	36	95	Police 1935–43
				–	–	83.2	Mentioned in Parsons (1955b)

cutive years of daily rainfall observations are shown for the Kerbala station. In two consecutive days (11 and 12 April 1971) more than 50% of the total rainfall for the 1970/71 season was recorded. In the next season (1971/72) more than 50% of the total rainfall again fell in three days only (January 21, March 15 and April 30).

For a number of stations the maximum rainfall recorded within 24 hours exceeds the absolute minimum total annual rainfall. For some stations with long records the frequency of daily rainfall has been worked out for various class

Table IV–5. Annual rainfall variability characteristics for selected stations in or near the desert areas with long observations records (based on UNESCO 1975).

Station	Rainfall (mm) Mean	Median	Max.	Min.	Standard deviation	Coefficient of variation	No. of years obs.
Mosul	390.2	372.1	603.5	208.2	140.0	0.36	30
Sinjar	403.2	–	669.6	199.7	151.1	0.37	22
Rutba	110.0	94.8	229.4	46.9	48.6	0.44	29
Baghdad	146.9	129.9	336.0	72.3	57.0	0.39	30
Diwaniya	119.0	111.2	270.3	47.1	43.6	0.37	30
Nasiriya	111.5	96.0	221.1	27.8	47.2	0.42	30
Basra	139.5	120.8	319.5	31.9	61.7	0.44	30

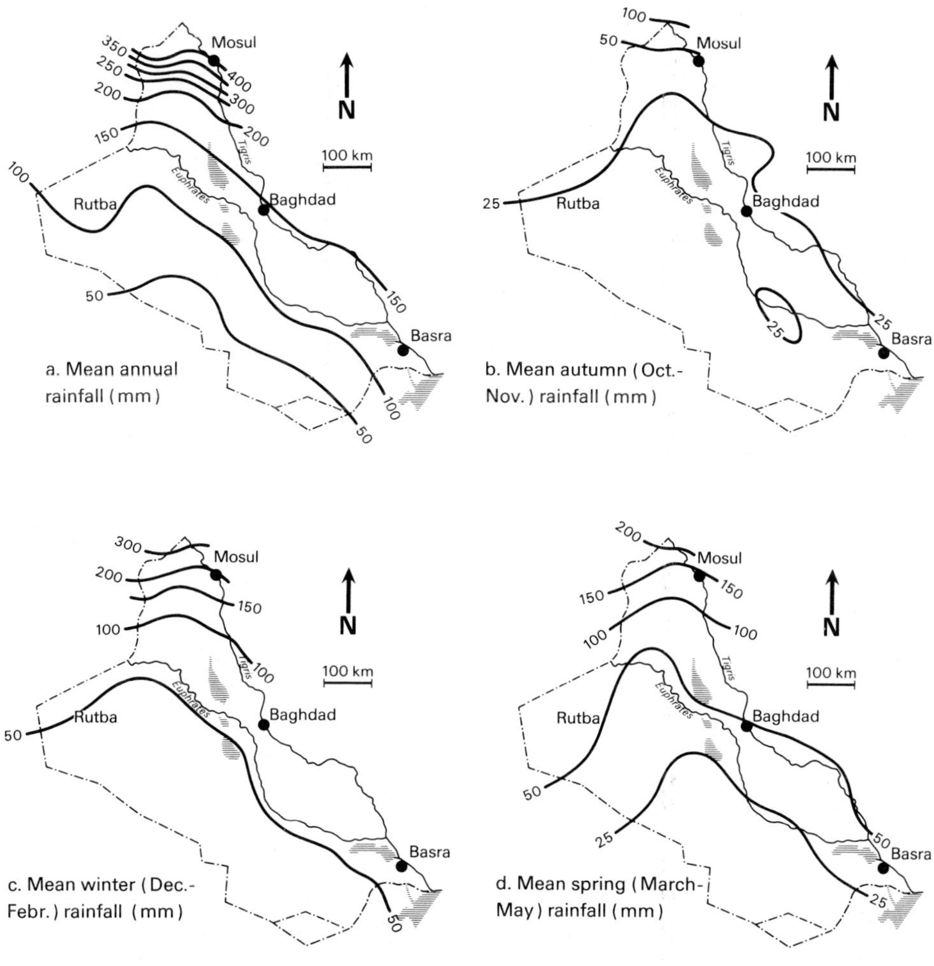

Figure IV–4. Mean annual and mean seasonal rainfall in the study area.

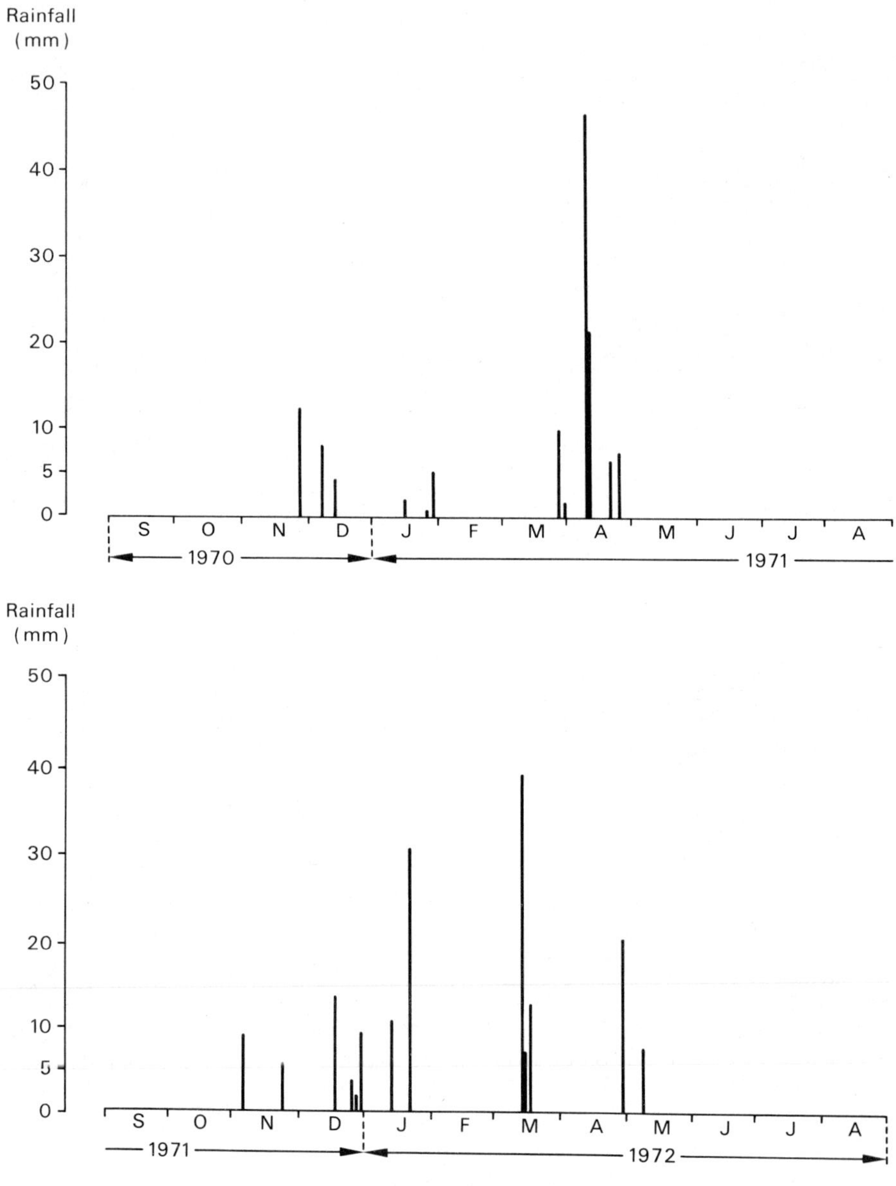

Figure IV–5. Two consecutive years of daily rainfall observations for Kerbala station.

intervals (table IV–6). The results show that the annual mean number of rainy days for the Jezira is from about 50 in the Lower Jezira to 100 in the Upper Jezira. For the Western Desert this is in the range of 25 to 50 and for the Southern Desert probably even less. Records do not allow more pertinent conclusions.

Table IV–6. Mean number of days per year with a rainfall in various class intervals for four stations in or near the desert areas (after UNESCO 1975).

Station	Days per year with rainfall per class (mm)					Days per year with any rain	No. of years obs.
	tr.–10	10.1–30	30.1–50	50.1–80	> 80		
Mosul	77.7	20.6	1.4	0.1	0.0	99.8	30
Rutba	55.6	3.5	0.2	0.0	0.0	59.3	29
Diwaniya	35.3	2.6	0.4	0.2	0.0	38.3	30
Basra	47.4	3.7	0.4	0.1	0.0	52.2	30

Snowfall in Iraq is virtually limited to the mountainous area in the north. Of the stations in and near the desert areas the average number of days annually with snowfall is: Rutba 0.4, Habbaniya 0.3, Diwaniya 0.2, Nasiriya 0.1, and Basra 0.0 (UNESCO 1975). The snow in these areas melts away quickly after reaching the ground.

IV.B.6. *Air humidity*

Relative humidity is a measure of the moisture in the atmosphere. The degree of saturation from the relation vapour pressure divided by saturation vapour pressure is expressed as a percentage. This relative air humidity can reach extremely low values in Iraq during the hot dry summer months. Mean monthly values can be less than 20% and values between 20% and 30% are normal in June, July and August. Table IV–7 gives details. Humidity is lowest for the desert areas in central and western Iraq (Baghdad, Rutba) and increases towards the Arabian Gulf in the southeast where for Basra relatively high values have been recorded. Mean monthly relative humidity there hardly drops below 50% (49%, 49% and 48% for June, July and August, respectively). Air humidity was measured hourly during two consecutive days in August 1972 in the Abu Dibbis-Ukhaidir area (Fig. IV–6). Between 1400 and 1700 hrs local time humidity dropped to 8% the first day and to 12% the second day. 26% was the highest humidity recorded between 500 and 700 hrs local time. The figure clearly shows the sinusoïdal shape of the curve. There is an inverse relation with the air temperature plotted in the same figure.

Table IV–7. Normal monthly and annual relative humidity for selected stations in or near the desert areas. Data kindly provided by Iraq Meteorological Department; period 1941–1970.

Station	Normal* relative air humidity (%)												
	Jan.	Febr.	Mar.	Apr.	May	June	July	Aug.	Sept.	Oct.	Nov.	Dec.	Annual
Mosul	82	76	70	62	44	30	26	29	34	49	67	83	54
Rutba	68	57	48	40	33	25	24	23	27	34	53	67	42
Habbaniya	73	62	55	47	35	26	25	27	30	39	61	73	46
Baghdad	71	62	52	44	31	23	23	24	27	36	56	71	43
Diwaniya	70	59	51	44	33	29	28	29	32	39	56	70	45
Nasiriya	66	58	51	44	36	33	30	28	27	35	53	66	44
Basra	78	70	64	58	53	49	49	48	50	56	69	78	60

* Mean of eight daily observational hours.

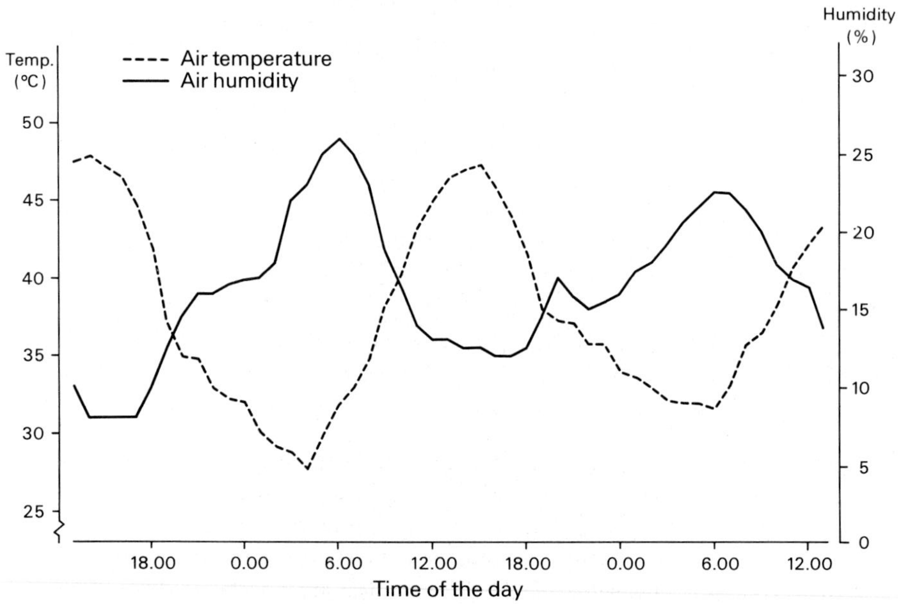

Figure IV–6. Air humidity and air temperature recorded over 48 hrs in mid-summer (6 – 8 August 1972) in the Ukhaidir – Abu Dibbis area.

IV.B.7. *Evaporation*

Evaporation in Iraq is measured by class-A pan. Preliminary data only, for a few stations, are available. As far as relevant for the present study data concerning the desert areas are given in table IV–8. In the wintermonths daily mean evaporation ranges between 1.5 mm and 3.7 mm. There is a rapid increase in spring and during the summer months evaporation is almost ten times as high as in winter. Autumn values are similar to those of early spring months, roughly between 3.0 mm and 9.0 mm per day. These values result in an annual total of over two metres. For Basra (high humidity, see above) the lowest total was recorded, 2044.2 mm. For Baghdad an amount of 3321.5 mm is given.

IV.B.8. *Subdivision and main climatic features*

Subdividing the study area into well defined territories with distinct climatic variants is not yet possible due to the fact that observations over longer periods are only available for very few stations within the study area. There is, however, a conspicuous difference between the climatic conditions in the Southern Desert and the Western Desert, compared to the Jezira. This can be seen from the quantity of rainfall (Fig. IV–4), the number of rainy days (table IV–6), extreme temperatures in summer and duration of the dry period (UNESCO 1975). The difference is clearly expressed in the climatic diagrams as constructed by Walter and Lieth (1960), which show the relationship of temperature with precipitation in a fixed proportion to each other: 10°C corresponding to a precipitation of 20 mm. By this relationship of temperature and precipitation unfavourable seasons due to water shortage are characterized. On the basis of the diagrams a climatic classification has been made. For Iraq the upper part of the Jezira is given as type IV (Mediterranean, winter rains), the rest of the study area as type III (subtropical, hot and arid). A more or less similar subdivision is found when other systems of climatological classification are used. Al-Kasab (1966) applied the aridity index of de Martonne (1926) and found the entire study area to be either arid or extremely arid with a transition in the Jezira to semi-arid. The same boundaries can be seen on the small scale vegetation map in Guest (1966) where from north to south a 'Moist Steppe zone', a 'Dry Steppe zone' and a 'Sub-Desert zone' are distinguished. It can also be seen in Zohary (1973) where the first two are combined in a climatic 'Steppe variant' and the third is called 'Desert variant'. With the present network, or rather lack of a network, of meteorological stations within the desert areas, all more detailed subdivisions must be regarded as highly speculative, based on unjustified extrapolations and interpolations.

The features of the climatic desert variant, *viz.* very low, erratic, rainfall (50–200 mm) a long, hot, dry summer with no rain at all and low relative air humidity (table IV–7), define very harsh conditions for plant growth. In the hot summer months surface water is lacking and with the dry air only the special adapted species can survive in this environment. They belong by and large to two groups, the ephemeral annuals and the deep rooted xerophytic, mostly woody, perennials. The first group germinates when the rains come, completes the life cycle in a few weeks only and survives the dry summer in a dormant stage as seeds. The second group has a number of adaptations, but most important is a deep root system enabling them to draw water from greater depths.

Table IV–8. Mean daily (per month) and mean annual evaporation for selected stations in or near the desert areas, measured by class A. pan evaporimeter (after UNESCO 1975).

Station	Evaporation (mm)														No. of years obs.
	Mean daily													Total annual	
	Jan.	Febr.	Mar.	Apr.	May	June	July	Aug.	Sept.	Oct.	Nov.	Dec.	Annual		
Mosul	2.1	3.0	4.3	6.5	9.7	14.5	21.1	17.7	11.7	6.0	3.6	2.4	8.6	3139.0	8
Rutba	2.1	3.7	5.7	8.6	11.3	15.7	18.1	17.0	13.7	8.5	4.3	2.6	9.3	3394.5	7
Ana	1.5	2.8	4.4	6.1	9.1	14.0	15.5	13.7	9.4	5.8	2.6	1.6	7.2	2628.0	5
Baghdad	2.4	3.6	6.0	8.7	11.9	16.4	18.6	15.3	11.7	7.4	4.3	2.3	9.1	3321.5	9
Diwaniya	2.0	2.4	4.0	5.2	7.4	9.9	11.0	10.3	8.2	5.2	3.0	2.0	5.9	2153.5	8
Nasiriya	1.8	2.7	4.5	6.2	8.2	10.7	11.2	11.0	8.9	5.7	3.0	1.9	6.3	2299.5	8
Basra	2.1	2.8	4.7	6.2	8.1	9.1	9.3	8.7	6.9	5.0	2.8	2.0	5.6	2044.2	8

Conditions in the remaining part of the study area, the northern half of the Jezira under the climatic steppe variant are more conducive for plant growth. Higher rainfall (200–500 mm), a rainy season of at least six months, a lower annual potential evapotranspiration (less than 1750 mm, UNESCO 1975), are reflected in an increased number of plant species in this area compared to the desert areas in the south. However, for the same reasons the Upper Jezira has become the centre of Iraqs extensive dry farming, whereby its value as rangeland has been reduced to almost nil (see chapter IX).

IV. C. Physiography

IV.C.1. *The overall structure*

The Southern Desert, Western Desert and Jezira have a rather simple structure. In the geological map reproduced in Guest (1966) the area is divided into six units of which two, Quaternary sanddune belt and Jurassic/Cretaceous expositions in the Qa'ara depression, occupy only a small area. The geological map, scale 1:1 000 000 published in 1960, shows a few more units, but a far more intricate pattern.

Descriptions of the geology of the desert areas can be found in Mitchell (1956), Parsons (1955 a, b, c) and Ahmed & Kraft (1972). The following information is mainly derived from these sources. Far more detailed data is available for selected areas as a result of intensive hydrological and oil exploration studies. This information is of little significance for the present study.

The Southern and Western Deserts are the Iraqi part of the Arabian shield, with relatively undisturbed, flat lying geological strata. They form a rather flat plain, gently sloping in northeastern direction towards the River Euphrates. The highest point is Jebel Anaiza about 915 m a.s.l. near the junction of the borders of Iraq with Jordan and Saudi Arabia. The lowest point, just below sea level, is found in a marshy area near Ur.

Although in general flat, locally a number of structures may be found like mesas, escarpments, large depressions and wadis. On the basis of lithology, structure, topography and drainage, several physiographic units have been distinguisted, first by Raisz (1951), later slightly modified by others (Parsons 1955 a, b, c). They are shown in Fig. IV–7 and briefly characterized below.

IV.C.2. *Description of the physiographic units*

IV.C.2.a. Dibdibba Plain

This unit, of which only a minor portion from Saudi Arabia and Kuwait extends

Figure IV–7. Map of the physiographic units in the study area.

into Iraq, has in general a gently rolling topography. It covers the southeastern part of the Southern Desert, reaching as far west as Busaiya, with elevations ranging from about 10 m to over 300 m a.s.l. The surface is sandy and gravelly (Fig. IV–8). There are some rounded hills and shallow depressions and the area is underlain by the sands and gravel and pebbly beds of the Dibdibba Formation, probably of Upper Miocene age. Rain seems to infiltrate quickly and the wadis are very wide and shallow. Drainage is on the one hand towards small, shallow, interior basins and on the other through very gently sloping small wadis towards the north to the flats bordering the Euphrates.

The Euphrates sanddune belt, discussed later, limits the northern part of the Dibdibba Plain. In the south, forming the border between Iraq and Kuwait, wadi

Figure IV-8 Example of an almost flat, sandy area in the Dibdibba Plain unit with a vegetation dominated by *Rhanterium epapposum*.

Figure IV-9 The most general landscape of the Al Hajara Stony Plain unit, a rolling to low hilly area, dissected by some wide vadis, with a rough stony or gravelly surface.

Batin is the marked feature. It is a depression of several kilometres wide, extending far into Arabia, with a fine textured bottom soil. At Al Haniya ridge, or perhaps slightly farther west, roughly along the line of deep wells Busaiya, Tuquiyid to the southern apex of the Neutral Zone, the desert changes in aspect. Here the next unit begins.

IV.C.2.b. Al Hajara Stony Plain

Named after the sharp angular fragments of limestones and chert which cover most of its surface, this unit extends as a 100–200 km wide zone from the Neutral Zone in a northwestern direction, roughly parallel to the Euphrates, to approximately 100 km from the Syrian border. It is a rather flat, slightly rolling desert, rising towards the west from less than 200 m to almost 500 m a.s.l. (Fig. IV–9). The underground is formed mainly by limestones and some marls of Eocene age. The many shallow depressions may have been formed by the collapse of roofs of caverns, resulting from groundwater dissolving the limestone. The depressions of Shabicha and As Salman are examples. Several large wadis cross the area in a northeastern direction: wadi Al-Khir, wadi Ubaiyidh, wadi Ghadef and wadi Hauran. They have their main catchment area in the Upper Wadian and either discharge into the River Euphrates or in depressions in the Lower Wadian unit. Most water, however, disappears before reaching this point. Drainage in this unit is mainly internal. To the southwest are escarpments, erosional, but perhaps also resulting from some folding along the Al-Batin. The sharp stony surface and the many depressions make this area of the whole Iraq desert the most difficult to traverse. The exposed stones are often coated with 'desert varnish', a veneer of mainly manganese dioxide and iron oxide, giving them a dark appearance (Engel & Sharp 1958).

IV.C.2.c. Upper Wadian

To the west of Al-Hajara extends an area with varied relief, due to a system of wadi branches, generally forming a dendritic pattern draining the area from a southwest to northeastern direction. The wadis have a different appearance, ranging from broad, shallow elongated valleys to steep-walled canyons, with cliffs up to 70 m high. Terraces are thought to be the result partly of regional uplift, partly of climatic fluctuations. The unit is by and large underlain by beds of limestone and sandstone of Cretaceous times. It forms the transition between the lower Euphrates area in the east and Syria and Jordan in the west. The surface of this area is hard and stony.

None of the wadis is perennial but water is often present at shallow depths beneath wadibeds. Wadi Hauran, coming from Jebel Anaiza, wadi Ghadaf with numerous tributaries from the plateau south of Rutba, and wadi Ubaiyidh with its tributary wadi Tibel are the most important wadi systems in the area. The Qa'ara depression, north of Rutba, and situated 80–100 m lower than the surroundings, is of particular interest from a geological point of view, because middle Triassic rocks are exposed there.

IV.C.2.d. Lower Wadian

In a strip of 40–90 km wide, this unit stretches between the Dibdibba Plain in the southeast, Al Hajara Stony Plain in the southwest, Upper Wadian and Syria in the northwest and the lower Jezira Plains in the northeast. It is a slightly rolling area, dissected by many shallow stream channels, which rise near the southwestern part of the unit and drain into the Euphrates River or the bordering flats and depressions. The depressions of Abu Dibbis and Habbaniya are the best known. Elevations go down to about 200 m in the northeast. Underlying strata dip in this direction and consist mainly of limestones with some marls and siltstones of Eocene to Middle Miocene age. Characteristic for this area are isolated remnants of hard resistant limestone, which withstand erosion better than the surrounding material.

IV.C.2.e. Al Hamad Plain

A small part of this Plain lies in Iraq, west of the Upper Wadian and extends into Saudi Arabia, Jordan and Syria. Elevations range from 700 m near the Syrian border to over 915 m a.s.l. at Jebel Anaiza. Apart from this mountain and the Al Walej and Hauran wadi systems, which drain the area, it is an almost featureless stony desert underlain by Eocene limestone.

IV.C.2.f. Euphrates Sand-belt and Ar Rahab

Special mention should be made of the only large sand dune area of the Iraqi deserts. Parallel to the River Euphrates from beyond Samawa in the northwest almost to wadi Batin in the southeast runs a dune belt of about 270 km long, ranging in width from a few kilometres to almost 40 km and in height from three to 30 m. In the northwest it fades away in low scattered dunes, in the southeast into a rather level sandbody. Apart from the crests of the highest dunes and some isolated dunes at the periphery the body as a whole seems rather well stabilized (Fig. IV–10 a, b).

A strip of rather fine textured, highly saline, flats known as Ar Rahab stretches between the sanddune belt and the physiographic units to the south. At some

Figure IV-10 The Euphrates sanddune belt is an impressive sandbody over vast areas, with (a) barren crests of ever moving sand and (b) slopes covered with a psammophytic vegetation as *Calligonum comosum* shown in this view.

locations, e.g. south of Ur and near Najaf it also extends north of the sanddunes, as a transition to the Lower Mesopotamian Plain.

IV.C.2.g. Jezira Plains

North of the Lower Wadian, from which it is separated by the drainage divide between Euphrates River and Jezira Desert, the Plains of the Jezira begin, extending from the Low Folds unit in the east and north (Jabel Hamrin and Jebel Makhel, Jebel Sinjar) into Syria. It is a low plateau sloping from about 500 m near Jebel Sinjar in the north and about 275 m near Abu Kemal in the west to less than 100 m a.s.l. in the east. No well defined structural features are found in this area. Alternating beds of sands, silts, clays, gypsum and limestones have a slight dip to the northeast and north. They belong for the majority to the Lower Fars Formation of the Middle Miocene. Minor areas in the northwest and central east have been classified as Upper Fars, dating back to the Upper Miocene. The most marked feature is the eastern, long, branched wadi Tharthar system of stream channels, emptying into the Umm Rahal north of Ramadi. Since it has a channel

Figure IV-11 The Jezira Plains with a flat to slightly rolling topography show the human impact most markedly. The area depicted has a rather compact silty soil, is devoid of most useful species known to have occurred here once, and is covered instead by *Artemisia scoparia* and *Peganum harmala,* two species useless for grazing, often found after cultivation and subsequent abandonment of an area (see IX. B.)

connection with the River Tigris near Samarra it is known as Tharthar depression and used as an overflow basin for Tigris water and a storage for water supply to the Euphrates with which it has been connected. Extensive shallow saline depressions are found in the central part (Sunaisala) and in the west along the Syrian border (Bouara, Barguth and Boghars, Fig. IV–16). The overall topography of the Jezira is flat to slightly rolling (Fig. IV–11). Low plateaus are mainly in the west, sometimes bordering the playa-lakes of the depressions. The surface material varies: limestone fragment scattered areas, compact clays and silts and hummock like low sanddunes, the latter especially in the northwestern part, were encountered.

IV.D. Soils

IV.D.1. *General characteristics of surface materials*

The surface materials and soils in the desert areas of Iraq are closely related to lithology, topography and drainage conditions. Chemical and biological soil forming and development are limited due to the small amounts of water available for the processes. Physical weathering is of great importance. Wind erosion has a tremendous impact on the surface materials, especially in areas with a sparse vegetation or no vegetation at all. Water erosion still plays an important role after the spring showers and has been even more important in the past as can be seen from the extensive deep wadisystems, from the gullies and fans (Buringh 1960).

The overall picture of the surficial deposits is one of mainly stony, gravelly and sandy materials, cut by wadis and with many depressions of fine textured, sometimes saline material, especially in the Western Desert and in the Jezira. Minor relief differences are often decisive for vegetation development and conditions along the wadis and in the many shallow, sometimes large depressions, greatly differ from the surrounding flat sandy or stony areas.

Large areas are stone covered, having a surface of *desert pavement* or *stone pavement*. Although such pavements often appear superficially to be similar, it has been found that the same form can be produced by several different mechanisms. Cooke & Warren (1973) name the following: (i) deflation of fine material by wind, (ii) removal of fines by water at the surface, and (iii) upwards migration of coarse particles by cycles of freezing and thawing and wetting and drying. Once the pavement is established, it protects the underlying material from erosion. This material can be of many kinds such as fragmented limestone, calcareous or gypsiferous sedimentary deposits with flints or limestone gravel, and coarse textured alluvium (Dregne 1967). Soils beneath pavements where coarse material is almost absent in the upper few decimetres to one metre of the profile have been recorded in the Middle East (Zohary 1973) and elsewhere (Cooke & Warren 1973).

Figure IV-12 A reg, consisting of alluvial, colluvial material transported from more elevated areas. The vegetation often occurs only along the runnels or small wadis, as shown in the upper right part of the photo.

The most pronounced desert pavements in Iraq are known as *regs*, stony surfaces of alluvial, colluvial material originating from more elevated sites (Fig. IV–12). Transported by wadis, other water courses and wind, from nearby areas, rounded but also often angular gravel, pebbles and larger size stones can cover extensive tracts of flat or undulating terrain. Vegetation in such areas is usually confined to the water courses, runnels or small wadis, fed by runoff water. water.

In contrast to regs an in Iraq widely distributed desert pavement named *hammada* is distinguished. This is a substrate of coarse stony material, boulders, formed in situ on flat hill tops and on slopes (Fig. IV–13; Zohary 1973). The rough surface impedes runoff and soil moisture is stored between the stones and in cracks, sometimes resulting in a sparse shrub cover. Hammada in Iraq is mostly derived from hard limestone and underlain by a stony soil. Hammada was also named *residual reg* as opposed to *transported reg* for the reg proper described above (Dregne 1967). The stones of the desert pavements are often covered by desert varnish, mentioned before.

Sands occur locally in Iraq over extensive stretches. Sand transport can be intensive and over vast distances in desert areas. Most of the sandy areas are rather stabilized although locally moving sands provide difficulties for traffic, e.g. near

Baiji on the Baghdad – Mosul road and at some places in the Southern Desert. Sand dune problems in the agricultural lands in or near the Mesopotamian Plain are mostly due to shifting pseudo sands (Dougrameji & Kaul 1971), soils consisting of loam or clay, cemented to sand-size grains (Buringh 1968).

Fine textured soils, *clays and fine silts*, are found in the lower situated often closed drainage areas. The easy pulverising of lime grains in transporting water is a factor here. Surface material is washed into the drainage basins during and shortly after the periods of heavy rainfall. The top soil of such areas dries out, shrinks and cracks in summer (Fig. IV–15). In certain areas an almost impermeable thin layer may be formed at the surface in the rainy season, resulting in the formation of rainpools, the so called 'barbak' (see IV.E.2.). In general the lime content of the desert soils is high. Altaie *et al.* (1969) found in their desert profiles 25–45%, except for the sands in the Dibdibba area and the gypsiferous soils, mainly in the Jezira. Gypsum content varies. High amounts are found all over the Jezira and locally in the Southern Desert, while in the Western Desert it is almost absent. Organic matter is low, varying from about 0.1% to 1.2% in the upper horizon.

Due to their limited agricultural potential the soils of the Southern Desert, Western Desert and Lower Jezira have been little studied as compared to those of other areas in Iraq. Buringh (1960) has compiled the knowledge on soils and soil conditions in Iraq to 1958. His 'Exploratory Soil Map of Iraq' (scale 1:1 000 000) depicts the whole country in 40 legend units. The desert is mapped in the following units: Mixed, gypsiferous desert land; Gypsum desert land; Stony desert land; Gravelly-desert and stony waste land, deeply eroded phase; Limestone

Figure IV-13 Example of hammada, a substrate of coarse stones formed in situ on a flat hill slope with an underlying stony soil.

desert land; Stony desert land; Sand dune land; Saline desert marshes, and Saline lake bottom land. The major boundaries follow the physiographic unit boundaries shown in Fig. IV–7. In the area under discussion little or no groundtruth for this map could at that time be collected (Buringh pers.comm.). Buringh's map has been the basis for a number of small scale, usually generalized versions shown in various publications (Guest 1966; Al-Kasab 1966). Some information has been added in the last decade. A survey and study has been made of soils of part of the Western Desert south of the Euphrates River, between Haditha and Al-Qaim, an area of 4875 km^2, by Yahia (1971). In Altaie (1968) and Altaie *et al.* (1969) the soil classification system of the seventh approximation (see USDA 1975) was applied to Iraq. Some 60 soil profiles, considered representative for the physiographic units of the country, were analysed and described in detail. About 15 profiles are located in the desert areas, but only half of these in the more central parts. It has resulted in a map of 'Soils and Soil Classifications of Iraq' (scale 1:1 000 000), with some 60 units in terms of the seventh approximation. This map has been virtually unchanged incorporated in the 1:5 000 000 FAO/UNESCO 'Soils map of the World', sheet VII.1. Altaie's map gives some more detail for the desert areas compared to Buringh's map and boundaries differ as a result of some reconnaissance traverses and study of new reference material (Altaie pers. comm.). Some additional soils data can be derived from investigations by consulting firms reporting on selected areas or problems (e.g. Parsons 1955 a, b, c; FMC Corporation 1964). A most useful account on soil types in the Middle East from the geobotanical point of view is given in Zohary (1973). In the framework of the present study samples from some 50 locations, distributed all over the area (Fig. V–1), were collected and analysed for a limited number of characteristics. This information is mainly incorporated in the next chapter, in relation to the shrub cover types.

IV.D.2. *Soil types and their distribution*

IV.D.2.a. Soil types

The already mentioned legend units used by Buringh (1960) in his 'Exploratory Soil Map' are, for the desert areas, mostly mixtures of two or more of the following Great Soil Groups: Desert Soil, Sierozem Soils, Red Desert Soils, Solonchaks, Regosols and Lithosols. New soil classification systems have been developed since and for reasons of easy comparison, in the following reference is made to the system of the 'seventh approximation' (USDA 1975). A detailed discussion on soil classification units in Iraq has been given by Yahia (1971). The soil types as distinguished in the seventh approximation, occurring in the desert areas of Iraq, are listed in table IV–9. The distribution of these units is shown in Fig. IV–14. This figure is mainly based on Altaie (1968) and Altaie *et al.* (1969). A brief description of the soil types and their distribution in the Southern Desert, Western Desert and Jezira is given below.

Table IV–9. Soil taxonomic units, according to the 'seventh approximation', described for the Lower Jezira, Western Desert and Southern Desert (derived from Altaie *et al.* 1969).

Order	Sub-order	Great group	Sub-group	Occurrence
ARIDISOLS	ORTHIDS	CALCIORTHIDS	TYPIC CALCIORTHIDS	Locally in the Jezira and on the middle and high gypsiferous river terraces.
			LITHIC CALCIORTHIDS	By far the most common soil type in the Western and Southern Deserts, occuring in different phases; stony, gravelly, eroded, and sandy.
		SALORTHIDS	TYPIC SALORTHIDS	Old alluvial deposits, river terraces, large depressions in the Jezira, transitions to the Lower Mesopotamian Plain.
	ARGIDS	PALEARGIDS	(not specified)	Western part of the Western Desert; in many other parts of the Western and also Southern Deserts, mainly the lithic phase, together with Calciorthids.
		HAPLARGIDS	(not specified)	Limited; an area in the northwestern part of the Western Desert.
ENTISOLS	ORTHENTS	TORRIORTHENTS	TYPIC TORRIORTHENTS	Limited; desert valleys Southern and Western Desert
	FLUVENTS	TORRIFLUVENTS	TYPIC TORRIFLUVENTS	Transition of Jezira to Euphrates River.
	PSAMMENTS	QUARTZIPSAMMENT	TYPIC QUARTZIPSAMMENT	Southeastern part of the Southern Desert.
		TORRIPSAMMENT	TYPIC TORRIPSAMMENT	Sanddune belt in the Southern Desert. Local dune fields in the Jezira.

IV.D.2.b. Soils of the Southern Desert

The southeastern part of the Southern Desert, largely the earlier mentioned Dibdibba Plains, is a gravelly and sandy desert. Coarse, gravelly sands of which the parent material for more than 95% is composed by quartz and other hard materials cover extensive stretches of the flat to undulating landscape. Altaie *et al.* (1969) have described a profile of a Quartzipsamment near Zubair. To 150 cm deep the texture was 80–92% sand with an organic matter content in the upper 20 cm of only 0.18%. Lime in this layer was 11.0–13.4% and about half that in the deeper layers. Some gypsum was found in these deeper layers. Buringh (1960) mentions the sandstone areas in Saudi Arabia as the origin of the sands. *Rhanterium epapposum* is here the typical dominant shrub species and a profile of the sandy soil under this species is given in table IV–10 (Al-Ani *et al.* 1974; Thalen 1974a).

The most eastern part of the Southern Desert, roughly between the sanddune belt and the Basra-Safwan road, is more gravelly to pebbly and gypsiferous. According to Altaie (1968) these are Lithic Calciorthids. Buringh (1960) mapped it, similar to the southeastern part, as pebbly and sandy desert land under Desert soils, with a high lime and very low organic matter content. The Euphrates sanddune belt, discussed before, borders most of the Southern Desert at the north side. Soil samples collected in this belt, south of Ur, showed the uniform, purely

Figure IV-15 Cracked heavy textured soils in the saline bottom of the Abu Dibbis depression with a sparse cover of *Cressa cretica*.

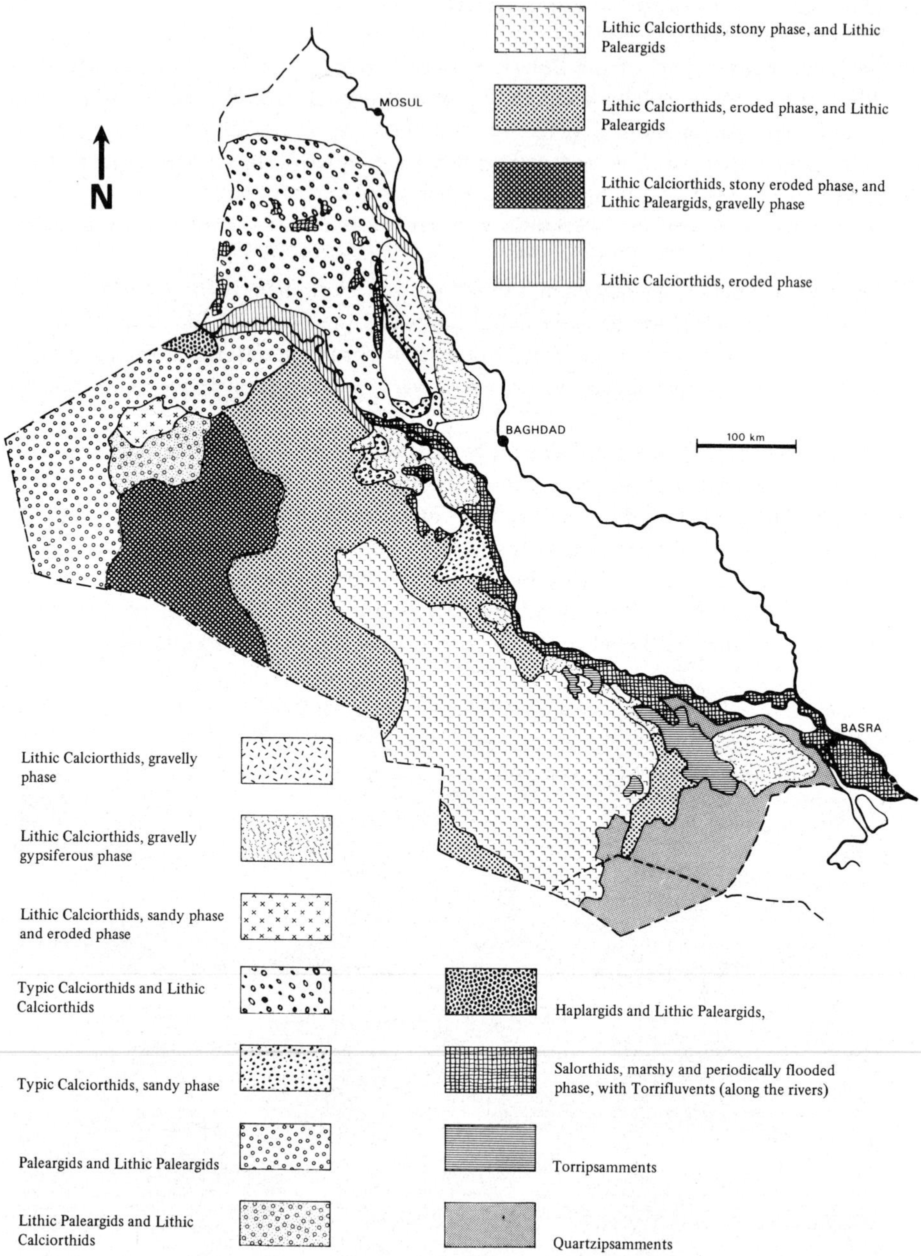

Figure IV-14 Soil map of the Southern Desert, the Western Desert and the Jezira.

Table IV–10. Physical and chemical characteristics of a soil profile under a good, almost monospecific, stand of *Rhanterium eppapposum*, sampled near Samah, Neutral Zone, 19 December 1971.

Depth (cm)	Texture				pH	ECe (mmho. cm^{-1})	Lime (%)	Gypsum (meq.$100g^{-1}$)	N (%)	P (ppm)	K (meq.$100g^{-1}$)
	Sand (%)	Silt (%)	Clay (%)	Class							
0– 5	90	6	4	sand	8.1	0.23	13.65	nil	0.013	4.60	0.18
5–15	82	12	6	loamy sand	7.8	0.20	16.75	nil	0.015	3.20	0.20
15–25	80	12	8	loamy sand	8.0	0.28	18.00	nil	0.009	1.13	0.32
25–55	76	14	10	loamy sand	7.2	2.44	23.00	16.8	0.005	1.73	0.12
>55	(hard compact layer)				7.6	2.50	27.25	55.0	–	1.93	0.07

Table IV–11. Characteristics of eight soil samples, collected from two different depths at four locations along the slope of the sand dune belt, near the crossing of the Ur-Busaiya track, 26 April 1973.

Sample number	Depth (cm)	Texture				Organic matter (%)	Gypsum (meq. $100g^{-1}$)	Lime (%)	ECe (mmho. cm^{-1})
		Sand (%)	Silt (%)	Clay (%)	Class				
1a	0–30	98	0	2	sand	0.06	nil	12.9	0.25
1b	30–60	97	1	2	sand	0.08	nil	12.4	0.59
2a	0–30	98	0	2	sand	0.05	nil	13.4	0.18
2b	30–60	98	0	2	sand	0.06	nil	16.3	0.24
3a	0–30	98	0	2	sand	0.01	nil	13.3	0.18
3b	30–60	98	0	2	sand	0.05	nil	14.1	0.57
4a	0–30	98	0	2	sand	0.06	nil	14.1	0.18
4b	30–60	97	1	2	sand	0.05	nil	15.2	0.24

sandy composition of the soils here (table IV–11). Only a few plant species can survive in this environment. Moist layers were found at a depth of only 25 cm, although no rainfall had been recorded during several weeks. Near the sanddune belt, in a transition to the Lower Mesopotamian Plain poorly drained, highly saline loam and clay soils are found. These soils, in relation to the halophytic vegetation, have been reported in Habib *et al.* (1970).

The major part of the Southern Desert, the central and western part is stony desert land, described before under the physiographic units Al Hajara Stony Plain and Lower Wadian Plain. The shallow soils, mostly sandy silts, are derived from the weathered limestone. The depressions are filled with clays and clay loams. A representative profile of such an area is given in Altaie *et al.* (1969). They describe a dark brown soil with a silty loam texture throughout the profile (clay 34–37%, silt 52–58%, sand 8–11%), almost no gypsum and a high lime content (24–30%). Organic matter ranges from 1.17% in the upper 20 cm to 0.18% in the 100–150 cm layer. Such depressions, often with a good vegetation, or at least a good potential, cover only a small percentage compared to the vast stony areas under regs and hammada.

IV.D.2.c. Soils of the Western Desert

The boundary between the Southern Desert and the Western Desert at Wadi Al-Khir is an artificial one and the limestone landscape with its partly filled depressions continues in the Western Desert. There is a gradual transition into the dissected landscape of the Upper Wadian. Altaie (1968) mapped all soils here as Lithic Calciorthids but in different phases; the Stony phase in the eastern part and the Stony Eroded phase and Eroded phase further to the west. He brings in another soil group here too, the Lithic Paleargids, and mapped the entire northwestern part of the Western Desert under these soil types. They have a low organic matter content, in a described profile ranging from 0.16% to 0.21%. A petrocalcic horizon has developed from the underlying bedrock after dissolution of the calcium carbonate and redeposition as a hard crust of secondary calcium carbonate. Lime content is very high. Figures of 30.7% (0–20 cm) to 47.0% (40–60 cm) are given from a profile 50 km west of station H–3.

In the wadibeds a number of different soil textural classes can be distinguished in layers. Most soils here contain subround cobbles and pebbles of limestone and siliceous rocks, mixed with calcareous sand, silt and clay (Fig. IV–16). Sorting of this material leads to differences in size of the fragments building the substratum. As a result a number of micro habitats can be recognized at certain distances from the wadi channel (Batanouny 1973).

In the interwadi areas different types of residual limestone deposits are scattered, usually forming desert pavements, covering shallow calcareous soils. At the northeastern part of the Western Desert, roughly in the line Najaf – Kerbala

Figure IV-16 Eroded bank of Wadi Ubaiyidh near Nukhaib, showing the different layers of mixtures of subround pebbles with soils of various textures, ranging from fine silt to coarse sand and gravel. In the wadibed a plant of *Citrullus colocynthis* with fruits and on the bank some shrubs of *Haloxylon salicornicum* and stumps of heavily grazed *Ephedra alata.*

Figure IV-17 The Bouara salt lake, a depression in the Jezira at the border with Syria. Most of the year a thick salt crust covers the soil. The salt is harvested by the Bedouins and sometimes sold in the villages and cities.

– Ramadi, some larger depressions are found of which the northern ones have been connected and are filled with diverted Euphrates water now forming large lakes: Habbaniya Lake and Abu Dibbis. Vegetation and soils of the saline depressions further south, near Najaf, were described by Abul Fatih (1975). In this region, bordering the Lower Mesopotamian Plain, we find locally deep fine textured soils with a high groundwater table and strongly saline. Evaporation and transpiration cause an upward movement of groundwater, containing in Iraq always some salts, to the surface. An accumulation of salts is the result. In areas with deep groundwater longtime irrigation combined with low precipitation can have the same effect. Details on these processes are given in Buringh (1960) and Yahia (1971). Common salts are chlorides and sulphates of sodium, calcium and magnesium, often found in mixtures. Sodium chloride (NaCl) is most common.

The best studied part of the Western Desert is the triangle south of the Euphrates between Haditha, pumping station H–1 and Al-Qaim, by Yahia (1971). He distinguished four land types as follows with their soil types:

- the plateau (appr. 400 m a.s.l.), an extension of the Al Hajara Stony Plain, with Calciorthids;
- an undulating area in the Lower Wadian with Lithic Calciorthids and Psamments, and Paleargids in the wadis;

IV.E. **Water resources**

IV.E.1. *Introduction*

The Iraqi desert areas have no perennial water courses, excepting the River Euphrates. Depressions with permanent small lakes or pools are only found at the fringe with the Mesopotamian Plain. The groundwater table, in general, is deep, too deep to be reached by the rootsystems. Plant life, therefore, has to rely upon the precipitation, sometimes forming pools in depressions or wadibeds, and for some time stored in the topsoil and subsoil. Over extensive areas no moisture at al will be within reach of plant roots for most of the year.

Water is often the decisive factor for the utilization of desert vegetation, not only as a limiting factor for plant development, but also for the survival of man and his livestock. Water supply for domestic purposes and livestock in many situations determines the accessibility of an area, and thereby the use of its vegetation. Considerable changes in this aspect of accessibility as related to availability of water have been taking place in the present century with a tremendous impact on the utilization of the desert rangelands. This is worked out in IV.E.3 and 4.

The water resources of the desert areas of Iraq have in the last decades increasingly been the subject of extensive and locally intensive investigations. Noble (1926), the first reporter, in his 'The Subsurface Water Resources of Iraq', dealt only briefly with the desert areas. From the early thirties onwards information became available through the activities related to oil exploitation and pipelines. MacFadyen (1938) worked on desert groundwater, giving details for the Nukhaib – Rutba area. From 1953 to 1955 the Parsons Company conducted a survey of the whole of Iraq, compiling all work previously done. This work is reported in several volumes under the general title 'Groundwater Resources of Iraq'. For the present study, the volumes 5 (Jezira Desert, Parsons 1955 a), 7 (Southern Desert, Parsons 1955 b), 8 (Northern Desert, Parsons 1955 c) and 9 (Selected areas, Parsons 1955 d) are important. Drilling of wells has taken place on an enormous scale and provided new data. H.T. Smith Inc. carried out a programme of 160 wells in Iraq during 1954–1959 of which 38 were in the desert area. It was followed by the work of the Yugoslavian Geoistrazivanja Company: a 110 (+2) wells programme in the Southern and Western Deserts and a 100 (+14) wells programme in the Mosul Liwa (INGRA 1960, 1964). In addition, work has been done by oil companies and governmental drilling teams. More recent complications of knowledge, including reports on studies in pilot areas, are given in van der Sluijs (1969) and in UNESCO (1975) for parts of Iraq, and by Ahmed & Kraft (1972) on the Southern and Western Deserts. For hydrological details reference should be made to the above sources.

IV.E.2. *Types of traditional watering places*

From olden times the following types of watering place have been used:

Springs

A zone of springs is found along the northeastern boundary of the desert, from Hit to Ain Hamud south of Nasiriya, over a total distance of about 450 km. The

Figure IV-19 Shitatha, the only real oasis in the desert areas of Iraq. The water of sulfur holding springs is guided by sluices in an irrigation channel system. With the help of this system date palms, alfalfa and other crops are grown.

source of the water is not yet known. The direction of the groundwater flow is from the Southern and Western Deserts towards the Euphrates river and water may come from greater depths along fault zones or through fractures. Alternative hypotheses suggest recharge from the Euphrates or even from groundwater in the upper Euphrates basin in the neighbouring countries (MacFadyen 1938; Parsons 1955 b; UNESCO 1975). Apart from a permanent domestic and stock water supply, the springs have a considerable value for irrigation of date gardens, fodder crops etc., for instance at Shitatha (Fig. IV–19).

Surface water

After the winter and spring rains, shallow pools may collect in wadis, depressions, etc. Many of these last only a few hours. In exceptional cases, depending mainly on the amount and intensity of the precipitation and the structure of the topsoil and underlying layers, they may last for months. They have always been an extremely important source of water for the Bedouins, illustrated by the different words in Arabic for the various types *e.g.*

'jalta' – pools formed in a rock basin, on elevated ground or left in rocky wadibeds after passing of a flood,

'barbak' – wide, shallow pools in loamy hollows (Fig. IV–20)

'thaqab' – pools in narrow clefts, usually long, narrow and rather deep, often surrounded by 'Sider'–bushes (*Zizyphus nummularia*), sometimes reflected in the name as in 'Shaib as Sidair' (30°10' N, 45°50' E) mentioned by MacFadyen (1938).

Masonry storage tanks are found along some ancient pilgrim-tracks to Saudi Arabia, of which the Darb Zubaida is best known. Increasing efforts have been made by the government authorities in charge of the desert to construct low dams in wadis, sometimes combined with dug storage basins, to collect water (Fig. IV–21).

Hand-dug shallow wells and water holes

Numerous shallow wells, intermittant or permanent, have been dug in the course of time, often grouped, as many as 50 together. They are located either in or near

Figure IV-20 A shallow pool of water in a wadibed, a 'barbak', in olden times about the only source of water for the Bedouins, apart from the few hand-dug wells.

wadis and other drainage lines, or in depressions. Water source is rainwater, infiltrated and collecting close to the surface. Depths vary roughly from one to about 30 metres and such wells sometimes pass through the alluvium into the bedrock. These wells are mostly in beds of sands and gravels and filled during the rainy season. The lower ones may be flooded temporarily, but most are dry by the end of the summer. The water quality is usually very good and for this reason and for the minor labor required to lift the water (Fig. IV–22), these wells are preferred by the Bedouins. Policeposts and other permanent settlements seem often to have developed at places where shallow wells provided a permanent water supply. As Salman, Shabicha, Nukhaib and Judaidat Arar are examples. At such places water of drilled wells has sometimes replaced that of the old shallow wells *e.g.* at Rutba and Nukhaib.

Deep hand–dug wells

A small number of ancient deep wells is known from the Southern Desert and Western Desert. Their origin is often unknown, but they may be several hundreds of years old, even dating back to the first millennium. MacFadyen (1938) argues that they may have been dug for strategic reasons, in view of the labor involved. If not started in the solid rock, they have a good masonry-lining at the top. Water

Figure IV-21 Low earthen dam across a wadi, constructed by the D. G. of Range Management; an attempt of water conservation. However, the unpredictable, often torrential waterstreams coming down the wadibed after rainfall easily destroy such structures and sofar they have proved of little use.

Figure IV-22 Shallow hand-dug well in the Jezira. Water is raised by hand, using a bucket or goatskin tied to a rope.

comes from great depths, some 15 to 100 m. It is raised in goatskin bags, with ropes pulled over inclined posts, often by camels or donkeys (Fig. IV–23). Many have radiating troughs for watering livestock and are known as 'star wells'.

IV.E.3. *Drilled wells*

During the last half a century or so a new category has been added, already mentioned before:

Drilled wells

After about 1930 the widely spaced and often temporary traditional watering points have been overgrown by a relatively dense network of drilled, permanent, deep wells (Fig. IV–24). The exact data on when and where deep wells have been

Figure IV-23 Very old, deep, hand-dug well near Ansab at the border with Saudi Arabia. A camel (or donkey) is used to pull up the water.

drilled are not easily available, if available at all. Combining the data of several sources clearly shows the trend.

The first attempts at drilling wells, by the Iraq Petroleum Company (IPC) along the Iraq – Mediterranean oil pipeline, included 49 wells at 12 localities of which nine wells at four localities were productive in 1935. The deepest was 359 m (MacFadyen 1938). The Iraq Ministry of Economics drilled at least 17 wells throughout the Southern Desert of which in 1955 seven were still in use, with new drillings being carried out (Parsons 1955b). Drilling by

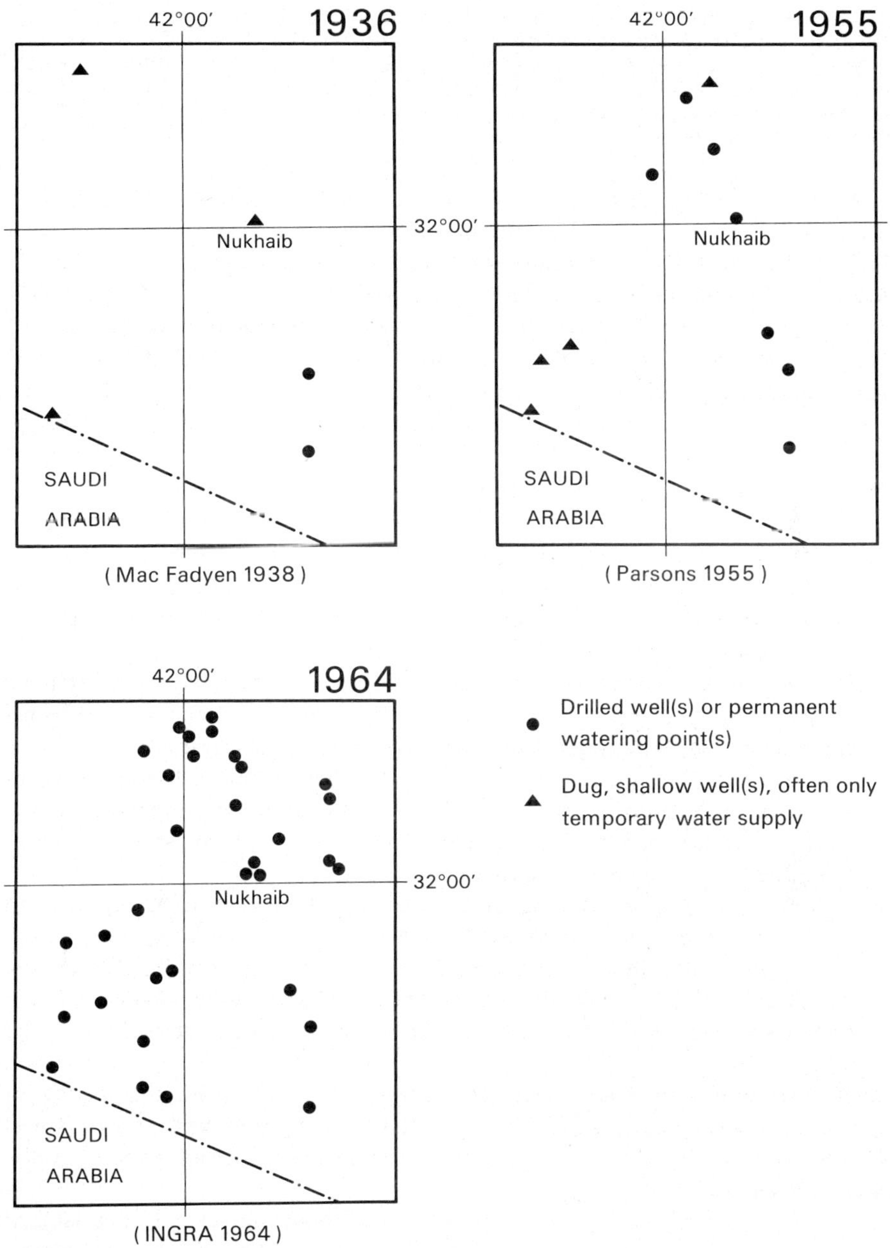

Figure IV-24 Increase in number of permanent watering points shown for a selected part of the Western Desert.

the Basra Petroleum Company in this area has not been well documented. For the Western Desert the Parsons report mentions drilled wells at Rutba, Habbariya, Nukhaib, Shinana, wadi Al-Walej (80 km northwest of Rutba), wadi Al-Auja (39 km northeast of Rutba, Ramla (135 km west of Ramadi), wadi Al-Muhammadi (67 km west of Ramadi), Sidr Kisra (50 km northeast of Nukhaib) and Shabwan (21 km southwest of Habbariya, and in addition IPC-drilled wells in the area north of Rutba (Parsons 1955c).

In about 10 years, between 1955 and 1965, at least another 150 wells were drilled in the Southern and Western Deserts by foreign companies. For this area a total of 500 drilled and hand-dug wells, and springs is mentioned in Ahmed & Kraft (1972). Fig. IV–24 illustrates the increase in watering points of a small part of the Western Desert after about 1930. The development of a network of perennial watering points had an enormous impact on the exploitation of the vegetation resources of the desert rangelands.

IV.E.4. *Water supply and vegetation utilization*

MacFadyen (1938), having traversed the 'Wadian' (an area of roughly 80 000 km^2, southwest of lat. 33°30' N and long. 43° E) in 1936 at the request of Shaikh Makruth al Hadhal to study the water supply, especially for the Anaiza Bedouins, wrote on the existing supplies:

> 'The part of the Wadian country that was seen is remarkable in the extreme paucity of the permanent water supply, Jadaidat at Al'ar being the only generally reliable supply between Rutba and the Barit and Machmi wells, a distance of some 300 km. Water is rarer that in any other part of Iraq with long distances between sources. Since most of these are only temporary, when they dry up the country becomes impossible to live in. There are no surface streams, except immediately after rain and no springs.'

For another area, after a discussion on the subject of new water supply with Shaikh al Arif of the Dhafir, the remark is made:

> 'He thought that if there were wells available, the area between Busaiya and the Batin would be inhabited. When traversed on 22 May, 1934 no Bedouin nor stock were seen for 93 km between 9 km West of Khabrat ibu Ghamain and Busaiya.'

That water supply for a long time has been one of the major limiting factors for grazing and other exploitation of the vegetation is by now well known. Guest (1966) based on his experience of desert travel in the midfifties, gives the following statement:

> 'Broadly speaking the check to grazing throughout the year is not lack of vegetation, except locally areas of edaphic or secondary desert, but lack of water points at which the animals may drink during the long dry summer season. Where well water is available flocks and herds can generally subsist in the desert region of Iraq throughout the year.'

Figure IV-25 a) Bedouin women with donkeys in search of the nearest water source. For many centuries a few hand-dug wells or some surface water after the rains collected in a wadi or depression were the only possibilities. Now it is usually one of the drilled wells as shown in b). b) Drilled well with pump, storage tank and tanker. Due to such provisions man can now move almost independently from water in the desert.

From the foregoing observations it may be concluded that until about half a century ago the lack of a permanent water supply was often the limiting factor for the movement of flocks for large parts of the desert, at least during part of the year. The drilling of hundreds of new wells, all over the desert, especially from 1950 onwards, has ended this restriction and has virtually 'opened up' the entire area the year round. In addition, tankers are being increasingly used, taking away almost all restrictions on stock movements with respect to water supply (Fig. IV–25 a, b).

IV.E.5. *Estimates of recharge versus exploitation*

Not all the wells once dug or drilled are still in permanent use. Increasingly poor quality water, often increased salinity, and improper maintenance of pumps and other installations are the most important reasons for a well to be abandoned. From the very beginning of groundwater exploration in Iraq, the recharge has been a point of major concern. It takes places at the outcrops of aquifers and in areas with permeable beds, in the rainy season. The existence of numerous wadis and local depressions where runoff water collects and the fractured limestone underlying most of the desert, are features to be mentioned in this respect. Although no detailed studies have yet been carried out, some recharge calculations have been made nevertheless.

Parsons (1955d) made calculations for four drainage areas in the Rutba-Nukhaib area. Assumed precipitation figures were 71 mm annually for Nukhaib as lowest and 119 mm for Rutba as highest. After applying correction factors for losses, mainly based on data from similar areas abroad, they arrive at annual recharge figures of 1.00 mm for Nukhaib area, 1.17 mm for Habbariya, 1.44 mm for Qa'ara depression and 1.80 mm for Rutba area. Total annual recharge is thereby computed at 52 million m^3 for a total area of 41 000 km^2. Probable actual figures are given at not less than 12.2 million m^3 and not more than 122 million m^3. Similar calculations have been made for the whole Western Desert and Southern Desert (Parsons 1955b, c). Ahmed & Kraft (1972) made estimates, using up-to-date wells data. They assumed an infiltration of 5%, compared to the 1–2% assumed by Parsons. The final figures, expressed in liters per second ($1.sec^{-1}$)are given in table IV–12.

The total capacity of the pumps of the wells installed under the '110 (+2) – programme' by Geoistrazivanja (105 working wells) was 236.5 $1.sec^{-1}$(INGRA 1964). It has been estimated that at the present time not more than 2000 $l.sec^{-1}$ is exploited (Ahmed & Kraft 1972). In Parsons (1955d) a figure of consumptive use being 60% of pumpage is given, the remaining 40% returning to the ground-

Table IV–12. Estimates of recharge to groundwater for the Western and Southern Deserts combined as given in, or calculated from, three different sources.

Source	Mean recharge ($l.sec^{-1}$)		
	Computed value	Actual value probably	
		not less than	not more than
Parsons (1955 b, c, d)	6 500	1 400	14 000
INGRA (1960)	–	15 000	25 000
Ahmed & Kraft (1972)	31 750	10 000	40 000

water. Based on such estimates it may be assumed that perhaps only 1000–1500 $l.sec^{-1}$ is withdrawn at present for the Western and Southern Deserts combined. This is ten times less than the lowest recharge estimates given in table IV–12.

There are so far no reasons to believe that a limited recharge to the groundwater will ever cause a reverse of the water availability as described, although reliable recharge figures are lacking. With a proper well maintenance service, already organized by the Government, water resources are, as mentioned above, no longer a limiting factor for the movements of man and livestock in the area.

To what extent the supposed good amount of renewable groundwater could be exploited for local small irrigation plots, e.g. for feedlots or as a help in early stages of afforestation, remains to be investigated. The stands of *Tamarix* spp. near Zubair provide an example of the possibilities. Van der Sluijs (1969), having studied the Dibdibba area along the border with Kuwait, suggests that the system of Wadi Batin, draining 375 000 km^2, alone might have a constant subsurface discharge of 12 000 $l.sec^{-1}$ enough to irrigate 24 000 ha at a rate of 0.5 $l.sec^{-1}$. One of the conclusions of INGRA (1964) as a result of the data provided by the 110 (+2) wells programme, is that

> 'there are regions particularly distinguished by the quantity and quality of water disposable, which could develop in to fertile oases . . .' etc.

Such statements, however, should be proved by detailed studies, perhaps combined with pilot projects, before larger scale implementation is undertaken.

IV.E.6. *The situation in the Jezira, water quality*

The Jezira can for hydrological work be divided into three distinct surface drainage areas: the wadi Tharthar area, the Central Interior Basins and the Western Interior Basins. The drainage is interior. Water flows to one of the many shallow depressions. Wadi Tharthar is the only extensive drainage system. The wadi flows north-south over a distance of several hundreds of kilometres, has numerous tributaries in the upper part of the Jezira and drains into the Tharthar depression. Since the connection with the Tigris the depression is continuously filled and forms a huge lake. The interior drainage and shallow depressions of the Jezira can be seen on the satellite image of Fig. IX–4. They form the basins where the runoff water collects and are often surrounded by distinct drainage lines. The Lower Fars Formation is the most important underlying formation. As a thick deposit of the Miocene, it consists of gypsum and anhydrite, limestones, clays and various other chemical precipitates. As a result, the groundwaters in the Jezira are high in dissolved salts (Parsons 1955a). Water quality is a greater problem here than water quantity and the potential for more intensive groundwater exploitation seems limited. The sources of water and the developments with respect to

Table IV–13. Indication of the quality of subsurface waters in the Southern Desert, the Western Desert and the Jezira, based on samples and analyses reported in Parsons (1955 a, b, c). Samples were taken from shallow and deep hand-dug wells, and drilled wells, all over the area in 1954 and 1955.

Area	Total number of samples analysed	Percentage of samples falling in indicated ppm – classes					Approximate average
		<500	500–1000	1001–3000	3001–5000	>5000	(ppm)
Southern Desert	70	18	15	30	23	13	3320 (2470*)
Western Desert	31	45	16	26	10	3	1290
Jezira	78	0	0	8	74	18	5070

* Excluding two extremes (32 600 and 32 000 ppm), only used for drilling water by the Basra Petroleum Company.

well drilling are not very different from the situation described for the Southern and Western Deserts, but high salt content imposes serious restrictions. Table IV–13 gives an indicative comparison in this respect for the Southern Desert, Western Desert and Jezira. All knowledge from chemical analyses has been compiled in Parsons (1955 a, b, c). The average ppm – values for the well waters sampled show a ratio of about 1:2:4 for the Western Desert, the Southern Desert and the Jezira, respectively. Over 90% of the Jezira samples had an amount of soluble salts exceeding 3000 pp,, against 13% for the Western Desert and 36% for the Southern Desert. One third of the Southern Desert samples had a soluble salt content under 1000 ppm, almost two thirds of the Western Desert samples had this quality, against almost none in the Jezira. In all Jezira groundwater samples the sulphate ion was the dominant, followed by chloride, nitrate and bicarbonate, while in general the calcium ion is the most common cation, followed by sodium and magnesium. The shallow hand-dug wells show the least mineralization and are, therefore, most intensively used. In general the amount of salts in the waters analysed falls within the tolerance limits of livestock, about 7000 ppm to 10 000 ppm for horses and cattle and 12 000 ppm for sheep, goats and camels (Vallentine 1971; Burdon 1971). It seems that waters may contain concentrations of salts that may cause serious physiological disturbances or make it otherwise undrinkable. Locally, water is pumped up for small scale irrigation, but this can only be continued under very good drainage conditions preventing the accumulation of salts in the topsoil.

Water quality in the Western Desert, with mainly gravels, sands, limestone and sandstone, is very satisfactory with a few exceptions. Maximum tolerance for domestic use in Iraq is 3500 ppm of soluble salts (Parsons 1955 c). Only in a few wells in the Western Desert was this amount exceeded in the 1955 data.

Far more variation is found in the Southern Desert waters. The majority in 1955 was found to be mixed salines, with a predominance of sulphates due to gypsiferous deposits. Locally, high chloride content has been observed. In the Dibdibba Formation in the southeast fresh froundwater is available only at relatively shallow depths. At greater depths it becomes increasingly saline, limiting the withdrawal. Less than one third of the surface waters analysed in 1955 exceeded the 3500 ppm tolerance rendering it less suitable for domestic use, while only very few sources would give problems for livestock watering.

IV.F. Animals

IV.F.1. *Large herbivores and carnivores*

Only a few taxonomic groups of animals occurring in the desert areas of Iraq have been studied. The documented knowledge on the mammals is easily available from the comprehensive treatise by Hatt (1959) and the monumental compilation in Harrison (1964, 1968, 1972). Only scanty records exist of the mammalian fauna of Iraq in former centuries. Hatt (1959) reviews some sources including the written and pictorial records from pre-Linnean times and some little known annotated lists published in the nineteenth century. The first comprehensive report in the twentieth century, based on collected material, is Cheesmans 'Report on the mammals of Mesopotamia, collected by members of the Mesopotamian Expeditionary Force, 1915 to 1919' (Cheesman 1920), later supplemented in Pitman (1922). A list of Vertebrates of Iraq was prepared by Mahdi & Georg (1969), and Allouse (1954–1968) compiled a bibliography on the vertebrate fauna of Iraq and neighbouring countries.

Before the permanent settlement of man in the area, big game must have been present in large numbers. The sources indicating that king Assurnasir-pal II of Assyria in the ninth century B.C. took 450 lions, and that at one feast the menu included over 1000 cattle, 15 000 sheep, 500 deer and 500 gazelles (Hatt 1959) may be exaggerating, but the figures mentioned suggest that during Assyrian times game and their predators were plentiful and that large numbers were hunted during the royal hunts. From horseback with lances and spears, later from cars with all sorts of firearms a potential valuable resource has been destroyed to the verge of total extinction in the second half of this century. As Mountfort (1965) put it:

> 'The senseless killing of exhausted animals which have been run to a stand-still by an automobile is a blot of shame on the Arab world; but it is chiefly by this method that the larger animals of the desert are now being exterminated'.

Two large herbivores, the onager or Syrian wild ass (*Equus hemionus hemippus* Geoffroy) and Arabian oryx (*Oryx leucoryx* Pallas) have become extinct in Iraq in the present century. Reference to onagers is made by many authors, briefly reviewed in Hatt (1959). The species was once found in good numbers in the Jezira. There are records from the Upper Jezira near Sinjar, but also from the Lower Jezira, near Ramadi and an area just north of Ana and Rawa. The last reports are from the period 1920–1930. The oryx once had a distribution extending from the Arabian peninsula north into Iraq. Although there are few documented records pertinent to Iraq, there is little doubt that this species once roamed the Southern and Western Deserts. Most of the foodplants for the oryx mentioned in Harrison (1968) for Arabia are also found in Iraq: *Cyperus* sp. (most likely *C. conglomeratus*), *Aristida plumosa* (=*Stipagrostis plumosa*), *Lasiurus hirsutus* and young shoots of *Tamarix* sp. and other shrubs. With its ability to live independently of free water for several months and its preference for sandy and gravelly terrain, the Southern Desert especially must once have provided the required living conditions. Nothing is known about numbers of oryx in Iraq. The last record of an animal shot is from 1914 (Hatt 1959).

The only large wild herbivores still occasionally found in the area under discussion are gazelles (*Gazella* spp.). The taxonomic place of the Iraqi gazelles is partly an open question. Taking into account the many thousands of animals hunted, remarkably little material (skulls, hides) has been secured for further study. It may be too late now to undertake such a study. Less than 25 years ago large

Figure IV-26 The large wild herbivore populations have almost disappeared from the Iraqi deserts. Gazelles, like this captive specimen kept by the Sheikh at Kesrah, were still seen in flocks of hundreds less than forty years ago, but are now rapidly vanishing.

groups of gazelles were still to be seen in the Jezira and locally in the Southern Desert and the Western Desert. During the several thousands of kilometres of desert traversed for the present study between April 1970 and June 1973, only once two gazelles were spotted at great distance and another animal that was seen, was kept in captivity by Sheikh Naif Mahruth al-Hadhal at Kesrah (Fig. IV–26). Hatt (1959) wrote:

> 'Gazelles are now abundant in only a few areas. One report of recent date was of a herd sighted between T–1 and T–2 which was so large that a driver had to stop his car and wait for the herd to complete its crossing. An area favored by some Baghdad hunters is the flat plain west of Ar Ramadi, where the animals are easily pursued by car'.

Guest (pers. comm.) found gazelles to be abundant along wadi Tharthar in the Jezira in April 1932. In 1952 he saw only six gazelles along the pipeline Haditha-Badji and in 1957 traversing the Southern Desert the only indication of gazelles was one spotted by the guide as a small far away dustcloud. The gazelles in the desert areas of Iraq most likely belong to one of the following taxa: *G. subgutturosa* Güldenstaedt (with the subgroups *G. subgutturosa subgutturosa* and *G. subgutturosa marica* Thomas), *G. dorcas* L. and *G. gazella* Pallas (Mahdi & Georg 1969; Harrison 1968). *G. subgutturosa* is considered the most common of the three. The available sources and information obtained from police guides and older Bedouins all indicate that until quite recent times gazelles were common and locally abundant all over the Iraqi desert areas. Unfortunately there can also be no doubt that the last remaining large game species have been reduced to a few local survivors due to an increased hunting pressure. Sheikh Naif Mahruth al Hadhal from Kesrah told in October 1971 how in the late fifties special hunting groups from the Gulf had hunted some 50 animals in about two hours. He himself had shot about 20 gazelles every spring between 1950 and 1960. He had seen six gazelles in 1970 and shot only two. Two types were mentioned: 'Afri' and 'Rhim', *G. dorcas* and *G. subgutturosa* respectively, according to the vernacular names given in Harrison (1968).

In the present situation the wild large herbivores are no factor of importance. Only re-introduction and protection might give them back their original place in the desert rangeland ecosystem.

A number of large carnivorous species too may still be encountered in the area. Most common are the foxes. *Vulpes vulpes* L. was regularly seen all over the area. Harrison (1968) regarded all Iraqi desert specimens as *V. vulpes arabica* (Thomas). Hatt (1959) collected a specimen of *Vulpes rüppelli sabaena* Pocock near pipeline station K–3 in the Lower Jezira. The other canidae of the Iraqi desert are the Asiatic jackal (*Canis aureus* L.) and the wolf (*Canis lupus* L.). The jackal is abundant in the thickets along the rivers, in cultivated lands and near settlements. Cheesman (1920), Hatt (1959) and Harrison (1968) do not mention any records from the desert areas. In conversations with policemen and villagers the occurrence of jackals near the major desert settlements was several timed indicated, e.g. for Rutba, As Salman and Shabicha. When camping at the small policepost of Mugr an Na'am at the border with Saudi Arabia some semiferal domestic dogs were fighting the jackals off during the night. *C. lupus* seems to be quite common all over Iraq. Many kills of sheep and goats during the night were attributed to wolves by the Bedouins and stories of wolves attacking man were often told.

Not less feared than the wolf in the desert is the striped hyena (*Hyaena hyaena* L.). Hatt (1959) gives a number of observations from the desert fringe, where the species may be most common. During a stay in Nukhaib in February 1971 the local police was joined one night in an unsuccessful attempt to shoot a hyena seen nearby by camping Bedouins. That damage can be done to the flocks by this species was often mentioned by the Bedouins.

A few species of Arabian felidae should be named as former or still existing predators in the desert: Asiatic lion (*Panthera leo persica* Meyer), cheetah (*Acinonyx jubatus* Schreber), caracal or desert lynx (*Caracal caracal schmitzi* Matschi) and the jungle cat (*Felis chaus* Güldenstaedt). That lions once were common can be deduced from a large number of records, reviewed in Hatt (1959). The last records are from about 1920. Whether the cheetah is also extinct in Iraq is an open question. Its disappearance may be well linked with the vanishing gazelle populations. Bedouins used to train cheetahs for the gazelle hunt in the previous century. The last record of a cheetah in Iraq in the literature is from 1948. Sheikh Metab Mahruth al-Hadhal from Habbariya told how he had shot a 'fahid' or 'nimr' in the Zarga area, rough rocky terrain about 40 km south of Habbariya, in the mid-fifties. A desert policeman told in February 1972 that he had shot a 'fahid' in 1966 in an area just north of Suq as Shuq near Nasiriya. These cases refer most likely to cheetahs, but the local names for large felidae in Arabic are often confusing. The caracal is a rare species, but known to occur still in the Western Desert and Southern Desert. In Thalen (1975) the earlier and some recent records of the species are discussed. Evidence is presented that the caracal is more common in Iraq than could be assumed from publications so far. *Felis chaus* is a not uncommon species of the irrigated areas and along the rivers. Its occurrence in some parts of the desert where surface water is available is not unlikely, but no such observations are documented.

The carnivora mentioned have no direct bearing on the exploitation of the desert rangeland vegetation and are of little importance in the present situation. However, in designing a future livestock grazing system the disturbing role that can be played by such predators should be seriously considered and protective measures incorporated. Losses due to large predators can be very high. Tigner & Larson (1977) found 24% of lamb deaths caused by predators and 18% of ewes deaths in a range area in Southern Wyoming, dominated by big sagebrush (*Artemisia tridentata*).

Moreover, the situation of the larger mammals in the area and the process of extermination by man with the help of modern technology is symptomatic of what is happening to the resources of the desert rangelands as a whole.

IV.F.2. *The smaller wild animals*

A fair number of small mammal species, birds and reptiles are found in the area under discussion. Little is known, however, about their relation to the vegetation. Most of them will depend on the vegetation for cover and food supply but more than this general remark can hardly be made.

The hare, in this area named as *Lepus europeaus, L. capensis* or *L. arabicus* and by Harrison (1972) considered to be one highly polymorphic species, is common all over the desert and regularly hunted for its meat by the inhabitants. Other common small mammals include jarboas (*Jaculus* spp.) and gerbils (*Gerbillus* spp.). Both these groups are mainly herbivorous and seeds and other plant parts are collected in the burrows. Their main impact on the perennial vegetation may be in the seed predation and seed distribution. Harrison (1956) excavated the tunnel system of *Gerbillus dasyurus mesopotamiae* Harrison and found that

> 'The burrows often combined blind side tunnels that were packed for a distance of six inches with stores of the dried seeds of the plant covering the mounds.'

The present knowledge of the birds of the desert areas of Iraq is insufficient to indicate their role as a factor influencing the vegetation. Some of the larger bird species are intensively hunted for their meat. The ostrich (*Struthio camelus* L.) has disappeared from this area. Two bustards, the great bustard (*Otis tarda* L.) and houbara (*Chlamydotis undulata* Jacquin) can still occasionally be seen. Hunting of houbaras by falcon, especially the saker falcon (*Falco cherrug* Gray) has been carried out so intensively that the species is rapidly disappearing now. Sandgrouse (*Pterocles* spp.) are every year shot by the thousand. They are hunted from cars mostly after an approach to a distance of some 20–30 m. Kills of five birds in one shot are no exception and stories are told of hunting parties shooting 2500 birds within a week. A good impression of the endangered position of the wild fauna of the desert areas in the Middle East has been given in Mountfort (1965). Only coordinated, intensive action may save what is left.

Of the reptiles two conspicuous species have to be mentioned, the desert monitor (*Varanus griseus* Daudin) and the spinytailed lizard (*Uromastix aegyptius* L.). The first species mainly feeds on smaller lizards and rodents, but *Uromastix* is a vegetable feeder. It reaches a length of about 50 cm (Fig. IV–27). Kevork & Al-Uthman (1972) studied its food habits in the area between Kerbala and Abu Dibbis and have listed the food plants. This list includes the shrubs *Anabasis setifera, Ephedra alata, Haloxylon salicornicum* and *Nitraria retusa*. The species can be abundant locally and is sometimes hunted by the Bedouins for its meat. Of the smaller reptiles the agamids (*Agama* spp.) and sand lizards (*Acanthodactylus* spp.) are most common. Snakes were occasionally seen and are unreasonably feared by the desert inhabitants, although some poisonous species do occur

Figure IV-27 The spinytailed lizard (*Uromastix aegyptius* Forskal) is one of the smaller plant feeding animals. Its eating of different types of shrubs has been reported. The species is occasionally caught by the Bedouins for its meat.

(Mahdi & Georg 1969). Sofar, little is known about the role of the birds and reptiles in the ecosystem of the desert shrublands of Iraq.

Figure IV-28 The two important livestock types for the Iraqi desert areas, the Middle East fat-tailed sheep and the onehumped camel or dromedary.

IV.F.3. *Domestic animals*

Apart from dog and cat, the domestic mammals in Iraq are the horse (*Equus caballus* L.), ass or donkey (*Equus asinus* L.) and mule (*Equus asinus* x *E. caballus*), the onehumped camel or dromedary (*Camelus dromedarius* L.), the domestic cattle (*Bos taurus* L. and *B. indicus* L.), the water buffalo (*Bubalus bubalis* L.), goat (*Capra hircus* L.) and the Middle East fat tailed sheep (*Ovis aries* L.). Camel and sheep, and to a lesser extent goat and donkey, are the species of importance for the desert shrub vegetation (Fig. IV–28). Horses, mules, cattle and buffaloes are mainly kept in the irrigated areas in the north. Three main breeds of sheep have been distinguished in Iraq: Awasi, Arabi and Kurdi, and two breeds of goats: Kurdi and Iraqi (see Hatt 1959). For the study of the utilization of the desert shrub rangelands estimates of the livestock population and trends in this population are of great importance.

Table IV–14. Estimated total livestock numbers of Iraq from available census data.

Year	Number of livestock (thousands)								Total animal units* (thousands)	Source
	Sheep	Goats	Camels	Donkeys	Mules	Horses	Cattle	Buffaloes		
1971	5 582	1 794	38	497	39	93	1 337	123	3 376	Central Statistical Organization (1973)
1965	11 040	1 846	202	542	72	122	1 455	235	5 037	Central Statistical Organization (1970)
1956	9 221	2 640	193	493	100	234	1 535	281	5 054	Haseeb (1964)
1950	7 490	1 754	279	326	67	184	1 035	201	3 886	International Bank (1952)
1938	5 525	2 224	82	–	–	–	250	60	–	Naval Intelligence Division (1944)

* One animal unit is defined as (the feed requirement of) a mature cow or the equivalent of a mature cow. The following conversion factors have been used, based on various sources e.g. Brown (1954), USDA (1976), le Houerou & Hoste (1977): 1 sheep = 0.20 animal unit, 1 goat = 0.15 animal unit, 1 camel = 1.25 animal unit, 1 donkey = 0.60 animal unit, 1 mule = 1.10 animal unit, 1 horse = 1.20 animal unit, 1 cow = 1.00 animal unit, 1 buffalo = 1.25 animal unit.

IV.F.4. *Changes in livestock population*

Amongst the oldest livestock population data available for the country are the figures obtained from taxes that had to be paid for livestock in the thirties. In the Geographical Handbook (Naval Intelligence Division 1944) the animals taxed in 1938 are given and a mention is made that after 1938 the method of taxation ceased to be any guide to the number of animals. Later livestock censuses for the country were carried out in November 1944, March 1946, June 1949, March 1956 and in 1965 (Haseeb 1964; Central Statistical Organization 1970). These so called censuses were carried out by Veterinary Department officials who visited the owners in various provinces, districts, and subdistricts and obtained from them estimates of their livestock.

A well-organized agricultural census was undertaken in autumn 1971 (Central Statistical Organization 1973). The whole country was covered, including the desert areas, where for the nomadic tribes an adapted enumeration procedure was followed. In addition, information can be obtained from the agricultural censuses of 1952/53 and 1958/59, which include figures on livestock in holdings only.

Shortcomings in the 1952–53 census in particular as compared to the later one have been pointed out and indicate the degree of reliability of the figures finally published (Haseeb 1964):
the census started in October and was only completed following summer, too long a period for comparable figures, especially livestock;
information was collected for the most part by untrained, part-time enumerators;
coverage of the census depended on number and extent of the interviews, as there was no ultimate basis in cadastral surveys. This makes it doubtful whether all agricultural holdings were traced and there is a belief that some enumerators rendered fictituous return;
there is likelyhood that some farmers have given erraneous information, on the one hand understating for fear of taxation, on the other overstating in the hope to secure land.

Apart from these censuses, in various publications rough estimates are given, which however, often in one way or the other, have been derived from earlier census figures (Ali 1955; Sarsam *et al.* 1956; FAO 1971; Al-Hashimi 1972). From the various sources table IV–14 was compiled.

Although these data are the best available, conclusions on trends should be drawn with caution only. The enormous drop in the number of camels for instance from 202 000 in 1965 to only 38 000 in 1971 cannot be explained satisfactorily and must be partially due to differences in census method. Mules, horses and cattle are easier to enumerate and their numbers for the various census years seem in agreement with what could be expected. Mules and horses show a continuing decrease after 1956, a phenomenon that could be ascribed to the enormous increase in number of motorized vehicles in Iraq since the early fifties. Cattle show a remarkable stability and the fluctuations shown in the number of donkeys seem realistic. For the present study in particular, the number of sheep is important. Although the absolute figures may be somewhat inaccurate, the trend is clear: an increase in number of sheep until the sixties, with roughly a doubling in 25 years and a sharp fall afterwards. It is the general trend shown in the last column, where all animals have been grouped in animal units. For the present study the important point is that the estimated total number of animal units increased until about 1965. This increase means on the one hand an added value to the national income, on the other hand a greater pressure on the vegetation of the rangelands because growing of fodder crops has by no means increased at the required rate. The sudden drop in livestock over the last decade or so (especially of sheep and camels) is one of the reasons for writing this account.

The available census data do not provide a direct estimate of the iivestock population of the desert areas at any season. Such figure being necessary for an assessment of the economic value of the desert shrub vegetation, this subject is discussed further in chapter XII. The impact on the desert shrub vegetation of grazing and browsing animals is treated in chapter VII and the role of livestock with respect to the development of man in the area is briefly described in the following section.

IV.G. Man

IV.G.1. *The oldest inhabitants*

Part of the desert areas of Iraq and especially the lands at the periphery of the desert have been inhabited by man for many thousands of years. An indication of the earliest presence of man in the desert areas of Iraq is found in artifacts, reported from various locations, and dating back to palaeolithic times. The desert has sofar been poorly explored for traces of inhabitation during the stone age but flints have been found at Jebel Anazeh in the extreme southwest, at wadi Hauran near Rutba and in the Qa'ara depression, north of Rutba. Other surface finds of palaeolithic times have been reported from sites in the Jezira (Solecki 1955). Voûte (1957) found an artifact in a boring at Razzaza near the Abu Dibbis depression. A few years earlier, palaeolithic flint scrapers, knives and points had been collected from several sites in the neighbourhood, near Shitathe, and Tar Kshaife on the road from Kerbala to Ukhaidir (Solecki 1954). There is ample evidence of an alternation of more humid, pluvial periods with extremely dry periods during the pleistocene, of which at least one fell in the palaeolithic period (Voûte 1957; Buringh 1960). Living conditions near the depressions must have been pleasanter than at present and early man may have temporarily settled in a savanna-like landscape near the lakes during the pluvial interphases, a view presented by various authors (Solecki 1954; Voûte 1957; Butzer 1958; Whyte 1961).

IV.G.2. *Domestication of animals and the first migrations*

From the fringe of the desert, in the northeastern foothills region, the evolution of man from hunter and food gatherer to sedentary food producer has been described *e.g.* by Braidwood (1952, 1954) after excavations at Jarmo and Solecki (1955) after his work in the Shanidar cave. First living in caves, later also in the open (by about 8000 B.C.), man may have been experimenting with potential domesticates of wheat and barley, sheep and goats. About 1000 years later, around 7000 B.C. village farmers in the Near East kept domesticated goats and sheep (Butzer 1964) and by 5000 B.C. neolithic farmers and cattle breeders inhabited parts of the foothills region (Roux 1964).

Prehistoric man was dependent on permanent shelter and food and water supply in his settlement sites. This must have kept him out of the deserts with the exception perhaps of some locations along permanent rivers and short seasonal movements into the area when watering points and food for the livestock were plentiful. The transition from early agricultural settlement to urban life and the rich historic sequence of civilizations in the Mesopotamian Plain are rather well documented (Roux 1964).

Relatively little is known about the parallel development of man's exploration, inhabitation and utilization of the desert areas in those times. Large groups of Semites left the Arabian Deserts to settle in the borderlands, Mesopotamia on the one side and Syria and Palestine on the other, *e.g.* the Akkadians (about 4000 B.C.), Amorites (2–3000 B.C.) and Aramaeans (about 1200 B.C.). It is certain, however, that the number of people and livestock migrating at any time through the desert was relatively small. With the exceptions of the desert fringes perhaps, near the agricultural lands, permanent pressure by man on the desert vegetation can be ruled out for these times.

IV.G.3. *Pastoral nomadism, the Bedouin*

The first real pressure on the desert vegetation started at the time – when, where and how is still an open question – that a type of land use now called 'pastoral nomadism' was developed. It has been suggested that it was 'derived from specialized hunting groups who from hunting a particularly abundant animal gradually pass to the control of its movements'.* A more plausible view is given by Johnson (1969):

> 'Historically, pastoral nomadism is best described as a specialized offshoot of agriculture that developed along the dry margins of rainfall cultivation. Here some groups, in order to exploit the nearby arid areas that were unusable from an agricultural point of view, gradually increased the numbers of their domesticated animals until concern for the flock began to take precedence over farming

* In Johnson (1969), presenting the point of view of Richard Thurnwald, described in 'L'économie primitieve' (Paris: Payot, 1937).

activities. Once the point was reacged a series of years in which rainfall totals were below normal would encourage abandonment of agricultural pursuits and concentration upon the flocks as the major subsistence course'.

When the ties with the cultivated lands of origin had been given up completely, pure nomadism was born. This process must have taken place in the second millennium B.C., perhaps even earlier. Some suggest the arid lands partly enclosed by the fertile crescent, others the steppes of interior Asia, as the place of origin (Krader 1959). For Arabia, including Iraq, use of the domesticated dromedary camel may have played a decisive role. The people involved in the region under discussion became known as 'Bedouins', those inhabiting the desert (Arabic: Badawy, pl. Badu, literally means pertaining to the Badieh or desert). A vast literature deals with the genealogy, culture and life of these true desert inhabitants. Their tribal organization, manners and customs, especially in relation to the harsh environment, have been well described. For Northern Arabia amongst others in Doughty (1908), Musil (1928), Raswan (1934) and Dickson (1951). In the context of the present study only some points are of direct relevance.

Pure nomadism, whatever its origin, can be characterized in a few points (cf. Peppelenbosch 1968):

- it is a non-sedentary form of animal husbandry, determined by the search for pastures*,
- with the entire human group accompanying the flocks and herds,
- with a seasonal nature of movements over winter and summer grazing grounds, following more or less fixed routes on tribal land,
- but without being self-sufficient.

Little is known about the numbers of the nomadic population inhabiting the Iraqi deserts in the early Bedouin years. What happened with the population in Iraq was closely linked with what happened in Arabia, in particular in the early islamic times, although there are controversial views. Roux (1964) in a review on the origin, movement and settlement of the various Semitic groups, writes:

'To consider the Syro-Arabian desert as the centre of diffusion of the Semites is out of the question. Only the Yemen, parts of Hadramaut and Oman, and a few oases in Arabia proper offer favourable living conditions, and it is extremely doubtful whether the great desert of central Arabia was inhabited at all between the Palaeolithic period – when it was not a desert but a savanna – and the first millennium B.C. Life in extensive desert areas presupposes long-range seasonal migrations in search of pastures, but only short-range migrations were possible before the widespread use of domestic camels in the Near East from the twelfth century B.C. onwards. Before that time the nomads, who rode on asses and practised sheep-rearing, were much more restricted in their movements than the bedouins of today and could not wander far beyond the limits of the grassy steppe which extends between the Tigris and the Euphrates and at the foot of the Zagros, the Taurus and the Lebanon'.

* 'and by watering points' should be added, a point not mentioned by Peppelenbosch.

IV.G.4. *Transitions of pure nomadism to a settled cultivation*

With their fast horses and riding camels, the migrating nomads had little difficulty in moving the farmers from their lands, *e.g.* in the Upper Jezira. Sometimes it was more a matter of shift in tribal areas from the desert to the cultivated lands that semi-nomadic tribes were forced to sedentary agriculture (Al-Kasab 1966). The population of Iraq as a whole is now of a great diversity, but virtually all people until recently living in the borderlands belonged to Bedouin tribes who were, or claimed to be, of pure Arabian descent.

Two processes have been indicated as possible causes of the expansion from Arabia to the north, (i) climatic, slight changes or fluctuations in the amount of precipitation, and (ii) political, decline of power of the ruling forces in the borderland, inducing migration and settling of nomads in or near agricultural areas.

Huzayyin (1956) discussed these points and gave an integrated picture. Evidence can be found from archaeology and historical documents for a change towards a greater aridity especially in the areas of the present Hadramout and Yemen. The author mentions that it has become proverbial amongst the Arabs that Yemen was the cradle-land of the Arab tribes, and that Iraq with its well-irrigated plains became their graveyard. The movement of people from the present Saudi Arabia, perhaps via stages of nomadism to a settled agricultural life in the borderlands, although particularly intensive during the early years of Islam has continued for many centuries. Information from sheikhs on, now at least partially, settled tribes in Iraq, published in Jamali (1934) provides an excellent record, as shown in the following example:

> '*Information secured from Sheikh Hassan Al Suhail regarding the tribe of Bani Tamim* (letter dated March 4, 1931)
>
> The tribe of Bani Tamim lived in Hota and Harij in the country of Nejd (Arabia), these being their villages, and their desert surrounds these villages. Then the nomadic Bedouin of Bani Tamim moved to Al Dahnaä and Tammam, in Nejd, and during the Caliphate of Omar Ibn al Khattab they immigrated to the desert around Basrah, and they were among the Moslem warriors whom the Caliph Omar Ibn al Khattab delegated to conquer Iraq. At that time they began migrating between Basrah, Kufeh, and Nejd seeking pasture for their camels. They maintained this status of life until the year 1150 A.H. when they came to the Hor of 'Aqarquf, which is our home at the present. Then they kept moving within Al Jazireh, the land between the Tigris and Euphrates, until the year 1180 A.H. when they had a year of drought and dearth of food. In brief, this year was one of the unpraiseworthy years in which the tribe were dispersed, as were the people of Saba, and the pasturage and cattle destroyed. A large portion of the tribe moved and settled in Hamrin Mountains, seeking for living and pasturage. At present they live in Diyala Liwa, which lies north-east of Baghdad. Some of them moved to the lands between Basrah and Samaweh and these are

now called Bani Malik and their chief is Badr al Rumaiyith, a well-reputed Iraqi leader. Those who lived in Hemrin Mountains are nearly ten thousand in number now and the Bani Malik are about twenty thousand in number. A small portion of these moved to the north of Baghdad and they at present live near Samarra, being about two thousand in number; and the rest live with me, and these are about six thousand in number. There are also smaller scattered tribes of Bani Tamim in the various quarters of Iraq and on Shatt al Arab. The bigger portion of these migrated into Persia during the days of Omar Ibn al Khattab, their migration being instigated by the Caliph for political reasons, and they are still in Persia to-day.

All these tribes are partly nomadic and partly settled; the nomadic tend the camels, sheep, and other cattle, while the settled engage themselves in agriculture.'

The events at the desert fringes were preceeded by shifts in tribal areas inside the desert. Raswan (1930) reports that a great northward shift of the Shammar and Aneza nomads, the two largest tribes in Iraq, took place during the first part of the seventeenth century, establishing roughly the later territorial boundaries. These boundaries were marked by waterholes, wadis, hills, particular variation in the desert surface and vegetation etc. Maps of the distribution and movement of the major nomadic tribes around 1930 have been given in Raswan (1930) and Naval Intelligence Division (1944), a later one in Al-Kasab (1966).

In the transformation from pure nomadism to permanent settled agriculture a number of stages can be distinguished, *viz.*

(i) the change to the possession and tending of sheep and goats and gradual elimination of the camel, living in tents;
(ii) cultivating some land (wheat and barley) for part of the year, combined with spring grazing in the desert (semi-nomadism), living still in tents, sometimes surrounded by reed screens or mud walls, or in sundried mud brick houses;
(iii) settling on the land and becoming a cultivator, living in houses build of reeds (in the marshy districts) or bricks.

For a long time the tribal population of Iraq could belong to any of these stages and even the same tribe or section of a tribe could consist of groups in various stages toward settlement as shown by the quoted letter above. In 1920 Gertrude Bell wrote:

'The tribal population of Mesopotamia exhibits every stage in the conversion of the Bedouin into the settled cultivator of the soil; there are tribes still purely nomadic who have never yet put their hand to the plough, others who are concerned solely with the care of their palm gardens, corn, and rice fields, others again who combine the occupation of that of an agriculturist with that of a shepherd, and yearly, with the coming of the winter rains, send half their number to pasture the sheep of the community in the desert . . .'*.

An understanding of this background of the recent and present users of the desert vegetation is essential for an assessment of the man-vegetation relation during the last half of a century or so when the living conditions changed so drastically.

IV.G.5. *Developments in the first half of the twentieth century*

From olden times the Bedouins had three main sources of income:

- their livestock, mainly through camel breeding,
- raids on enemy tribes, trade caravans, oases and agricultural settlements at the desert fringe,
- services related to trade and travel parties, *e.g.* guiding caravans and protection *i.e.* receiving money for obstaining from raids**.

In the course of a few decades in the present century, in particular the first two after World War I all three sources were undermined completely. The main causes were the arrival of the motorcar, the establishment of the new state boundaries and an increased governmental control over the former Bedouin domain, all interrelated. With the coming of the motorcar as a strong, fast and easy to feed means of transport, and also with the introduction of mechanical pumps in the irrigated areas, the demand for camels dropped drastically in a few years. A journey which previously took two weeks by camel could be covered in 24 hours by car. In the process of fixing the borders of Iraq with the surrounding countries, after World War I, thought had been given to tribal areas and movements. An abstract from the agreement between the representatives of the King of Iraq and the Sultan of Najd and the British High Commissioner in Iraq, of 5 May 1922 illustrates this:

'The wells and lands used from old times by the Iraq tribes shall belong to Iraq and the wells and lands used from old times by the Shammar Najd shall belong to Najd. In order to determine the location of these lands and wells and to fix a boundary line in accordance with this principle a committee shall be formed . . .'

and . . . 'whereas many of the wells fall within the Iraq boundaries and the Najd side is deprived of them, the Iraq Government pledges itself not to interfere with those Najd tribes living in the vicinity of the border should it be necessary for them to resort to the neighbouring Iraq wells for water, provided that these wells are nearer to them than those within the Najd boundaries' (cf. Gehrke & Kuhn 1963).

The new boundaries, however, had a great impact, and clashes were unavoidable. What happened at the Iraq-Arabian border between 1927 and 1929, illustrated

* From: Review of Civil Administration of Mesopotamia, 1920, p. 21, quoted in Jamali (1934).

** For the role of the desert in north-south, east-west connections, see Holt (1923).

'. . . all too typically, the effects of confronting the proud but primitive and fanatical bedouin of north-eastern Arabia with a territorial frontier, which they could neither locate nor comprehend, as opposed to one tribally indicated; with an attempt at national discipline in their own country, which they resisted with savage individualism; and with modern weapons, which they resented as evil and unjust' (Longrigg 1953).

With fast terraincars, airplanes and well organized armed forces, a new order was established and the people, who had proved their flexibility over centuries, had to adapt to the changed situation, in the desert as well as in the areas of rainfed and irrigated agriculture. A decrease in the pure nomadic population was the result and an increasingly sedentary life.

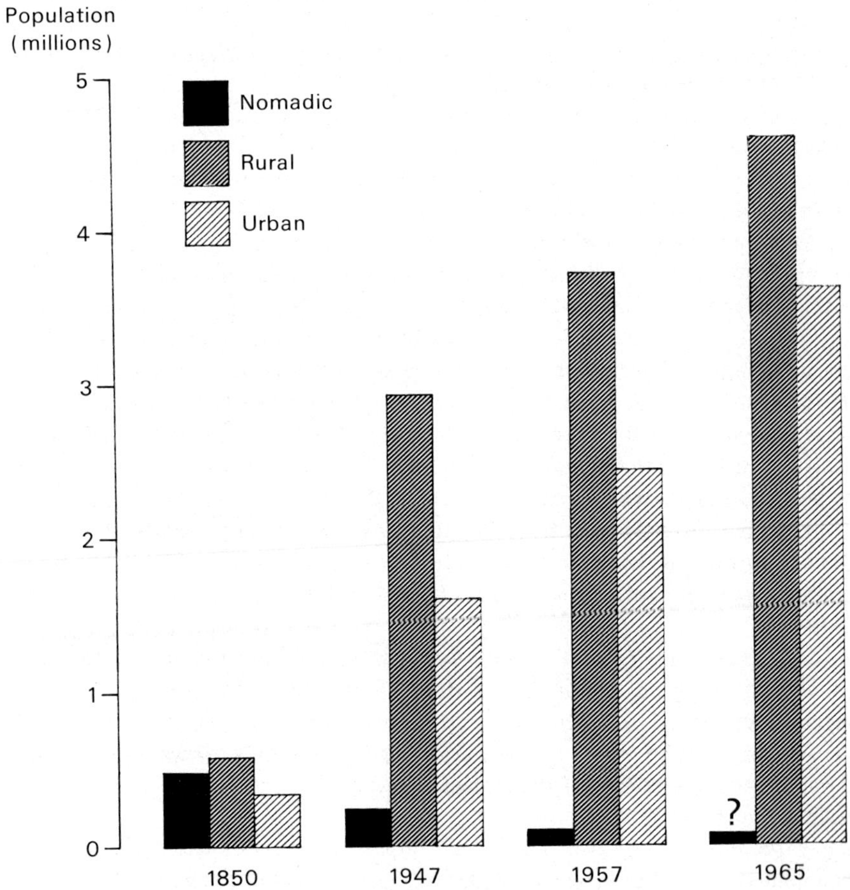

Figure IV-29 Estimated nomadic, rural and urban population in Iraq in 1850, 1947, 1957 and 1965 (from data given in Lawless 1972).

All statistics on the nomadic population of Iraq are of questionable reliability. First the pure nomads as a group are not easily defined against the population who had settled temporarily and moved out seasonally into the desert with the livestock. Second, at the time no highly reliable census methods could be applied while the required maps and enumerators were lacking and aerial census methods not yet developed. As a result there are only rough estimates, perhaps better 'guesstimates'. The figures encoutered in going through the various sources are presented in table IV–15. The best possible sources indicate a gradual decrease in the nomadic population, at least over the last fifty years, against a great increase in the rural and urban population (Fig. IV–29).

Table IV-15 Estimated nomadic population for selected years in relation to total population of the country. From various sources, as indicated.

Year	Nomadic population (thousands)	Total population of the country (thousands)	Nomadic as percentage of the total population	Author and source
1867	500	1 280	39.1	Hasan (1958)
1890	no data	1 726	–	Hasan (1958)
1900	no data	2 000 - 2 500	–	Longrigg (1953); estimated, partly based on Turkish census figures (?)
1905	no data	2 250	–	Hasan (1958)
1919	no data	2 848	–	Hasan (1958)
1935	no data	3 605	–	Hasan (1958)
1938	600	3 871	15.5	Al-Kasab (1966); quoted from a 1961-report of the Min. Agr. Refrom
1947	170	4 750	3.6	Lebon (1953); taken from 1947 census
1947	250	4 861*)	5.2	Al-Kasab (1966); quoted from a 1961-report of the Min. Agr. Reform
1947	250	4 861*)	5.2	Lawless (1972); based on 1947 census
1952	200 - 500	5 720*)	3.5 – 8.7	Int. Bank (1952); no source mentioned
1957	270	6 340*)	4.3	Al-Kasab (1966); quoted from a 1961-report of the Min. Agr. Reform
1957	65	6 340*)	1.0	Lawless (1972); based on 1957 census
1962	200 - 250	7 480*)	2.7 – 3.3	Yacoub (1971); quoted from a 1962-report of the Min. Agr. Reform
1965	272	8 262*)	3.3	Yacoub (1971); quoted from a report-by A.J. 'Oraim (1965)
1970	200 - 270	9 440*)	2.1 – 2.9	Yacoub (1971); estimate, based on some of the above mentioned sources.

* From Annual Abstract of Statistics (Central Statistical Organization), several volumes. All figures based on the 1947, 1957 and 1965 census data.

The 500 000 nomads estimated for the mid-nineteenth century decreased to an estimated ten per cent one hundred years later, while over the same period the total population of Iraq increased from 1.3 million to over 5 million.

IV.G.6. *Recent developments and the present situation*

Reduction of the pure nomadic population, however, has in no way meant a reduced pressure on the resources of the desert. An ever increasing number of wells and the fading tribal rights made a seasonal movement with large numbers of livestock relatively easy. Permanent water supply, improved communications, full governmental control and facilities provided by the various development projects have caused great changes. At the edge of the desert, villages and even large towns have developed. Zubair, Nasiriya, Samawa, Najaf, Kerbala and Ramadi are examples. Inside the desert areas an ever growing system of marked tracks, and even of tar roads is developing especially under the influence of oil exploration and exploitation, and this, together with a system of policeposts, equipped with wireless connections and fast vehicles, guarantees adequate communication and full governmental control.

Villages and small towns have grown near policeposts and wells, often along old desert routes, e.g. for the hadj to Saudi Arabia (Busaiya, As Salman, Shabicha, Nukhaib), to the Mediterranean (Rutba) and along the Euphrates across the Jezira (Ramadi, Haditha, Hit, Ana and Rawa). Other fixed points are the policeposts for control of the borders with Kuwait (Khedr-el-Mai), in the Neutral Zone (Samah, Ansab), with Saudi Arabia (Jumaima, Aidaha, Judaidat Arar, Mugr an Naam) and with the western neighbours (Traifawi).

Moreover, an increased governmental policy of settlement of the remaining nomadic population is pursued by virtually all governments, not only in the Middle East, and Iraq is no exception. Contrary to the spontaneous sedentarisation described before, this may be called induced sedentarization. In view of the developments taking place in the area as a result of technological development and oil revenues, the action taken will perhaps only speed up a natural development. A number of reasons can be given to defend this policy: social and administrative, political and economic. Education, health services and general integration with the rest of the population are the arguments most often mentioned in the numerous documents dealing with this matter (e.g. United Nations 1970; FAO 1970; Yacoub 1971; FAO 1972). A major problem faced is the attraction of the cities causing an urbanization rather than a dispersed settlement pattern of the rural areas, including the desert. Measures are being taken to counteract this development. Their effectiveness can only be judged in the future.

Apart from the drilling of wells, in Iraq schools and health centers have been established in the larger settlements in the desert areas. More recently programmes have been launched and are now executed for road construction and the building of new villages and expansion of existing ones in order to improve the living conditions of the nomads and have them settled permanently. A scheme for villages, each with 30 to 50 houses, school, market, health services and an agricultural unit and good water supply, described in Al-Thaura of 17 July 1975, serves as an example. The ecological consequences of such measures, *e.g.* in terms of resource exploitation, should not be overlooked and must be a matter of serious concern which will be touched upon in later chapters.

IV.H. Summary of developments, especially with respect to man

For the purpose of the present study, three phases in the human occupation of the desert areas of Iraq may be roughly distinguished:

(i) From palaeolithic times to the second millennium B.C.

An occasional local occupation originally perhaps only during pluvial periods near depressions or lakes, in historical times most likely seasonal and only permanent at a few points along the Euphrates and near natural springs.

(ii) The second millennium B.C. to the early twentieth century.

An increasing occupation by the nomadic pasturalists and establishment of trade routes, all governed by the Bedouin tribes with their own culture. Fluctuating numbers, mainly as a result of irregular expanding tribal movements from the Arabian peninsula into the present Iraq and a continuous shifting pattern of occupation.

(iii) Present century, especially after World War I.

Strong increase of communications and permanent settlements, at the periphery a growth of large towns, inside the desert new villages, policeposts, etc., all this combined with (semi-)nomadic movements from the periphery, a development induced by facilities provided by modern technology.

Human impact on the vegetation, fauna and other resources must have been very limited at first, because of the relatively small number of people and the restrictions imposed by the few permanent watering points and limited means of transport. At times of expansion of the population from Arabia into Iraq and the centuries following, the impact may have increased locally along established migration lines and trade routes. However, limited water supply facilities and established tribal grazing rights were the regulating factors and there is no evidence for marked changes in fauna, vegetation, and environment. Taking the gazelles (*Gazella* spp.), some larger birds as bustards and ostrich (*Otis tarda, Chlamydotis undulata* and *Struthio camelus*), and some of their predators the hyena (*Hyaena hyaena*), cheetah (*Acinomyx jubatus*) and desert lynx (*Caracal caracal*), as indicators, most of these species have been reported well into the present century. This picture of human inhabitation of the area and especially man's utilization of its resources has changed dramatically over the last fifty years or so. These changes had been going on for some time but became apparent with the political changes and technical innovations after the first World War.

In general, at least for Iraq, the number of livestock that was built up in an area under extensive seasonal grazing was governed by a few factors only, *viz.*

- forage supply,
- water supply,

– number of people required to herd (guide and protect) the animals,
– consumer demand of the livestock products,
– communication and means of transport (strongly related to the above points).

In the fifteen years following the first World War, the present boundaries of Iraq were established and control of the area by a central government was ensured. It worked mainly through the products of technological development: wireless communication, fast terrain cars, sometimes even airplanes (landing grounds were easily established), used by an organized and disciplined army and police. Because of the lack of security, raiding had prevented breeding of large flocks by small communities (Longrigg 1925; Raswan 1934; Sweet 1965). Raids were stopped under increased governmental control.

Stock watering points had always been the limiting factor for any livestock movement in the desert. There were a few permanent watering points and the seasonal ones often did not last long. In fact lack of stock water supply rather than forage supply was in many cases the cause of moving on. The large number of wells drilled and constructed during the last decades was discussed before. They have made a large part of the desert permanently accessible for man and his livestock. The parts still remote from water supply can be served by tankers, owned or hired by a group of families (Fig. IV–25).

The demand for camels with their variety of services and products has decreased. It has been replaced by an increasing demand for mutton and other products of sheep, due to an increasing population with a higher per capita income, to a great extent accounted for by the oil revenues (Ahmed 1968). Fast movements of flocks is now possible by using trucks for transporting animals in times of emergency or for marketing purposes.

This left forage supply as virtually the only limiting factor in building up large numbers of livestock in the area. It often was not limiting, at least not in the early phases. Newly drilled wells and increased transport facilities opened up new areas for an almost permanent inhabitation and facilitated the seasonal migration into the area from the periphery by semi-nomadic herdsman with large flocks of sheep. The impact of these changes on the natural large herbivore populations and their predators has already been mentioned. The impact as found in pattern and process in the vegetation is presented in the following chapters.

CHAPTER V

RANGE COVER TYPES

V. RANGE COVER TYPES

V.A. Introduction

Two types of information on the vegetation of the desert areas of Iraq are available in the literature, *viz.*

(i) detailed descriptions of the vegetation in selected, small areas *e.g.* Ukhaidir area (Nasserula 1971), Baiji sand dunes (Al-Hilli 1970), saline swampy areas near Najaf (Abul-Fatih 1970, 1975), area along the Baghdad-Husaiba road (Batanouny & Sheikh 1972), the Ghurfa desert (Batanouny & Hilli 1973) and Jabal Sanam (Weinert & Al-Hilli 1975);

(ii) general descriptions of the vegetation types of the area as a whole, usually resulting from a number of reconnaissance trips (Guest 1966; Monakhov 1972; Al-Ani *et al.* 1970b, 1974; Kaul & Thalen 1971b; Zohary 1940, 1973). These descriptions characterize the types in terms of the one or two dominant species only. Their relation to the environment is only mentioned as resulting from impressions recorded in the field.

For the present study a more detailed picture of at least the most important communities* of the desert rangelands as a whole was desired. A study was undertaken with the main objective being an identification of community types, to be described in terms of the contributing species with the prevailing environmental conditions.

A classification and description of vegetation types for an area as large as the study area in Iraq has to be based on an analysis of selected subareas or stands. A complete (100%) survey of the selected parts is usually impractical or impossible and within the stands a subsampling scheme has to be applied. In this respect four steps can be distinguished in vegetation sampling (see Mueller-Dombois & Ellenberg 1974), *viz.* (i) segmentation of the vegetation cover or recognition of entities, (ii) selection of samples in the recognized segments, (iii) the decision on size and shape of the sample, and (iv) the decision on what to record within the sample units or plots. These points are discussed below, followed by an outline with discussion of the classification procedure followed. Finally a description is given of the community types arrived at, including their habitat characteristics.

---

* The term 'community' is often used confusingly in an abstract as well as concrete sense. In the concrete sense it may be considered identical to stand or phytocoenose. In the abstract sense a classification unit is meant, more or less similar to community type or nodum, for which the term phytocoenon has been proposed. For a detailed discussion see Westhoff & van der Maarel (1973). In the present study the term stand has been most commonly used for the concrete vegetation segment.

V.B. Sampling

V.B.1. *Selection of the stands*

As a result of reconnaissance survey trips over about 4000 km along different tracks, an overall knowledge had been gained on the general features of the vegetation in the desert rangelands. The major species constituting the stands covering extensive areas or formed in particular habitats had been identified. A brief characterization of cover types by their dominant species could be given (Al-Ani *et al.* 1970b; 1974; Kaul & Thalen 1971b). These types were an abstraction from the overall impressions from the field and not based on any planned field sampling. They nevertheless provided the necessary basis for the field sampling.

When extensive areas have to be dealt with, the importance of a good reconnaissance knowledge of the area as a whole can hardly be overestimated (Mueller-Dombois & Ellenberg 1974). The stands that were subsequently sampled in detail were not taken at random or systematically spaced over the study area, but chosen by a procedure of subjective selection. In the early phase of the sampling, criteria for the selection of a stand were either a supposed high representativeness for a widely distributed and often also common vegetation type or the obvious occurrence of a particular set of environmental factors. An example of the last category are the stands sampled inside the exclosures at Rutba and Khedr-el-Mai. In a later phase stands were sometimes selected on the basis of an 'uncommon' combination of dominant and co-dominant species, especially when these species had already been encountered earlier as important constituents in species combinations of other stands. It may, therefore well be that stands with some 'unusual' species combinations are relatively overrepresented in the material. However, such situations can sometimes be highly informative. Subjective sample site selection as carried out for this study had great advantages. The knowledge of vegetation variation before the sampling was initiated better guaranteed the incorporation of the two most interesting types of stands *viz.* on the one hand the most common and widely distributed species and species combinations, on the other hand the species combinations under habitat situations known to be extreme in one way or another. Pure random or systematic sampling could in view of the large area concerned and the limited time available easily lead to an oversampling of the first category and an undersampling of the second. Moreover, organization in terms of logistics makes these procedures unpractical and costly.

In following a procedure of subjective selection of stands there can be the danger of a circularity, strengthening itself. This happens when in the selection emphasis is laid on the sampling of situations fitting in a preconceived abstract vegetation typification. In that case community types are not objectively derived from a sample set, but created and continuously reinforced by a circular reasoning. When the sampling for the present study was carried out, the available knowledge did not allow such a prejudice about the community types to be expected.

V.B.2. *Location of the sampled stands*

By July 1973, when the sampling programme had to be concluded, detailed data had been collected on 54 stands, distributed over the Southern Desert (19), the Western Desert (26) and Lower Jezira (9), as shown in Fig. V–1. A certain geographical clustering can be observed, *e.g.* near Samah, Kehdr-el-Mai, Ukhaidir, Nukhaib, Rutba and Hatra. This is mainly due to organisational and logistic pro-

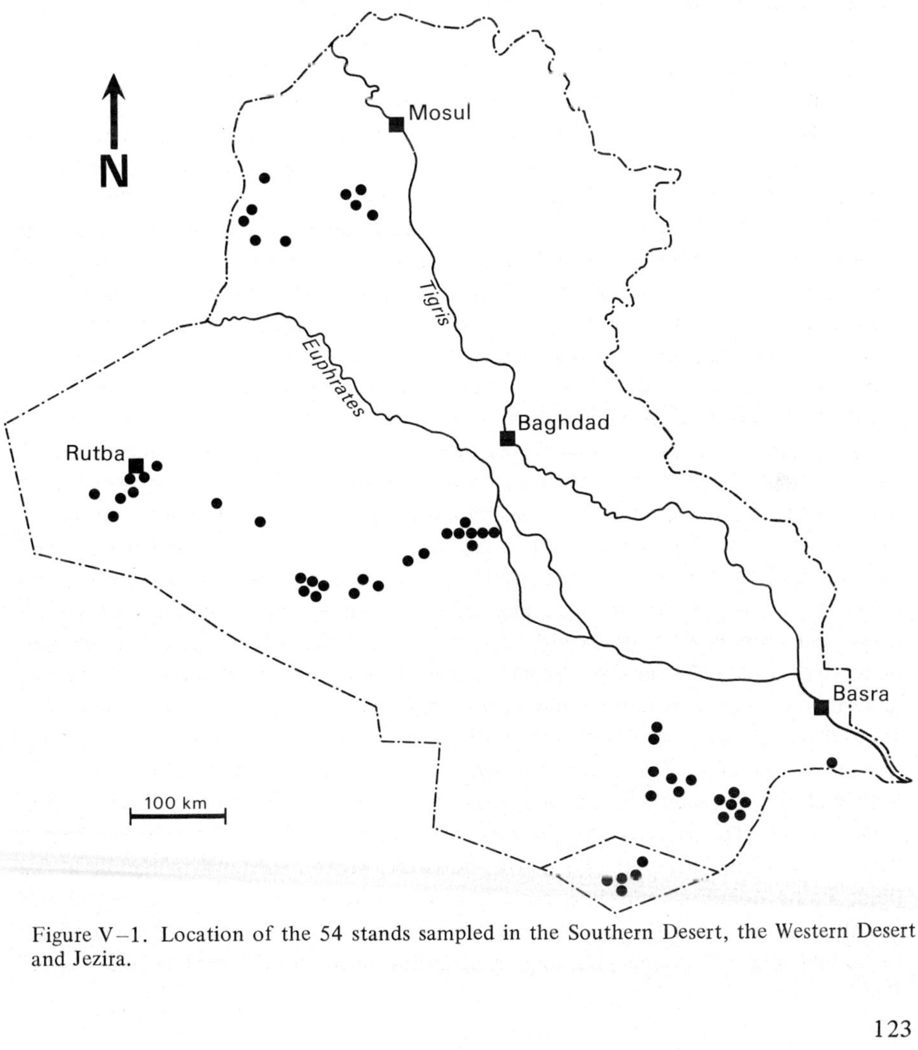

Figure V–1. Location of the 54 stands sampled in the Southern Desert, the Western Desert and Jezira.

blems. A significant influence on the results has not been found, except that the spatial distribution of certain types seems more limited now than it is in reality. A more intensive sampling programme should reveal this. The Jezira is, compared to the Western and Southern Deserts, undersampled.

V.B.3. *Size, shape and number of sample plots*

Taking into account (i) the heterogeneity of many of the stands to be sampled, often related to some minor local relief or surface soil differences, and (ii) the openness (low cover) of almost all stands, it was obvious that a relevé method, using one sample plot as representative of a stand would either require an unworkable large sample plot size or be liable to introduce large errors. Moreover, it had been decided that a number of vegetation parameters would have to be quantified through actual measurement, to avoid as far as possible personal bias resulting from observations and estimates by different observers. A method of subsampling with a large number of small sample plots was therefore adopted.

Shape, size and number of sample plots cannot be seen independently when a defined sampling precision for a certain vegetation parameter is the objective. Another factor to be considered is that when stands are very open (and often also heterogeneous in their spatial pattern) a much larger total sample plot area will be required, compared to a dense and homogeneous stand, to reach the same accuracy. The sampling unit can theoretically be any shape, but in actual practice a choice has to be made between circle, square, rectangle, point and line. The point can be considered the smallest form of the first three, the line as an extreme rectangle. All five shapes mentioned have their advantages and disadvantages (Brown 1954; Perry 1962; Mueller-Dombois & Ellenberg 1974). For the present study a square of 4 m^2 was used as sampling unit. The shape was determined by the fact that such a quadrat was easily laid out in the field when using fixed wooden frames. Circular sampling units of the same area would have been more difficult to lay out or construct and transport under the prevailing field conditions. The only advantage would have been less 'edge-effect'. The two by two meter was about the smallest practical size in relation to the average shrub size and proved well adapted to the general spacing of the plants. This had been worked out in field trials with different quadrat sizes. A smaller size would have introduced an additional error through a higher edge-effect. A rectangle would have had the considerable advantage of incorporating in many cases more variation compared to a square with the same area (Bormann 1953; Greig-Smith 1964). Any frame with sides over two metre, however, proved difficult to handle in the field. Point quadrats and line transects were not applied in the present stand sampling because certain parameters related to area (*e.g.* density and biomass) had to be incorporated in the sampling. A sampling unit size of 4 m^2 has been suggested for frequency determinations in low shrub and tall herb vegetation by Cain & Castro (1959) on empirical grounds.

Total sample size, or the number of times a given sampling unit (sample plot, quadrat) should be repeated, is often arbitrarily determined in vegetation studies (Mueller-Dombois & Ellenberg 1974). What constitutes an adequate sample size is difficult to determine on statistical grounds when dealing with a variety of attributes usually to be determined on a number of species, differing in growthform, density and distribution pattern. As a practical method the plotting of the running mean during the quadrat sampling has been suggested (Kershaw 1964). This and other methods, e.g. the ratio of the standard error of the mean to the mean when assuming a random distribution of individuals (Greig-Smith 1964), may be applicable for detailed studies in relatively small areas. For practical work involving routine sampling of large areas, such methods are usually too time consuming and the sample is established empirically after a few tests in stands representative of the situations to be tackled. Such tests showed for the present study that a very large number of quadrats would be required to reach a high precision. A sample size of 50 quadrats was finally adopted for all samples. This constancy in sample size for all stands greatly facilitated data analysis and comparison of stands in a later phase. It implied a relative oversampling in the few homogeneous stands with a high cover and a relative undersampling in the more open, heterogeneous vegetation. Aspects of the resulting precision are discussed under V.B.6.

The 50 quadrats of 4 m^2 were laid out in five sets of 10, each randomly distributed over an area of 1000 m^2 (20 m x 50 m). This means for each stand a sampling intensity of four per cent for an area of half a hectare. This sampling has great similarity to quantitative studies of other workers. Bannink *et al.* (1974) sampled in vegetation of weeds on arable fields 10 circles of 0.2 m^2 in an area of 50 m^2. This works out to the same sampling intensity (4%) in an also very open type of vegetation. Whittaker (1973), referring to earlier work, makes a mention of his 'tenth-hectare strip quadrat' in which small shrubs and herbs were recorded in a number of subquadrats either randomly or systematically spaced along a 50 m tape. Sobolev & Utekhin (1973), in a description of the Russian (Ramensky) approaches to community systematization, also mentioned 1000 m^2 for desert communities and wrote: 'The shape of the test area can be different, and instead of one large plot a series of smaller ones can be used. Intentional selection of 'typical or average' conditions is accepted'. Westhoff & van der Maarel (1973) give 25–100 m^2 as the minimal area for shrub communities and 50–100 m^2 for steppe communities. Werger (1973) using the relevé method for a syntaxonomical study in South Africa used 50 m^2 or 100 m^2 quadrats for dwarf shrub communities of different kinds. Batanouny & Abu El-Souod (1972) in a study on the vegetation in the Libyan Desert along the Cairo-Alexandria road used a sample area ranging from 200 m^2 to 475 m^2 in the different stands. The 475 m^2 was used for a very open *Aristida plumosa** community occurring in patches of 4–25 m^2 at distances

* Stipagrostis plumosa

of one or more kilometres apart. This pattern is more extreme than found in any of the stands sampled in Iraq.

The 200 m^2 resulting from the 50 quadrats of 4 m^2 for desert shrub vegetation compares well therefore with the above values.

V.B. 4. *Vegetation information recorded*

The first decision to be made was whether to include all plant species in the study, or to make a selection. In view of the limited time available sampling had to continue during all seasons. In order to make the data comparable it was decided to record only the species that could be seen during all seasons. The annual, ephemeral, species and geophytes were thereby ruled out. Moreover, these species form synusia of their own, react in their own way to spatial and temporal changes in environmental conditions and, therefore, warrant a separate study. After a heavy rainfall in the wintermonths they may put extensive areas under a dense cover, while in a dry year they may not appear at all. In desert vegetation annuals can make up 60–80% of the total species number (Guest 1966; Zohary 1973).

> 'The total number has for the desert areas of Iraq been estimated as comparatively small – not more than 250–300 species at the most' (Guest 1966).

This leaves about 100 species or less to be dealt with, including a number of rare and uncommon ones. In the sampling all perennial species encountered were recorded and also a few annual or biennial species recognizable in the field the year round, *e.g. Artemisia scoparia* and *Cornulaca* and *Salsola* spp.

For each sample plot height and crown diameters of each individual plant unit were measured and recorded per species on standard forms. Combining these data from the 50 quadrats per stand provided for complete calculation of frequency, density and crown cover (by diameter method, see III.B.3.d.) per species. In addition, all measured quadrats were clipped at surface level and the aerial biomass (fresh and dry, see III.B.3.f.) determined per five quadrats combined. Particular features of interest, *e.g.* species being grazed selectively, were separately noted.

Although a relatively small number of species had to be dealt with, the proper species name could not be established in all cases because of unavoidable damage done to collected specimens or the absence of parts vital to identification. Another difficulty was that plants recorded under one species in the beginning later proved to encompass more than one species of the same genus. Incorporating all knowledge on species, identified or not, was considered more correct than leaving out the information for unidentified species. Wherever identification problems existed, the generic name with sp. (for one species) or spp. (for more than one species) has been used.

V.B.5. *Habitat characteristics*

Habitat characteristics were recorded for each sampled stand. The aim of this part of the sampling was to allow a proper description of the environmental conditions for the stands and wherever possible indicate the relation between community types and environmental conditions. The following physiographic, edaphic and biotic information was collected:

Main terrain characteristics, including soil surface features and the situation of the stand with respect to the surrounding area, were noted.

Soil samples were taken at two sites within the stand, for two depths, 0–30 cm and 30–60 cm. The soil samples were analysed in the laboratory of the Institute for Applied Research on Natural Resources, Abu Ghraib for the following factors: soil texture (% sand, % silt, % clay), organic carbon, gypsum, lime, electrical conductivity (EC), and pH (not for all samples). The samples were collected by digging small pits. Particular features exposed in the profile (horizons, depth of gypsum nodules, stony layers, etc.) were also recorded.

Factors related to utilization by man and animals such as degree of grazing, occurrence of shrub cutting for fuel, recent ploughing, etc. were recorded.

Climatic factors could only be generally assessed, based on the location of the stands, by interpolation and extrapolation of records from the few meteorological stations.

V.B.6. *Precision of the sampling*

The often unavoidable arbitrary nature of the sample size in many descriptive vegetation studies has already been mentioned. When the minimal area concept is applied the representativeness of the sampled area is usually taken for granted once its size has been established from the species number – sample size relation as worked out for a similar vegetation. A check on the minimal area would in most cases be more time consuming than the actual sampling itself and is therefore omitted.

When a particular vegetation attribute is sampled through a number of plots, the required number of plots for an estimate with a defined precision and probability can be calculated. The number of plots required depends not only on the precision and confidence or probability level, but also on the variation among plots. This, in turn, is largely determined by plot size, plot shape and the distribution pattern of the plants within the stand. The number of plots necessary to sample with a desired precision and confidence can be found with the following formula (Milner & Elfyn Hughes 1968; Kinsinger 1977):

$$N = \left(\frac{t.s}{p.\bar{x}}\right)^2$$

where N = number of plots necessary to sample within certain prescribed precision and confidence;

t = confidence or level of probability, obtained from standard statistical tables;

s = standard error*, calculated from a group of values of trial plots;

p = sampling precision or desired level of accuracy, expressed as a decimal;

$\bar{x}$ = mean value, calculated from a group of values of trial plots.

With n sample plot values available, the obtained precision (P) for a certain probability level (t) can be worked out from:

$$P = \frac{t.s}{\sqrt{n}.\bar{x}}$$

This has been done for the data of part of the sampled stands with crown cover (cc) and aerial biomass as measured attributes. Evaluation of the precision when working from the separate quadrats (n = 50) was too time consuming for the crown cover measurements and impossible for the aerial biomass determinations.

Table V–1. Sampling precision of the total crown cover estimates for different confidence levels for 16 stands. The stands were selected systematically according to their place in the rank order of crown covers from highest to lowest.

Stand number	Percentage crown cover as measured in 10 subsamples of 5 times 4 m² each		Calculated sampling precision (% deviation from the mean) for different probability levels			
	$\bar{x}$	s	60%	70%	80%	90%
41	39.0	12.8	9	11	14	18
25	28.8	10.1	10	12	15	20
5	20.2	5.5	8	10	12	16
34	19.1	10.7	16	20	25	33
53	16.8	9.9	16	21	26	35
21	16.1	6.9	12	15	19	26
52	13.7	9.3	19	24	30	40
28	10.7	4.3	11	14	18	24
44	10.1	9.8	27	34	43	57
29	8.7	5.5	18	22	28	37
11	7.8	2.9	10	13	16	22
14	7.4	3.9	14	18	22	29
4	6.8	4.3	18	22	28	37
13	3.8	2.8	21	26	32	43
37	3.3	2.7	22	29	36	48
16	2.0	2.4	33	42	52	70

* The standard error in itself is an estimate and the number of trial plots should not be less than ten (cf. Milner & Elfyn Hughes 1968).

The values for five quadrats combined were, therefore, treated as one sample, giving a total of 10 samples per stand.

The 54 stands sampled were ranked according to their crown cover and for 16 systematically selected stands a calculation was made for sampling precision. These stands represent covers from the highest (cc = 39.0%) to almost the lowest (cc = 2.0%). The sampling precision has been worked out for four probability levels: 60, 70, 80 and 90% (table V–1). The three stands with a cover exceeding 20% showed a precision of ± 10% of the mean at 60% confidence and ± 20% of the mean at 90% confidence. For six stands with estimated mean covers between 10% and 20%, the adopted sampling resulted in a precision ranging from ± 11% to ± 27% of the mean at 60% confidence and from ± 24% to ± 57% at 90% confidence. For the very low covers, cc below 10%, precision varied even more, from ± 10% to ± 33% at 60% confidence and from ± 22% to ± 70% at 90% confidence.

A comparison of the precision of the sampling for the low cover stands with the high cover stands shows a two to three times higher precision for the high cover ones. This means that the variation among plots in the low cover stand is much higher. It may be due to the quadrat size in relation to the distribution pattern of the vegetation. Under extreme arid conditions, especially when there is in addition a heavy biotic influence, the distribution tends towards a clustering of the plants in the few microhabitats where survival is possible.

Sampling precision for the aerial biomass estimates at 60% and 90% confidence levels is shown in Fig. V–2 for all stands. The stands with an estimated aerial biomass of 50 $g.m^{-2}$ or more all have a ± 25% precision or higher at 60% confidence level. For the stands with a biomass of 120 $g.m^{-2}$ or more this precision is even reached at 90% confidence level, with one exception. For the stands with a lower aerial biomass, between 5 and 50 $g.m^{-2}$, at 60% confidence level the precision ranges from ± 7 to ± 42%, at 90% from ± 14 to ± 88%.

In evaluating the minimum precision to be tolerated, it must be kept in mind that for most practical applications a possible relative large error for low biomass stands is less serious than a small error in a high biomass stand. Fig. V–2 shows that the adopted sampling scheme had the advantage that the stands with a higher cover and aerial biomass have been measured with a relatively higher precision. Increasing the precision to levels well above the ones shown in Fig. V–2 would have required an unworkable large sample size. To increase the precision of the stand with the highest biomass (286 $g.m^{-2}$) at the 90% confidence level from the ± 17% achieved to ± 10%, would have required a sample size of 30 times 5 quadrats of 4 m^2! This is a three times higher sampling intensity than was adopted, 150 quadrats to be measured and clipped, instead of 50. Such precisions are, at the present state of affairs, for management reasons not required for the quantification of the desert shrub vegetation. Table V–1 and Fig. V–2 indicate that for detailed studies, particularly of the open shrub cover types, a large sample size

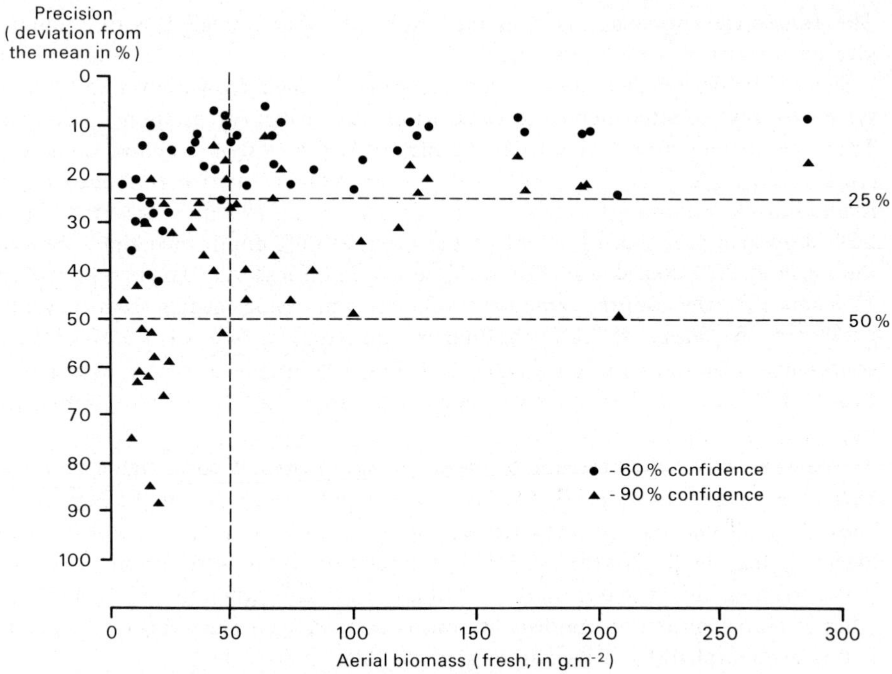

Figure V–2. Sampling precision for the estimated aerial biomass values, at 60% and 90% confidence level.

may be required to achieve the desired precision level. This aspect has so far been largely neglected in such studies.

For the present survey of range cover types the adopted sampling of 50 quadrats of 4 m^2 in each stand proved a reasonable compromise between the precision desired and the limitations set by the time involved.

V.C. Classification

V.C.1. *The approach, selection of characteristics and guiding principles*

In vegetation classification observations on different concrete segments of the vegetation are grouped into classes. Observations sharing particular characteristics are brought together in the same unit. Such a unit is then described and characterized in a way that it can be communicated to others. A community type may be the result, an abstract unit derived from concrete situations sampled in the field (Whittaker 1973). The 'characteristics' used for determination and recognition of the classification unit should be properties of the item to be classified. In vegetation classification the properties most commonly used are taken from (i) physiognomy and structure, or (ii) floristics. Properties outside the vegetation are sometimes used, such as site factors (climatic, edaphic, anthropogenic) or geographical factors (for a listing see Mueller-Dombois & Ellenberg 1974). A vegetation classification, however, should preferably only use properties of the vegetation itself as characteristics and the nomenclature of the classification units should be designed accordingly. Only then a concrete vegetation segment (stand) can be recognized and given its place in that classification system (Zonneveld 1974). Environmental factors, e.g. based on climate, soils or management practices, can be brought into the classification as 'guiding principles' guiding the selection (abstraction) of characteristics and the arrangement of the classification units into a meaningful order, with or without a hierarchy (Zonneveld 1974).

It may be clear that the same set of observations can be classified in as many ways as there are properties or combinations of properties that can be selected as characteristics for working out the similarity between observations. The final classification system is a reflection of the selected characteristics and guiding principles which the researcher considered most relevant for a distinction and description of the classification units. This aspect of choice has lead to a number of distinctly different approaches and as many 'schools' in vegetation classification (see Shimwell 1971 and Whittaker 1973).

For a classification of the vegetation cover of the desert shrub rangelands of Iraq for the present study a number of possible properties were considered to serve as characteristics for the classification.

Physiognomy and structure. A useful classification, based on physiognomy, for desert vegetation in the Middle East, has been given in Kassas & Girgis (1970) in a study on the plant cover of a large part of the Egyptian desert. In the present study emphasis was on shrub vegetation types. Ephemeral vegetation was excluded from the sampling and open forest types were not encountered. This leaves only a few physiognomically different types possibly to be distinguished, certainly too low a number for a meaningful classification. The main advantage of a vegetation classification based on physiognomic criteria seems the possibility it provides of comparing the vegetation of geographically widely separated areas (Beard 1973). It is thereby the best basis for worldwide vegetation classification (Fosberg 1967; UNESCO 1973).

Floristics. Florestic properties have been most commonly used in vegetation classification. Units have been defined by (i) dominant species only, (ii) selected diagnostic species, and (iii) full floristic composition. The first approach has been accepted particularly in American, British and Russian work related to range and forest management (Whittaker 1973). The types are dominance types but have often been indicated as associations. This last term is confusing while the association is the basic unit of another, the floristic – sociological, approach often referred to as the Braun-Blanquet school (Braun-Blanquet 1928, 1964; Westhoff & van der Maarel 1973). The dominance types approach was adopted for the Iraqi desert ranges during the reconnaissance phase. A major disadvantage was soon realized. Relatively few species are the dominants or co-dominants for most of the area (Guest 1966; Al-Ani *et al.* 1970 b, 1974), while a few other dominants have only a very restricted distribution (Guest 1966). It seemed that only the extreme environmental conditions were sufficiently reflected in the dominance of certain species, *viz.* species with a relatively narrow distribution range in terms of environmental conditions. In other words the major and economically most important part of the desert shrub rangelands would fall under a few dominance types only. These types are defined by species with a wide environmental range and thereby of little ecological significance for a practical management oriented classification, however important they may be from the range resource point of view.

This leads to the Braun-Blanquet approach as promising alternative, where the full species composition is taken as the best expression for the relation between vegetation types on the one hand and for the vegetation type – environment relation on the other (Westhoff & van der Maarel 1973). It has been found by some workers that the vegetation of desert areas poses a number of difficulties for a straightforward application of the floristic – sociological approach (Batanouny pers. comm.; le Houérou pers. comm.; Zohary 1973). Some important problems are:

a) The large number of annuals and the ubiquity of these species causing high floristic overlaps between communities and variation within communities (Zohary 1973). A great number of these annuals can be found in a variety of habitats. Zohary (1973) named them an unstable part or a kind of 'filling', not integrally connected with the permanent matrix formed by the chamaephytes. A minor and often temporary difference in the conditions of the topsoil or a difference in rainfall may have an enormous effect on the distribution and growth of the annuals, while hardly affecting the perennial species.

b) The temporary high cover of ephemeral species and the relatively low permanent cover of the perennials. Batanouny & Hilli (1973), in their study of the vegetation of the Ghurfa desert, included the annuals in their sampling. The average cover for the 46 stands mentioned is 58%. Not less than 17 stands had a cover between 70% and 100%. These figures are in contrast with the average of 11.2% cover recorded for the 54 stands in the present study, where only perennials were included. More than half of the sampled stands had covers below 10%. Including the ephemerals could mean a masking of the perennials. Visual estimates in terms of cover/abundance and sociability would at least require an adapted scale and a high accuracy at such low covers might be difficult to achieve, certainly when different observers take part in the sampling.

c) The variation in (spatial) pattern and (temporal) process in environmental condition. Over large tracts vegetation is confined to narrow wadis, runnels and small depressions. In such areas there is a great variation in vegetation, related to soil type, soil depth and microclimatic conditions. Such areas form an intricate pattern of microhabitats (Batanouny 1973). Sampling and classification of samples into meaningful classes is highly problematic (Zohary 1973).

Changes in environmental conditions, particularly related to the surface soil, are common. Zohary (1973) mentions sand onblow on vegetation growing on hammada or loess soil which may be removed after some time. Another example is a temporary usually fine textured soil deposited after wadi flooding, gradually being removed by wind erosion. Results of such processes can be a temporary 'mixture' of species in many possible gradations, difficult to interpret as a species combination related to a particular set of environmental conditions.

d) The low number of perennial species per stand; one of the main problems encountered. Virtually monospecific stands are not uncommon. The average number of perennials found per sampled stand was 5.7 (range 2–12) Batanouny & Hilli (1973) found an average of 4 perennials (range 1–10) for about 50 stands sampled in the Ghurfa desert.

Vegetation in a large part of the Iraqi desert shrub rangelands is at the threshold of its existence. Surviving species must not only have a resistance against the extremely harsh climatic conditions, but also against the continuous effects of utilization (see chapters VII–XI). Under the prevailing conditions heavy grazing does not result in a change in floristic composition in the sense that species disappear and others take their place; the 'decreasers' and 'invaders' as defined by Dijksterhuis (1949). The present vegetation of the desert rangelands is by and large composed of species that are 'increasers' and 'invaders'. Continued overuse will only lead to a change in relative contribution of species to the stand, followed by a lower species number prior to total denudation. Many stands are, thereby, probably fragmentary and modified relicts. Such stands can provide difficulties in phytosociological studies. Werger (1973) in his study on the phytosociology of the Upper Orange River Valley in South Africa writes (p. 96):

> 'When specific relevés with reference to the total table appeared to be clearly heterogeneous, or their total number of species was so low that it evidently showed their fragmentary nature, these relevés were rejected for the table after having been interpreted' and 'Twenty further relevés were of such a fragmentary nature in floristic composition, obviously owing to extreme overgrazing and trampling, that, although they were ecologically interpreted, they were not represented in the tables'.

The present study was not primarily aimed at a complete description of well defined syntaxonomical units in their full floristic composition. The objective was rather to describe the current vegetation in terms of recognizable community types, whether representing a degradation stage or not. Heavily overused stands or otherwise poorly developed stands had, therefore, to be included. They are more representative of the present vegetation of the desert rangelands of Iraq than the relatively little disturbed or well developed stands.

The four points raised above do not imply that the Braun-Blanquet approach would be unsuitable. They rather point at some desired modifications for a successful application, *viz.*

(i) to treat the ephemeral and perennial species not on the same basis and, preferably, separately;

(ii) to apply a sampling procedure within the stands that caters for their often heterogeneous and open (low cover) nature;

(iii) to add to the species one or another quantitative value, allowing a comparison of the importance of that species in relation to other species in the same stand or the same species in other stands.

A justification of points (i) and (ii) as incorporated in the present study has been given under V.B. The adopted approach for point (iii) is discussed under V.C.3.

Numerical processing

The subjectivity in any classification procedure has already been mentioned. Reducing this subjective element is continuously attempted. Numerical methods are receiving a great deal of attention, especially during the last 25 years or so, partly triggered by the easy access and applicability of digital computers to handle the timing-consuming calculations involved. Reviews of numerical classification in vegetation have been given amongst others in Gounot (1969), Goodall (1970, 1973b) and Williams (1976).

For the purpose of a numerical approach, the stands can be considered points in a multidimensional space. The axes in this space represent the different variables specifying the stands. The variables are usually floristic; presence and absence of species or one or another measure of species quantity are commonly used. Where stands are highly related in terms of the variables used, the points representing the stands in the multidimensional space should appear close together. A number of groups of related stands should form clusters. With a separation of these clusters the classes are defined. The arbitrary nature of these procedures lies in the choice of variables, the measure of quantity per variable used and class delimitation wherever no distinct discontinuities are found.

Clustering procedures usually start with a matrix of similarities between all pairs of samples representing stands. Based on these similarities in a number of steps the samples can be arranged in a hierarchy depicted in a dendrogram. The dendrogram shows the pattern of relationship of the samples, which can be used to distinguish the desired classes (Williams 1971; Goodall 1973b). Such numerical procedures have been applied to the data set of the desert shrub rangeland. They are presented under V.C.4. and V.C.5.

Environmental factors as guiding principles

In the adopted classification procedure environmental factors have been used only in the later phases. The classification was done on the basis of characteristics derived from the full floristic (perennials) composition of the sampled stands. Once the stands had been classified, the environmental factors (mainly edaphic and anthropogenic site factors) were incorporated, particularly to rearrange the distinguished classification units in an ecologically meaningful sequence. The different steps and the results are presented below.

V.C.2. *Basic data processing*

The density, crown cover, and frequency for each species in each stand were calculated from the raw field data (2700 quadrats sampled in the 54 stands) that had been entered on standard forms. These data were used not only for the cover type classification envisaged but also for description of particular situations in relation to utilization (see chapters VII–XI).

Using only a presence or absence notation at first, the sample data of the stands were entered into a matrix, the columns for the 54 stands, the rows for the species.

V.C.3. *Relative importance percentage, a basis for comparison*

For an indication of the relative contribution of a species to a stand a parameter had to be worked out that should allow a quantitative comparison between physiognomically different species and make stands with a similar floristic composition comparable, irrespective of their total cover. For such situations 'importance values' have been successfully applied (Mueller-Dombois & Ellenberg 1974). For the present study a new parameter was developed, the Relative Importance Percentage (RIMP). It is very similar to a parameter used by Risser & Rice (1971) in their studies of Oklahoma Upland forest vegetation. It combines density, crown cover and frequency as follows for a species (a):

$$RIMP_{(a)} = \frac{100}{3} \left(\frac{d_{(a)}}{d_{(\text{all species})}} + \frac{cc_{(a)}}{cc_{(\text{all species})}} + \frac{f_{(a)}}{f_{(\text{all species})}} \right)$$

where: d = density (number of plant units per unit area)
cc = crown cover (calculated by diameter method, see III.B.3.d.)
f = frequency (presence or absense in 50 quadrats).

The expression as a percentage does allow an easy comparison within and between species. The percentages of all species in a stand together add up to hundred. One advantage of the parameter is that species, and thereby stands, are easily compared. A species contributing to the stand with many individuals with a low cover will score high by its high density and frequency. A species with few individuals, but high cover can also score high by its high cover and frequency. Both species could be dominants. The Relative Importance Percentage expresses this quantitatively on a comparable basis.

Relative values are used. This makes the species contribution to a heavily overgrazed stand directly comparable to the contribution to a protected stand. For a first classification of stands affected by regular heavy utilization this is a great advantage. A disadvantage is that a value in itself does not give direct information on the nature of occurrence of the species in the stand.

The similarity between cover-abundance scales combined with an indication of sociability, as used in the Braun-Blanquet sampling, and the Relative Importance Percentage may be noted. The first one is in absolute terms, the second in relative, but both incorporate density, cover and distribution pattern.

V.C.4. *Numerical classification based on species presence and absence*

In working out the similarity between the samples of two stands the presence of a species in both could be used as an indication of similarity, the occurrence of a species in one, but not in the other as an indicator of dissimilarity. Indices of similarity have been developed on this principle. Among the oldest and best known are the index of similarity of Jaccard (1902) and the index of similarity of Sörensen (1948).

When:

a = number of species present in both samples considered,
b = number of species present in the first but absent in the second,
c = number of species absent in the first but present in the second,
d = number of species absent in both samples considered, but present in others,

the index of similarity of Jaccard (IS_J) can be written as:

$$IS_J = \frac{a}{a + b + c}$$

and the index of similarity of Sörensen (IS_S) as:

$$IS_S = \frac{2a}{2a + b + c}$$

The main difference between the two is that IS_S gives greater weight than IS_J to the species common to the two samples than to those in only one.

One could also argue that a species occurring in a sample of a third stand but absent in the sample of the two others (the d, above) should be positively valued for the similarity between these last two stands. This has been incorporated in the matching coefficient of Sokal & Michener (1958), IS_{SM}, which can be written as:

$$IS_{SM} = \frac{a + d}{a + b + c + d}$$

Many other, more complicated, indices have been developed (see Goodall 1973 a). For the present study there was no easy access to automatic data processing equipment where such indices could be used. Moreover, it was found that with the facilities at hand a satisfactory classification could be achieved, as will be shown below.

The Iraq data were processed by computer for the production of a similarity matrix of the stands, followed by an agglomerative clustering, using centroïd sorting, finally presented in a dendrogram. Two different similarity indices were applied, IS_J and IS_{SM}. The resulting dendrograms are shown in Fig. V–3 and Fig. V–4. For computer technical reasons the stands no. 1, 2, 3 and 4 were left out and the species *Atriplex leucoclada, Alhagi mannifera* and *Andrachne telephioides**. Fig. V–3 and Fig. V–4 indicate that certain stands are highly similar (5 and 6, 22 and 23, 34 and 35, 26 and 27), while others only fuse at a very low level of similarity with other groups (10, 18, 39, 51). At the same time can be seen that two 'objective' classifications can produce quite different results from the same data set. This aspect has been discussed in detail in the literature (Williams 1971, 1976). Dendrograms of the kind shown have, therefore, to be read with caution and it must be realized that the fusions at higher levels may be of little significance (Clifford 1976). High similarities between groups of stands can, however, be observed in both the dendrograms and some of these groups are of the same composition, e.g. (22, 23, 24, 25) and (46, 47, 48, 49).

V.C.5. *Numerical classification incorporating relative importance percentages*

Because of the relatively low species number and the sometimes great differences in the relative contribution of species to the stands, a numerical classification based on a measure of species quantity was also carried out. The Relative Importance Percentage (RIMP) was used to quantify the species contribution to the stand, the Euclidian Distance to work out the relative similarity between the stands. Using a centroïd sorting method, another dendrogram was constructed from which relative similarity between stands and a preliminary grouping of stands was derived.

V.C.6. *Final sample grouping and distinction of types*

In the Braun-Blanquet approach the species x relevé matrix is rearranged until

* The stands were selected because of their known high similarity with two other stands (5 and 6) sampled nearby. The species were selected because of their low presence, all occurring only once.

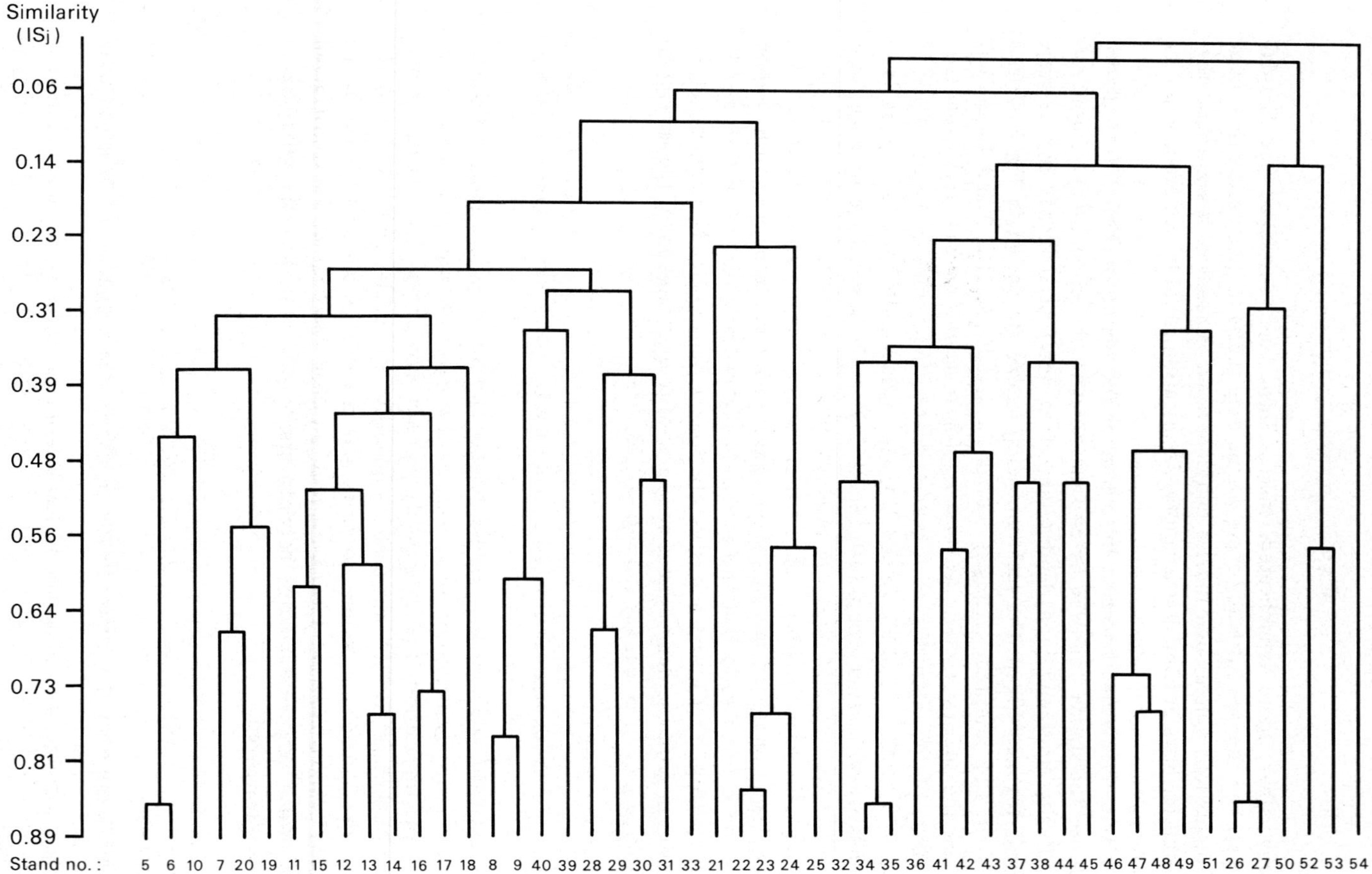

Figure V–3. Dendrogram of stands-similarity, using IS_J for the similarity calculations and a centroïd sorting technique to construct the dendrogram.

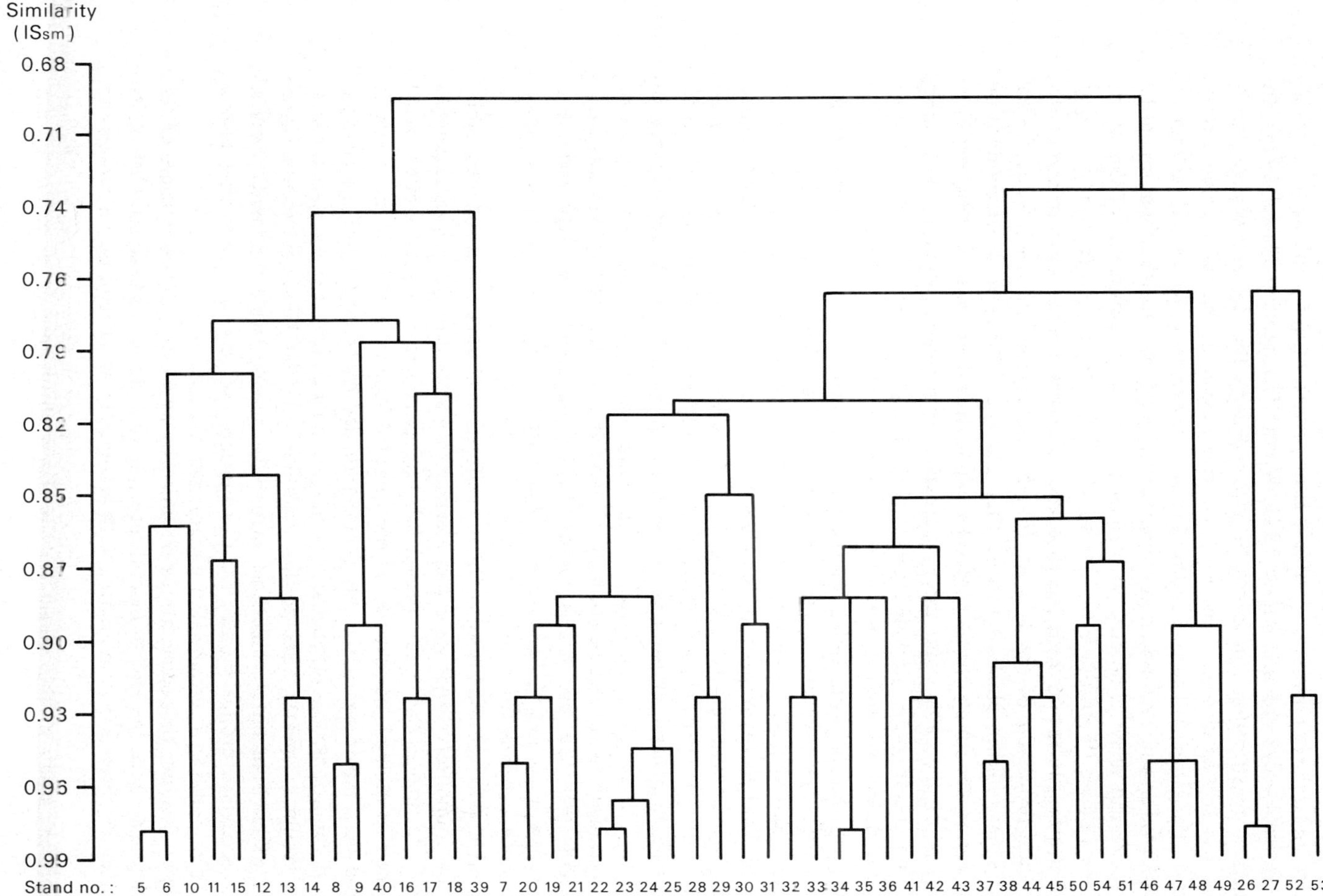

Figure V–4. Dendrogram of stands-similarity, using IS_{SM} for the similarity calculations and a centroïd sorting technique to constuct the dendrogram.

species and relevés (samples) exhibit a clear pattern of noda, mutually discriminant species-relevé groups (Werger 1973; Westhoff & van der Maarel 1973). In numerical classification stands can be grouped according to their sampled attributes, but also species association can be worked out on the basis of their occurrence in the samples. The two approaches are not exclusive. Numerical techniques can be used in the Braun-Blanquet approach to work out distributional similarity of species and relevé similarity can be calculated, using coefficients of similarity, as shown above (Westhoff & van der Maarel 1973). In the present study manual table re-arrangement by ordering taxa and relevés was done hand in hand with numerical processing of the data set. For the manual rearrangement a board with small square plastic blocks was used that allows a shifting of entire columns or rows in a rather simple manner, thereby eliminating the time-consuming multiple rewriting of the table. The three alternative numerical procedures with their resulting dendrograms provided preliminary groupings. Table V–2 shows how these groupings are related to the final classification. Each dendrogram provided some new information, useful for the rearrangement. Table V–2 shows for instance that the exclusive position of stand 21 was clear from the IS_J – dendrogram, but the 32 – 33 link from the other two dendrograms. Some stands were difficult to fit properly in the classification on the basis of the dendrograms alone, e.g. the numbers 10, 17, 18 and 54. They were not rejected. After comparison with already established groups, they could be placed and were found to represent, in general, transitions or remnants. As such they should be incorporated for a more complete picture of the range cover types. In the last phase of the rearrangement a certain (subjective) weighing for selected species was used. Emphasis was laid on species known to be dominant over large areas, e.g. *Haloxylon salicornicum, Ephedra alata, Rhanterium epapposum, Artemisia herba-alba* and *A. scoparia*.

In table V–3 the final result of the classification is shown as a synoptic table. Fifteen community types were distinguished on the basis of their floristic composition. The species have been arranged in fourteen socio-ecological groups, based on their combined occurrence in the sampled stands. Although more stands should be sampled and preferably a more detailed analysis made of the environmental conditions, this preliminary species grouping could serve as a start for indicator groups of species. Such groups of species will often be a more accurate indicator of environmental conditions than only one or two (dominant) species. Socio-ecological groups should, for this purpose, be sets of species that have a similar response to a certain set of environmental conditions.

Table V–3 shows that some groups are mutually exclusive, while others are partly overlapping. Once the environmental spectrum of a group has been established, its presence or absence in combination with presence or absence of others groups can be a highly effective indicator of the prevailing conditions. The contribution of the present study to this aspect for the desert shrub rangelands is worked out below in V.C.7.

Table V–2. Final sample grouping in relation to the sample groupings in the dendrograms obtained by different ways of processing.

	Community type	Sample numbers fusing at relatively low levels in the dendrograms			Sample numbers as grouped in the final classification
		Dendrogram based on IS_J (Fig. V–3)	Dendrogram based on IS_{SM} (Fig. V–4)	Dendrogram based on Eucl. Distance and RIMP-values	
I.	*Calligonum comosum – Cornulaca leucacantha*	26 – 27 50	26 – 27	50 – 26	50 – 26 – 27
II.	*Haloxylon salicornicum – Halogeton alopecuroïdes*	5 – 6	5 – 6	1-2-3-4-5-6	1-2-3-4-5-6
III.	*Rhanterium epapposum – Cyperus conglomeratus*	21			21
IV.	*Ephedra alata – Rhanterium epapposum*	(7-) 19 – 20	(7-) 19 – 20	19 – 20	24 – 19 – 20
V.	*Rhanterium epapposum – Helianthemum lippii*	22 – 23 (-24) 25	22 – 23 (-24) 25	22 – 25 23	7 – 25 – 22 – 23
VI.	*Ephedra alata – Haloxylon salicornicum*			10 – 16	10 – 16 – 31
VII.	*Zygophyllum coccineum – Cornulaca leucacantha*	11 – 15 12 – 13 – 14	11 – 15 12 – 13 – 14	11 – 15 12 – 14	11-12-13-14-15
VIII.	*Haloxylon salicornicum – Astragalus spinosus*		32 – 33	32 – 33	32 – 33
IX.	*Haloxylon salicornicum – Artemisia scoparia*	8 – 9	8 – 9	8 – 9	8 – 9 – 17 – 18
X.	*Haloxylon salicornicum – Artemisia herba-alba*	(32-) 34 – 35 36	34 – 35 36		34 – 35 – 36
XI.	*Artemisia herba-alba – Helianthemum lippii*	28 – 29 (31-) 30	28 – 29 (31-) 30	28 – 39 29, 30	28 – 29 – 30 – 39
XII.	*Artemisia herba-alba – Salsola spp.*	37 – 38 44 – 45	37-38-44-45	37 – 38 – 44 45	37 – 38 – 44 – 45
XIII.	*Achillea fragrantissima – Artemisia herba-alba*	41-42-43	41-42-43	42-43	40-41-42-43-54
XIV.	*Artemisia scoparia – Peganum harmala*	46-47-48-49 51	46-47-48 49	47-48-51	46-47-48-49-51
XV.	*Tamarix* sp. – *Aeluropus lagopoïdes*	52 – 53	52 – 53		52 – 53

Following the Braun-Blanquet approach, after the analytical phase (sampling) and the synthetical phase (preliminary classification), a third one, the syntaxonomical phase, should follow. In this phase the distinguished types are entered in a formal hierarchy (Westhoff & van der Maarel 1973). Other synoptic syntaxon tables are consulted and character- and differential taxa recognized. Syntaxa at different levels have not yet been established for Iraq except as far as incorporated in Zohary (1973). The typification in this extensive work is one based on dominance mainly and not documented with tabular data. In the last decade several studies have been made where plant communities have been described in phytosociological terms (Al-Hilli 1970; Abul-Faith 1970, 1975; Nasserula 1971; Batanouny & Sheikh 1972; Batanouny & Hilli 1973; Weintert & Al-Hilli 1975). Guest

(1966) in his section on 'Phytosociology' gives a description of 'Associations' of the Desert, of the Steppe, of the Forest, and of the Alpine and Sub-alpine areas. These 'Associations', however, are a listing of dominant species with their common associates, as observed during reconnaissance travel and reported in the literature. The author uses the term 'Association' only in the heading and further mainly refers to 'Community'. In its concrete sense, not meant as an abstract classification unit or phytocoenon, this seems the more appropriate term. With the material at hand a modest start could now perhaps be made for a syntaxonomical treatise of Iraq. This was considered beyond the scope of the present study and in its initial stage is bound to be highly speculative. For an improvement in the management and use of the vegetation of the desert shrub rangelands of Iraq a characterization of the vegetation types in proper syntaxonomical terms could be useful, but is not a prerequisite. In the present study, where only a relatively low number of stands (54) could be sampled and an equally low number of (perennial) species was encountered (42), the classification presented must be regarded as a preliminary one and as a starting point for a more comprehensive classification of the shrub cover types on a detailed floristic basis. A characterization beyond community type or phytocoenon has, therefore, not been attempted.

V.D. Description

V.D.1. *Ecological basis for the types*

V.D.1.a. Edaphic factors

The different community types established on the basis of properties derived from the vegetation itself must be related to differences in environmental conditions. The ecological meaning of a type can only be established when the coincidence of occurrence of the type and a distinct set of environmental conditions has been demonstrated. In the present study the factors investigated are related to soil, human activity and, to a lesser extent, climate.

The possibility, demonstrated in table V–3, to arrange the types and species in a diagonal order suggests that there may be one factor or a set of related factors increasingly different when comparing the habitats going from one side of the table to the other, of the stands where the sampling was done. The two extremes are the community type found in high sand dune areas on the one side and the community type found in somewhat saline depressions on the other. The factor clearly different in the two habitat types, apart from elevation, is soil texture. The percentages of sand, silt and clay found in the soil samples were, therefore, arranged per community type (Fig. V–5). Although only a few samples per type could be analysed, a clear trend was found from pure sand in the stands of the

Table V–3. Synoptic table, showing the occurence of the species in the community types. The values are mean RIMP – values (see V.C.3), obtained by dividing the total RIMP – value of a species in the samples of a community type by the number of samples (sampled stands) in that type. Underlined figures indicate occurence of the species in all samples in the type, figures marked with an asterisk (*) occurence in only one sample.

Number of stands sampled	3	6	1	3	4	3	5	2	4	3	4	4	5	5	2
Community type / Species, arranged in socio-ecological groups	I *Calligonum comosum – Cornulaca leucacantha*	II *Haloxylon salicornicum – Halogeton alopecuroïdes*	III *Rhanterium epapposum – Cyperus conglomeratus*	IV *Ephedra alata – Rhanterium epapposum*	V *Rhanterium epapposum – Helianthemum lippii*	VI *Ephedra alata – Haloxylon salicornicum*	VII *Zygophyllum coccineum – Cornulaca leucacantha*	VIII *Haloxylon salicornicum – Astragalus spinosus*	IX *Haloxylon salicornicum – Artemisia scoparia*	X *Haloxylon salicornicum – Artemisia herba-alba*	XI *Artemisia herba-alba – Helianthemum lippii*	XII *Artemisia herba-alba – Salsola spp.*	XIII *Achillea fragrantissima – Artemisia herba-alba*	XIV *Artemisia scoparia – Peganum harmala*	XV *Tamarix sp. – Aeluropus lagopoïdes*
(i) *Calligonum comosum*	$\underline{30}$														
Cornulaca monacantha	7														
Schanginia aegyptiaca	1														
(ii) *Halogeton alopecuroïdes*		$\underline{19}$				0*	0*		0*						
Anabasis setifera		1													
(iii) *Cyperus conglomeratus*	0*		$\underline{37}$*												
(iv) *Rhanterium epapposum*			$\underline{22}$*	$\underline{23}$	$\underline{51}$				5						
(v) *Ephedra alata*				$\underline{25}$	0	$\underline{26}$	5		0*	11	1*				
(vi) *Moltkiopsis ciliata*				0		6*	9		1*						
Lycium barbarum						0*	3		4*						
Zygophyllum coccineum	2						$\underline{22}$								2*
Aellenia subaphylla							4		5*						3*
Salsola baryosma							2								
(vii) *Haloxylon salicornicum*		$\underline{27}$		6	7*	$\underline{20}$	25	$\underline{90}$	$\underline{45}$	$\underline{53}$	19		5		19*
(viii) *Stipagrostis* spp.	23	$\underline{28}$	$\underline{41}$*	$\underline{29}$	12	19	$\underline{9}$		10		6	0*	1*		
Helianthemum lippii		$\underline{20}$		15	$\underline{25}$	$\underline{22}$	$\underline{3}$		4		$\underline{11}$				
Cornulaca leucacantha	$\underline{36}$	$\underline{1}$					$\underline{15}$		5				1*	1	8
Heliotropium bacciferum		1			0*	0*					1*				
Anvillea garcini		0*									2				
Reseda muricata						0*			0*						
Haplophyllum tuberculatum											4				
(ix) *Astragalus spinosus*					1	$\underline{6}$	2	$\underline{2}$	2	17	$\underline{4}$	0*	4		
Cousinia spp.		0*					0		2		0*	0*	2		
(x) *Artemisia herba-alba*				2*	4					$\underline{15}$	$\underline{49}$	$\underline{86}$	$\underline{26}$		
Salsola vermiculata										6*					
Atriplex sp.													1*		
(xi) *Achillea fragrantissima*						4*					0*	0*	$\underline{28}$		
Zilla spinosa								1*		0*	1*		1*		
Haloxylon articulatum								5*					4		
(xii) *Artemisia scoparia*						0		3*	$\underline{16}$	$\underline{5}$	1*	6	16	$\underline{25}$	
Peganum harmala							0*			3			7	$\underline{37}$	
Noea mucronata									1		1*	1*	5*		
Salsola spp.											0	$\underline{7}$		16	
(xiii) *Lagonychium farctum*														14	
Andrachne telephioides														3*	
Achillea conferta														2	
Teucrium polium														1	
Scrophularia deserti														1	
Alhagi mannifera														0*	
(xiv) *Aeluropus lagopoïdes*															$\underline{16}$
Bienertia cycloptera							3								$\underline{47}$
Tamarix sp.															$\underline{6}$

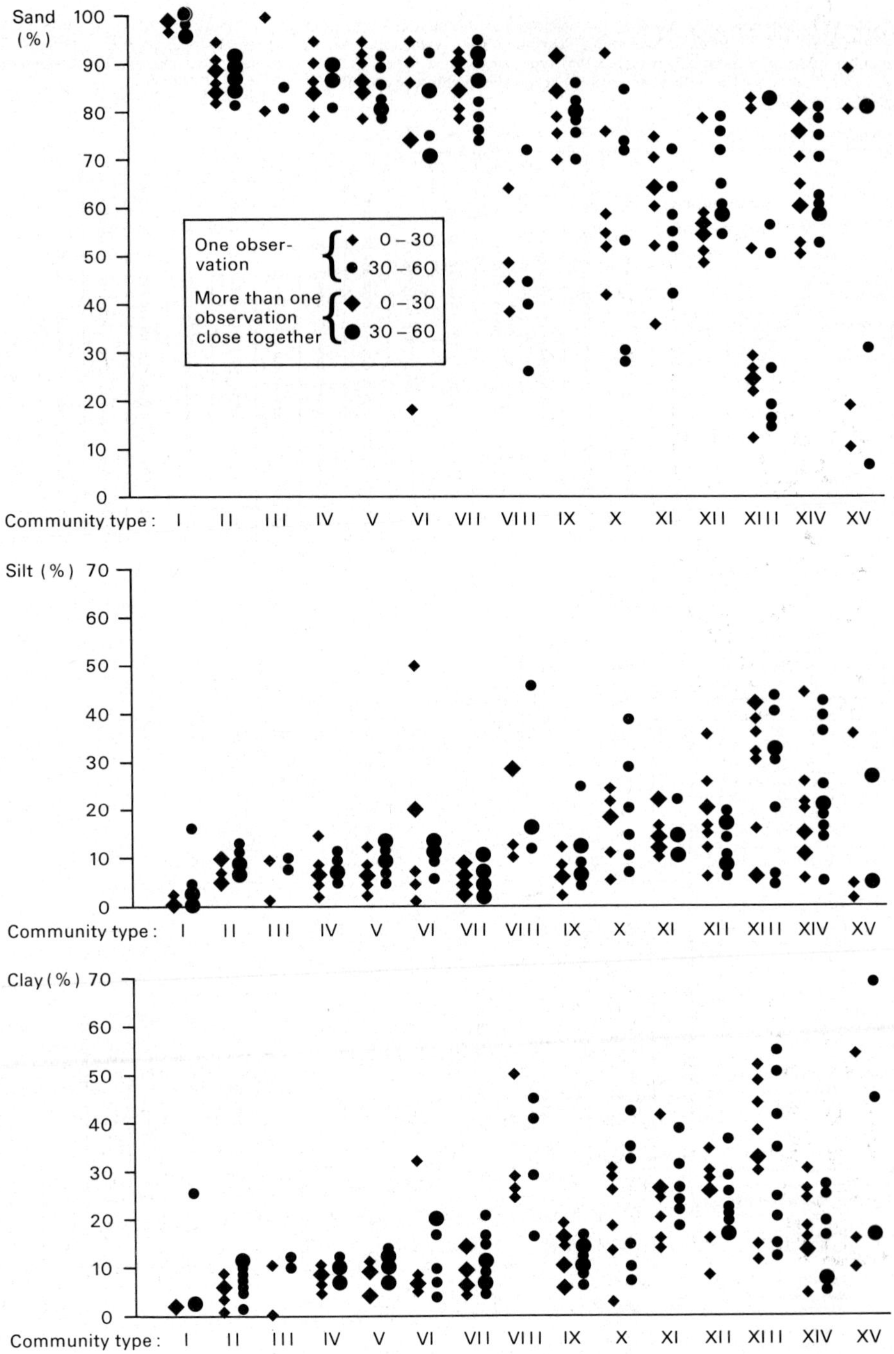

Figure V–5. Granular composition of the soils under the 15 distinguished community types.

first types to clay loam, silty clay loam and even clay soils under the stands grouped in the last community types. Soil texture itself has little or no direct influence on plant development. Only in extreme cases, e.g. where large stones become obstacles for root development, one might speak of a direct influence. Its indirect influence, however, must be enormous while soil texture is one of the main properties determining the depth and amount of available soil moisture, the overruling factor for plant development in the arid areas. The relation is not a straightforward one. Infiltration and permeability is higher in coarse textured soils, resulting in water storage at greater depth. Upward movement of water may be quite slow. Fine textured soils with a higher conductivity when dry, can continue loosing water from considerable depth (Dregne 1969). In the Iraqi deserts the fine textured soils are mostly found in the lower situated areas which receive run off from the surrounding areas. Fieldwork in the Negev has shown that loessial plains with deep saline soils had the least amount of water available to plants annually, rocky slopes somewhat more, sands even more than rocky slopes and loessial wadis with deep soils by far the most (Tadmor *et al.* 1962). Batanouny & Hilli (1973) found for the Ghurfa Desert in central Iraq the distribution of communities to be greatly affected by the physiography and the physical soil properties, especially soil depth and soil texture. A similar statement had been made before when investigating the plant communities along the Baghdad – Husaiba road in the Western Desert (Batanouny & Sheikh 1972). For the desert rangelands in Jordan, Poore & Robertson (1963) formulated this aspect as follows:

> 'The accumulation of run-off may mean that some parts have effectively many times the mean annual rainfall, other parts have much less. Because a layer of coarse material prevents evaporation from the surface of the soil, gravels usually afford a more favourable environment than sand, sand than silt, and silt than clay. Salinity may also be a problem on heavy soils. The water balance of the vegetation varies therefore with topography and soil texture.'

Kassas & Girgis (1970) in their account on habitat and plant communities in the Egyptian desert made similar observations and their quotation of Shreve (1951) is worthwhile to be repeated here, while it seems to hold equally well for the Iraqi deserts:

> 'The physical texture of the soil, its depth and the nature of its surface are equally important . . . The profound influence of the soil upon desert vegetation is to be attributed to its strong control of the amount, availability and continuity of the water supply. This fundamental requisite of plants is the most effective single factor in the differentiation of desert communities!'

The data on soil texture shown in Fig. V–5 provide the evidence for Iraq. A systematic difference between soil textures of the topsoil (0–30 cm) and the subsoil (30–60 cm) was not found. Both exhibit the same trend, from coarse to finer texture in relation to the community type arrangement as shown in table V–3.

Electrical conductivity measured in the soil samples varied from 0.2 to 14.7.

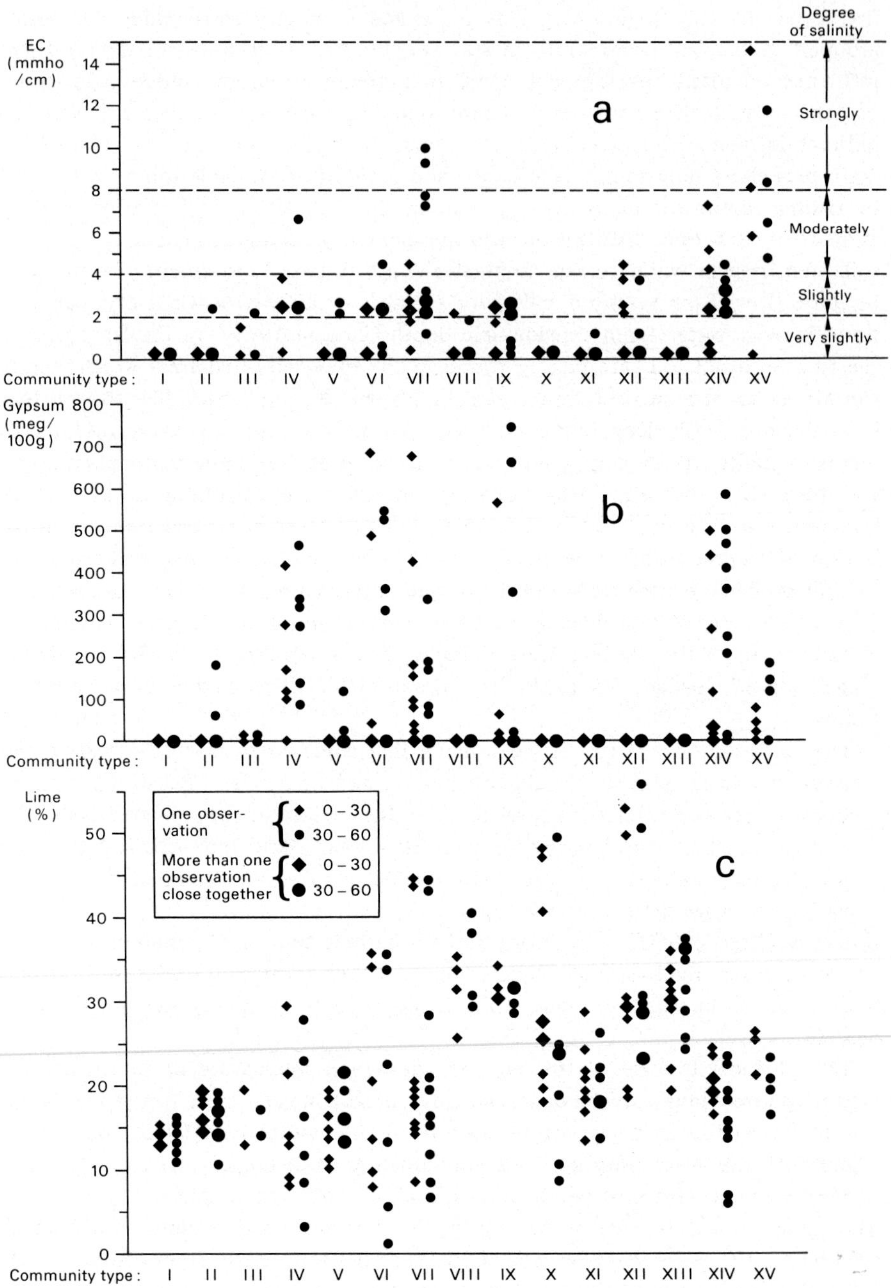

Figure V–6. EC, gypsum and lime in the soils under the 15 distinguished community types.

mmho.cm^{-1} (Fig. V–6a). No chemical analysis has been carried out to determine the type of salts, but the soluble salts for the region are known to be mostly NaCl, and some $CaCl_2$, $MgCl_2$, Na_2SO_4 and $MgSO_4$, and sodium chloride and gypsum salines are most common in the area (Yahia 1971; Zohary 1973). A detailed classification of soil salinity for Iraq is given in Yahia (1971), based on three principles: salinity degree, salinity type and distribution of salts throughout the profile. His classification of salinity degree is incorporated in Fig. V–6a. For only one community type (XV), the *Tamarix* sp. – *Aeluropus lagopoïdes* type can the measured soil salinity be called strongly saline. It is most likely the main differentiating factor for this type. For another type (VII), the *Zygophyllum coccineum Cornulaca leucacantha* type, some subsoil samples showed a strongly saline level. Weinert & Al-Hilli (1975) in their description of the vegetation of Jabal Sanam near the Iraq – Kuwait border distinguished a *Zygophyllum propinquum* community. Regarding the species name they mention that *Zygophyllum propinquum* represents as a northern Arabo-Irano-Afghano-Sindian shrub the eastern part of the *Zygophyllum coccineum* group. *Z. propinquum* was not distinguished by Guest (1966) and Rechinger (1964) and must, therefore, be the same as their *Z. coccineum*. As a habitat for this community Weinert & Al-Hilli give a sand dune habitat over saline loam, which seems in agreement with the observations in Fig. V–6a. For four more community types one or a few samples reached a moderately saline level, *viz.* type IV *Ephedra alata* – *Rhanterium epapposum*, type VI *Ephedra alata* – *Haloxylon salicornicum*, type XII *Artemisia herba-alba* – *Salsola* spp., and type XIV *Artemisia scoparia* – *Peganum harmala*. All nine remaining types had either very low salinity levels or were virtually salt free.

The other soil properties analysed are gypsum and lime content (Fig. V–6 b,c). The soil samples of community types I, III, VIII, X, XI, XII and XIII had no gypsum at all, while high amounts were found in the profiles of the soils of the stands constituting the types IV, VI, VII, IX and XIV. The types X, XI, XII and XIII where no gypsum was found are all characterized by an extremely high presence of *Artemisia herba-alba*. High amounts of gypsum were found in the types where *Ephedra alata* occurs abundantly (IV, VI and VII). Lime content of all samples was high (Fig. V–6c), ranging from less than 15% to over 50%, the first in the dunes, the later in wide shallow wadibranches with the *Artemisia herba-alba* – *Salsola* spp. community type. No consistent difference between lime content of the topsoil and subsoil (only to 60 cm depth) was found. Yahia (1971) found gypsum and lime accumulations in all soil horizons in the desert soils and desert colluvium soils and states that the distribution is probably influenced by the underlying parent materials and rocks. Under the gypsiferous desert soils these are primary gypsum, anhydrite, limestone and locally mudstone and sandstone. Under the stony desert soils these are gypsum and limestone interbedded with marls. Where gypsum-containing parent rocks are present, gypsum accumulations increase with depth. Lime accumulations increase similarly in those stony soils which overlie limestone. The direct influence of gypsum and lime content on plant development is limited due to the low solubility (Buringh 1960). No direct relation with the community types was found for the lime content, except that in the in table V–3 centrally placed types (VI – XIII) lime content was generally higher and showed more variance compared to the types I – V and XIII and XIV.

V.D.1.b. Anthropogenic factors

Results of the observations on anthropogenic factors are summarized in table V–4. Grazing must be considered a factor of importance for virtually all types. All sampled stands showed signs of moderate or heavy grazing, except the ones in exclosures and higher upslope on the sand dunes. Heavy grazing was found for all except one stand grouped in type V (*Rhanterium epapposum – Helianthemum lippii*) and type VII (*Zygophyllum coccineum – Cornulaca leucacantha*). Of these two, the type VII shrubs also showed signs of cutting (with one exception) and crown cover in none of these sands exceeded 10%, while for type V values of 3, 14, 16 and 29% crown cover were measured. This difference illustrates the essential difference between grazing and cutting, which is further discussed in the chapters on utilization (chapters VII – XI). This also becomes clear when comparing the crown cover of stands where cutting was observed (total 22) with those without cutting (total 32, including some where nothing is left to be cut): an average of 6.5 (± 4.4)% against 14.8 (± 9.0)%. The impact of cutting is far more severe and leads to a quick elimination of the species desired for their fuel value. Cutting therefore has definitely contributed to the observed differences between community types. A separation of past and present impact is impossible and highly degraded stands where no signs of recent cutting were observed may well be partially the result of this factor. A clear relationship between observed cutting and perticular community types could, therefore, hardly be expected and was not found. Ploughing, on the contrary, can be shown as a likely factor in determining the types XIV (*Artemisia scoparia – Peganum harmala*) and to a lesser extent XIII (*Achillea fragrantissima – Artemisia herba-alba*). The process is dealt with in chapter IX. Signs of ploughing were only found in stands of these two community types.

V.D.1.c. Climatic factors

The final factor to be considered, climate, can be only generally assessed because of the lack of detailed climatic data for the desert areas. All stands later classified in the types I, II, III, IV and V are from the Southern Desert. The stands later grouped in community types VI, VII, VIII, IX, X and XI are, with two exceptions from the northern part of the Southern Desert and the Western Desert.

Table V–4. Selected anthropogenic factors and crown cover as observed in the stands sampled per community type.

	Sampled stands	Community types														
		I	II	III	IV	V	VI	VII	VIII	IX	X	XI	XII	XIII	XIV	XV
Grazing (1)	1	–	++	+	++	++	+	++	+	++	+	++	++	+	–	+
	2	–	++		+	++	+	++	++	+	++	+	+	++	+	+
	3	+	++		+	++	++	++		+	+	+	+	+	+	
	4		–			+		++		+		+	++	++	+	
	5		–					+						++	+	
	6		–													
Cutting (2)	1	–	+	–	–	–	+	++	–	+	++	+	–	–	–	–
	2	+	+		–	–	–	++	++	++	–	–	–	++	–	+
	3	+	+		+	–	+	+		+	–	–	–	–	–	
	4		–			–		+		–		–	–	++	–	
	5		–					–						+	–	
	6		–													
Ploughing (3)	1	–	–	–	–	–	–	–	–	–	–	–	–	–	–	–
	2	–	–		–	–	–	–	–	–	–	–	–	+	++	–
	3	–	–		–	–	–	–		–	–	–	–	–	++	
	4		–			–		–		–		–	–	–	+	
	5		–					–						++	+	
	6		–													
Crown cover (%)	1	(.1)	7	16	16	14	8	4	22	17	5	7	10	39	8	14
	2	4	10		19	16	2	8	3	2	5	11	10	2	3	17
	3	9	7		1	3	2	9		8	19	9	3	19	8	
	4		24			29		8		12		26	30	2	12	
	5		20					9						2	14	
	6		20													

(1) Grazing:
- – no grazing observed in the area and no signs of grazing found
- \+ moderate grazing; area known to be grazed and/or plants found which showed signs of recent grazing
- ++ heavy grazing; plants show signs of severe grazing and area known to be under heavy grazing pressure

(2) Cutting:
- – no signs of cutting
- \+ some plants affected by cutting (see chapter VIII)
- ++ many plants cut and/or area known to be continuously used for fuel collection

(3) Ploughing:
- – no signs of ploughing
- \+ some signs of ploughing, probably several years ago (see chapter IX)
- ++ area relatively recently ploughed, furrows and destroyed shrub cover often clearly visible.

Community types XII, XIII and XIV, are derived from stands sampled from Rutba north into the Jezira to Hatra with two stands sampled near Nukhaib as the only exceptions. The two stands later classified in the last type XV, were sampled near the Abu Dibbis depression and form a special category with clear links to community types found further south. These observations provide evidence that the arrangement as given in table V–3 also reflects a broad southeast – northwest climatic gradient, going together with an increase in rainfall, more rainy days, somewhat lower temperatures and increased water availability (see chapter IV.B). It is the transition from the true desert conditions to sub-desert and steppe (Guest 1966; UNESCO 1975). Such a difference might be expected from the distribution patterns already known of certain dominant species in the types.
Rhanterium epapposum which is widely distributed and common in the Southern Desert from only a few locations in the Western Desert, and absent in the Jezira;
Ephedra alata which is not known from the Jezira;
Haloxylon salicornicum which is the most common species of large parts of the Southern Desert, common in the Western Desert, known from the Lower Jezira and absent in the Upper Jezira, etc.
Zohary (1940, 1973) has worked this out for a phytogeographical subdivision and shows that in the study area two regions meet, the Irano-Turanian region in the north and the Saharo-Arabian region (or Saharasindian region sensu Eig) in the south. Far more community sampling is needed for a detailed picture but the overall climatic influence is clear even from these very general observations.

V.D.2. *Brief characterization of the types*

I. *Calligonum comosum – Cornulaca leucacantha* type

Species: The rather tall shrub *Calligonum comosum* is the dominant species, in combination with the low, often prostrate *Cornulaca* spp. A few annual grasses and forbs were also found, but could not be identified. The indicated occurrence of *Schanginia aegyptiaca* (table V–3) is misleading, as this species was only found at the lowest part of the dune slope, where the shallow sand is underlain by a saline, fine-textured soil. In general the community type is extremely poor in species. A related community *(Calligonum comosum – Haloxylon salicornicum)* is described in Dougrameji & Kaul (1971) for a low sand dune area near Tel Lachm, southeast of Nasiriya. In this community *Cornulaca* spp. were also the common associates. In the dune heels and the interdune hollows the grass *Stipagrostis plumosa* becomes important. Here also numerous ephemerals can be found after the rains (see Weinert & Al-Hilli (1975) for Jabal Sanam and Al-Hilli (1970) for the Baiji sand dunes).
Ecology and distribution: The distribution of this type in Iraq is limited to sand dune areas. The type was distinguished from stands sampled in the Euphrates sand

dune belt, between Ur and Busaiya (see IV.C.2.f. and Fig. IV–10b). A stand of this type is shown in Fig. VIII–6 and results of analyses of the pure sandy soils are given in table IV–11. Zohary (1940) mentioned a *'Calligonetum comosi'*, observed between Basra and Ur. Guest (1966) used the same 'community' name and gave *C. comosum* as an important associate of *Haloxylon ammodendron* on sand dunes in the Southern Desert and as the most abundant associate of *Haloxylon salicornicum* in the Lower Jezira near the Tharthar depression in a vegetation type named after this species. *Calligonum comosum* communities are well known from neighbouring territories. The species is found from Iran to Morocco (Mobayen & Tregubov 1970; Zohary 1973; Batanouny 1974).
Utilization: The economic importance of this type is limited to camel grazing (Serkahia 1956 for Iraq, Batanouny 1974 for Egypt) and, more important, fuel cutting (see VIII.B.3.).

II. *Haloxylon salicornicum – Halogeton alopecuroïdes* type

Species: Halogeton alopecuroïdes (= Agathophora alopecuroïdes) and *Anabasis setifera* are the two species most distinct for this type, in combination with *Haloxylon salicornicum* and some species of the *Stipagrostis – Helianthemum lippii* group (table V–3). The ephemeral cover can be extremely rich in species in this type (see table VII–14).

Ecology and distribution: The distribution is limited. The type is based on six sampled stands in the Khedr-el-Mai and adjacent area, which is discussed in detail in chapter X. Soil characteristics are given in table X–1. Guest (1966) distinguished the *'Agathophoretum alopecuroïdi'* and mentioned for its distribution the Southern Desert and the eastern end of the Western Desert to within 40 km of Nukhaib and not further north than near Ramadi. Kernick (1966) mentioned a similar species combination for Kuwait.

Utilization: The type is important for grazing as well as fuel. The stands provide good grazing, especially for camels, but also for sheep and goats in spring when the ephemeral cover is well developed and the shrubs provide additional forage.

III. *Rhanterium epapposum – Cyperus conglomeratus* type

Species: This community type was distinguished on the basis of one sampled stand only, near Um Quasr at the border with Kuwait (Fig. V–7). Only three perennials were recorded, the two after which the community was named and *Stipagrostis plumosa,* covering together some 16% and with an aerial biomass fresh of 486 $kg.ha^{-1}$. In the vicinity Weinert & Al-Hilli (1975) carried out a more thorough sampling on the floristic aspects and gave a long list of species, mostly annuals. Regarding their data this type could only represent a subtype of a *Rhanterium epapposum – Haloxylon salicornicum* type which they distinguished.

Ecology and distribution: This community type is most likely limited to deep

Figure V–7. The *Rhanterium epapposum* – *Cyperus conglomeratus* community type, as found near Um Quasr, close to the Kuwaiti border.

sandy flats, sometimes overlain by fine gravel. The type occurs in the southeastern part of the Southern Desert, in the Jebel Sanam – Um Quasr area from where it extends into Kuwait, Kernick (1966) writes about the community in this country under the heading *Cyperus* steppe: '....It would appear that *Rhanterium epapposum* was also prevalent in this region at one time, but over the years through being pulled up for firewood, it has largely disappeared.... Normally the division between one major plant community and another is fairly abrupt in most parts of the country, but between Suleibya and Abdulyah the *Cyperus conglomeratus* community changes gradually to a pure community of *Rhanterium epapposum* over a transition zone of about 10 to 12 kilometers.'

Utilization: The community is used for fuel collection *(Rhanterium epapposum)*, as seen above, but is also grazed. When the stand near Um Quasr was being sampled heavy grazing by goats and sheep was going on (see Fig. VII–5).

IV. *Ephedra alata – Rhanterium epapposum* type

Species: Ephedra alata, Rhanterium epapposum, Stipagrostis plumosa and *Helianthemum lippii* are the most common species in this community type. The ephemeral species appeared to be relatively scarce.

Ecology and districution: This species combination was recorded from Joë Liban (Fig. V–8), some 25 to 30 km south of Busaiya along the track to Khedr-el-Mai in

Figure V–8. Stand of the *Ephedra alata* – *Rhanterium epapposum* community type sampled near Joë Liban, approximately 25 km south of Busaiya.

a flat to slightly rolling area with rather shallow sandy soils, but also on a gravel plain near Samah in the Neutral Zone. In all three situations the gypsum content of the soil, especially of the subsoil (30 – 60 cm) was high . Although the two leading species occur in many different species combinations, the combination of the two as dominants seems a special situation which could have a high indicator value. The stands near Joë Liban had a good cover (16% and 19%) and were only slightly grazed, probably due to the absence of drinking water in this area, while *Ephedra* and *Rhanterium* are often found very heavily grazed.

Utilization: The type has a good grazing value but the danger for overgrazing is great, for all species found in this type are highly palatable and nutritious. *Rhanterium* can be cut for firewood.

V. *Rhanterium epapposum* – *Helianthemum lippii* type

Species: This type respresents in fact the often pure *Rhanterium epapposum* type, found in the Southern Desert with a very dense cover (Fig. V–9), but also in stands with a cover of less than five per cent. *Helianthemum lippii* is the common associate and was found in the samples of all four stands grouped in this type. A good number of other species can be found in this type, such as *Ephedra alata, Haloxylon salicornicum, Astragalus spinosus, Stipagrostis plumosa* and *Cornulaca* spp. (see also Guest 1966), but these never occur in abundance. The number of geophytes and ephemeral annual species can be very high in open stands of *Rhanterium epapposum*. In this respect especially *Plantago* spp., *Iris sisyrinchium* and *Gagea reticulata* should be mentioned.

Ecology and distribution: The type is found on pure sand, sandy loam or loamy sand, which can be compact and overlain by a gravel layer. The statement of Guest (1966) that the *'Rhanterietum epapposi'* is found on soils 'with underlying rock at a greater depth than where the *'Haloxyletum'* flourishes', was confirmed during attempts to dig up the rootsystems of these two species. The distribution of the type is mainly in the Dibdibba area of the Southern Desert (see IV.C) but the community type can be encountered locally in the entire area where *Rhanterium* is found in Iraq (see the distribution map of Fig. VI–7).

Utilization: This type must be regarded as one of the best shrubtypes for grazing, especially by camels and is recognized as such by many authors, not only in Iraq

Figure V–9. Very dense stand of the *Rhanterium epapposum* – *Helianthemum lippii* community type near Samah in the Southern Desert.

Figure V–10. Example of an open stand of the *Ephedra alata – Haloxylon salicornicum* community type near Judaidat Arar in stony terrain, being grazed by camels.

but also in Kuwait (Kernick 1966) and Saudi Arabia (Vesey Fitzgerald 1957b). *Rhanterium* is also extensively used for fuel (see chapter VIII). Aerial biomass (fresh) for the four stands sampled was 99, 490, 608, and 1274 $kg.ha^{-1}$

VI. *Ephedra alata – Haloxylon salicornicum* type

Species: Ephedra alata and *Haloxylon salicornicum* form an often observed species combination usually in open stands in undulating stony terrain, where in the lower areas, with soil accumulation, the situation is conducive for plant growth (Fig. V–10), or on flat areas with a shallow soil. Because of the variety of (micro)-habitats in such areas, many other species can be found in this community type. Guest (1966) provided a list when he discussed the *'Ephedretum alatae'*. The species found in the samples for the present study can be seen in table V–3. They range from *Moltkiopsis ciliata,* a species of sandy soils, to *Achillea fragrantissima,* a species of fine textured soils.

Ecology and distribution: The type occurs in the Southern Desert and the eastern half of the Western Desert in habitats as indicated above. Often there is gypsum in high concentrations near the surface. Good examples are found in the Busaiya area. The type was found as far west as Nukhaib. It rarely occurs over large distances and its cover is low, 1.6%, 2.0% and 7.5% for the stands on which this type was based. The species combination is also known from Saudi Arabia (Zohary 1973).

Utilization: Camels like stands of this type particularly (Fig. V–10). They were often seen moving from one patch of vegetation to the other, eating the *Ephedra* with preference. The grazing value is rather low for sheep and goats, while in the stony habitat of this type not only the shrub cover is low, but also the grasses and forbs (annual and perennial species) are poorly represented. It is likely that the type represents a degradation stage of a once more productive cover.

VII. *Zygophyllum coccineum – Cornulaca leucacantha* type

Species: Table V–3 gives the (14) perennial species found in the five stands together grouped in this type. Weinert & Al-Hilli (1975) list 53 species of which 35 are annuals in their description of the synonimous *Zygophyllum propinquum* community. The perennials in their list have some halophytic affinity.

Ecology and distribution: This community type occurs in areas where sandy soils are overlying a saline, often fine-textured subsoil (see Fig. V–6a). Such areas are found in a belt from Zubair in the southeast to about Rutba in the west. Kernick (1966) describes the type for the seashore in Kuwait. Communities with *Zygophyllum coccineum* are widespread in the Middle East (Zohary 1973, Batanouny 1974) but for a higher order phytosociological arrangement more field data should be available which can be directly compared.

Utilization: The use of this type is limited to some salt grazing by camels.

VIII. *Haloxylon salicornicum – Astragalus spinosus* type

Species: This type is based on two extreme stands which were sampled in the Nukhaib area, representing probably a most degenerated and very well-developed aspect (see Fig. V–11) of the type. The covers are 3.4% and 22.4%, respectively. *Haloxylon salicornicum* is the dominant and often the only perennial species, but *Astragalus spinosus* is the common associate, while occasionally *Zilla spinosa, Haloxylon articulatum* and *Artemisia scoparia* were found.

Ecology and distribution: This community type is one of the commonest of the Southern Desert, the Western Desert and Lower Jezira. Where the soils are not too fine textured or too coarse sandy, and there is a regular use-pressure on the vegetation, this type can be found. It is composed of relatively less palatable and unpalatable species except for *Haloxylon salicornicum*, but this species can withstand not too heavy grazing and fuel cutting very well, as will be demonstrated later. This community type, together with type IX, discussed below, seem to reflect a certain degradation as a result of the continual use of the vegetation. In the western half of the Southern Desert and the eastern half of the Western Desert, types VIII and IX cover most of the area.

Utilization: The present use of this type is virtually limited to the possibilities of

Figure V–11. Well developed shrubs of *Haloxylon salicornicum* in the *Haloxylon salicornicum – Astragalus spinosus* community type as coccurring near wadi Ubaiyidh at Faidhat al-Jeez.

grazing *Haloxylon salicornicum* and cutting the shrubs of this species for fuel. All other species can hardly contribute as far as these factors are concerned.

IX. *Haloxylon salicornicum – Artemisia scoparia* type

Species: The relatively high number of perennials (15, of eight different socio-ecological groups) is shown in table V–3. *Haloxylon salicornicum* and *Artemisia scoparia* are the only species found in all four stands comprising this type.

Ecology and distribution: This type has a very wide distribution and was recorded from the Lower Jezira, Western Desert and especially the Southern Desert, where it was found on sand-filled, shallow depressions and gravelly plains. Together with the previous community type it forms what Guest (1966) described as the *'Haloxyletum salicornici'*, the 'most important and characteristic association of this territory'. A good deal of the work reported in the following chapters has been carried out in stands of the types VIII and IX.

Utilization: See type VIII above.

X. *Haloxylon salicornicum – Artemisia herba-alba* type

Species: This community type (see Fig. VI–2) clearly represents the transition of the *Haloxylon salicornicum* dominated types to the *Artemisia herba-alba* dominated ones, as shown in table V–3. Associates of both these species can be found in this type, but of the three sampled stands comprising this type the total number was low. *Artemisia scoparia* was well represented and also *Astragalus spinosus.* Two extreme species in terms of environmental requirements, *Ephedra alata* and *Zilla spinosa,* are found together in this type, illustrating the relative inhomogeneity in this type.

Ecology and distribution: Stands of this type extend over large areas at places along the Ramadi-Rutba road; they can also be found along the Nukhaib-Rutba track and Ukhaidir-Nukhaib track. It roughly forms the belt in the middle part of the study area, where two phytogeographical units meet, the Irano-Turanean with *Artemisia herba-alba* coming from the north and the Saharo-Arabian with a Sudanian influence of which *Haloxylon salicornicum* is a representative, from the south (see Zohary 1973 for a lengthy and detailed discussion). The edaphic conditions such as soil texture (Fig. V–5) and lime content (Fig. V–6c) of the different stands as well as within the stands themselves vary widely. Textural classes range from sand to clay, but mostly fall in the sandy loams and sandy clay loams.

Utilization: The type has a good grazing value for camel and sheep and can also provide fuel.

XI. *Artemisia herba-alba – Helianthemum lippii* type

Species: Somewhat similar to type X, this type also forms a transition towards the *Artemisia herba-alba* dominated stands, but with a higher dominance of this species. *Herlianthemum lippii* and a number of other species of the same socio-ecological group (table V–3) are the common associates. *Achillea fragrantissima* and *Zilla spinosa,* the species, found in this type together with *Artemisia herba-alba,* are also found together in other types.

Ecology and distribution: The type is found on rather fine textured, non-saline soils with a gravelly top layer (sandy loams, sandy clay loams and clay loams). The cover is usually sparse (Fig. V–12) and forms a patchy pattern. The type is quite common between 'Mia sittien' and Rutba and between Nukhaib and Rutba. It was also seen in the Lower Jezira. Guest (1966) mentioned the same species combination.

Utilization: Although relatively low in production, the type is often heavily grazed by sheep and, to a minor extent, by camels.

XII. *Artemisia herba-alba – Salsola* spp. type

Species: This type represents the often almost pure *Artemisia herba-alba* stands. The only species commonly found together with *Artemisia herba-alba* in this type are *Salsola* spp. *Artemisia* and *Salsola* both urgently require a detailed study on the species, subspecies and ecotype level. For further ecological and syntaxono-

Figure V–12. Stand of the *Artemisia herba-alba – Helianthemum lippii* community type, sampled a few kilometres south of Rutba.

mical work the species are extremely important. In this type they have been grouped because efforts to separate more types failed due to taxonomical problems. *Noea mucronata* is a common species in stands of this type and also *Artemisia scoparia.* The ephemeral cover is usually very well developed in between the shrubs. *Poa sinaica* (Bor 1968) is often well represented too.

Ecology and distribution: The type is found on sandy clay loams and clay loams, mostly in wadibeds and wide shallow depressions receiving runoff water. It has an extensive geographical distribution in Iraq in the Western Desert and Jezira. The area around Rutba provides good examples. Work on *Artemisia herba-alba* reported in the following chapters was mostly carried out in stands of this type.

Utilization: Sheep grazing and to a minor extent goat and camel grazing. In places cover and biomass are very high. In a degraded stand of this type 3.3% cover and

43 kg.ha^{-1} fresh aerial biomass was found; in a well developed stand this was 30.3% and 528 kg.ha^{-1}. Because of their location in wadibeds and depressions these stands have often been destroyed or degraded by ploughing for cereal cultivation (see chapter IX).

XIII. *Achillea fragrantissima – Artemisia herba-alba* type

Species: This type is formed by a very distinct species combination: *Achillea fragrantissima, Artemisia herba-alba* and *Haloxylon articulatum.* A good example is found in the Rutba fenced area (Fig. V–13; see chapter X). *Artemisia* is usually the dominant species. Many other perennials are occasionally found such as *Zilla spinosa, Peganum harmala, Artemisia scoparia* and *Atriplex (leucoclada?).*

Ecology and distribution: The type occurs all over the study area but is relatively best developed in the Western Desert. In the Jezira it occurs in degraded forms with a very low cover while in the Southern Desert it is limited to shallow depressions filled with a fine textured soil. Clay loams and loams are the common textural classes under this type, but a high variation is possible as can be seen from Fig. V–5. (Guest (1966) discussed this type under two community names, '*Achilletum fragrantissimae*' and '*Haloxyletum articulati*'. He gave the first as very

Figure V–13. The *Achillea fragrantissima – Artemisia herba-alba* community type in the Rutba fenced area.

common for the Southern Desert and the second as common for the Syrian Desert. It is quite possible that subtypes can be distinguished at a later date when more field data will become available. The stands grouped in this type were found as far south as Nukhaib and as far north as Baaj.

Utilization: As mentioned for the previous type, stands of this type are commonly destroyed by ploughing. The grazing value is good for sheep and goats, the fuel value is low.

XIV. *Artemisia scoparia –Peganum harmala* type

Species: By these two species a type is characterized that was found to be most common over large areas of the Central and Upper Jezira *e.g.* in combination with *Lagonychium farctum, Achillea conferta* and *Alhagi mannifera.* In the 'ground layer' the P.-C.-R. group of Eig (1935) was sometimes found: *Poa sinaica, Carex stenophylla* and *Ranunculus asiaticus.* These species were not included in the sampling. Guest (1966) describes them separately under the *'Poetum sinaicae'.*

Ecology and distribution: The type forms the most common species combination in the northern half of the Jezira, especially on abandoned formerly cultivated fields. It is discussed in chapter IX. In the Western Desert it is found at some locations in formerly ploughed areas.

Utilization: The type, although being grazed by livestock, has a very low forage value and no fuel value.

XV. *Tamarix sp. – Aeluropus lagopoïdes* type

Species: The tall shrub *Tamarix* and the grass *Aeluropus lagopoïdes* together constitute a community type at the fringe of saline depressions. At the lower side also *Cressa cretica* was found, while somewhat higher a species group comes in that is shown in table V–3. This is a transition to the *Haloxylon* community types at higher grounds. The type is only given as a link to the many halophytic community types that can be distinguished in Iraq when more detailed and extensive sampling has been carried out. For the desert shrub vegetation the saline areas are of minor importance.

Ecology and distribution: The saline conditions under which this community type was found are shown in Fig. V–6a. The soil texture under this type ranged from silty loam to clay. The stands of this type were sampled in the Abu-Dibbis area. Studies as reported in Abul-Fatih (1975) should in due time give the proper data base for a syntaxonomy of the halophytic community types in Iraq.

Utilization: The type is being grazed by sheep and camels. *Tamarix* is sometimes cut for fuel.

The community types described above and the socio-ecological groups as shown in table V–3 should only be regarded as a preliminary result of an exploratory phase of species grouping and ecological characterization, probably at some time to be developed into a complete syntaxonomy of this area when more fielddata become available. Such a study, which is badly needed, would require an approach different from the one adopted for the present study. For an example see Werger (1973).

Table V–3 shows that three species, *Haloxylon salicornicum, Rhanterium epapposum* and *Artemisia herba-alba,* come out with very high RIMP-values and constancies in a number of types. Moreover, they are to a great extent mutually exclusive. These three species were, in an early phase of the study, selected for detailed autecological work. The results are presented in the next chapter.

CHAPTER VI

ECOLOGY OF THREE KEY SPECIES

VI. ECOLOGY OF THREE KEY SPECIES

VI.A. Introduction

Of the shrub species encountered in the desert rangelands of Iraq, three species were found to be dominant over extensive areas, *Haloxylon salicornicum, Rhanterium epapposum* and *Artemisia herba-alba*, either as the leading associate in a mixed shrub cover or, commonly, in virtually monospecific stands. In earlier publications the importance of these species had been already outlined. Zohary (1940) considered the *Haloxyletum salicornici* probably the most important and most characteristic association of the Saharo-Sindian territory of the Syrian Desert. Guest (1966) added that it is perhaps the most widespread community in the Southern Desert of Iraq, covering also considerable areas both in the Western Desert and the Lower Jezira. The value of *Haloxylon salicornicum* and *Rhanterium epapposum*, as a source of fodder and as fuel for the inhabitants of the desert, is also indicated in the place names Rumaitha and Arfajiya; from RIMTH and ARFAJ, the Arabic names for *H. salicornicum* and *R. epapposum*, respectively. This value has been indicated by most authors dealing with the desert shrub vegetation of Iraq, e.g. Gillett (1948a), Serkahia (1956, 1972), Guest (1966).
Artemisia herba-alba is in the Western Desert and the Lower Jezira, although in the latter often in degraded stands, as widespread as *Haloxylon salicornicum* and *Rhanterium epappossum* are in the Southern Desert. According to Eig (1935) and Zohary (1940, 1941, 1973) the species gives true expression to the Irano-Turanian conditions for this part of Iraq. Its economic value, especially through sheep grazing, is beyond doubt and is indicated amongst others by Serkahia (1956, 1972), van der Veen (1960 b) and Guest (1966). As for *Haloxylon salicornicum* and *Rhanterium epapposum, Artemisia herba-alba* has also given its name to a location in the desert: Ash-Shihiyat, a plain some 50 km southwest of Busaiya, after the Arabic name SHIH for the species.

In view of their extensive distribution in Iraq and important place in many community types (chapter V), the differences in environmental conditions in their habitats and apparent economic importance, it seemed well justified to examine the ecology of these three species in greater detail. In the chapters on utilization therefore (VII–XI) a good deal of the analytical and experimental work is reported for especially these species. In the present chapter the distribution, habitat, reproduction and development for the three species are discussed.

VI.B. Brief description of the species

VI.B.1. *Haloxylon salicornicum* (Moq.) Bge.

Haloxylon salicornicum is a leafless, succulent dwarf shrub of the family Cheno-

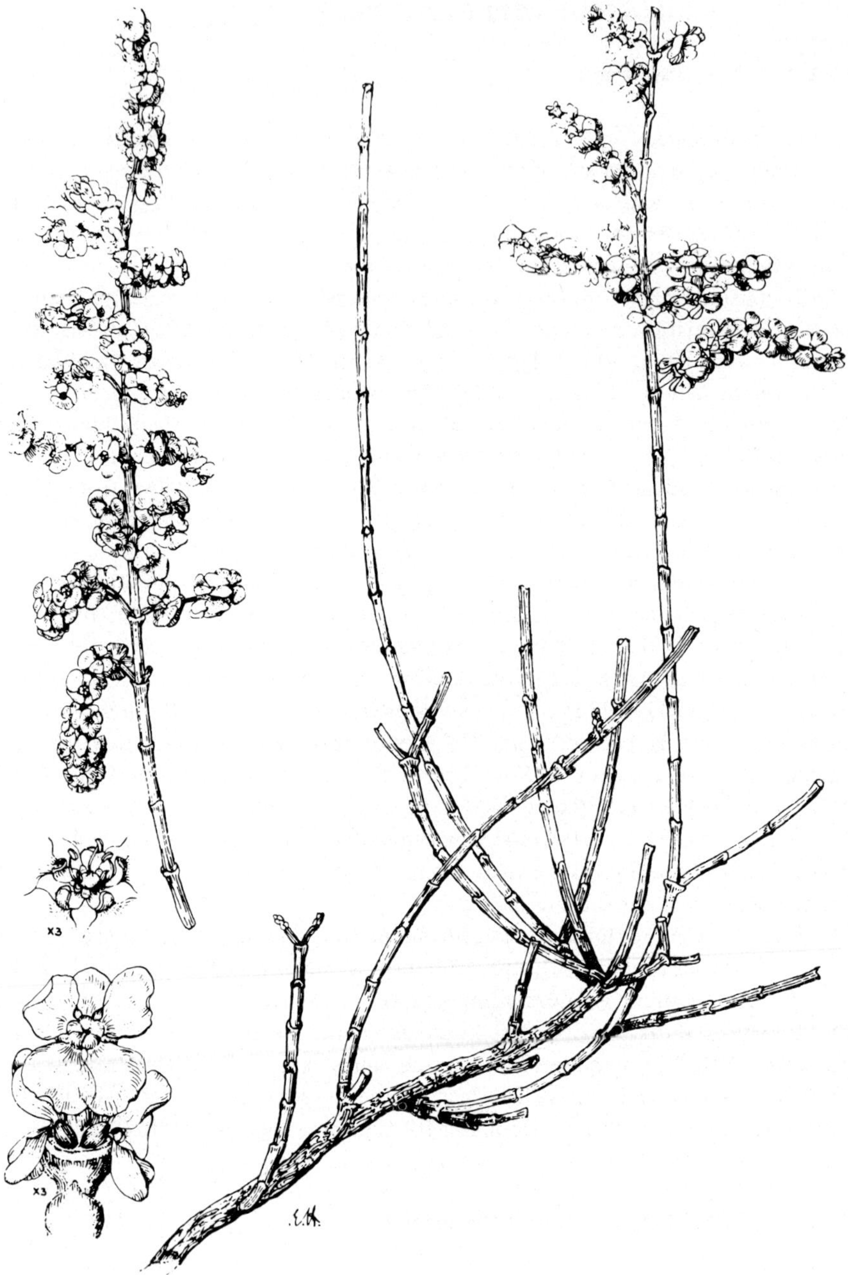

Figure VI–1. *Haloxylon salicornicum* (Moq.) Bge. (1/1).

podiaceae. At the base the woody, old branches may reach a diameter of five centimetres. The woody part is often densely branched and partly covered by a hummock. From the branches erect succulent green twigs develop. These jointed twigs later carry the flowers, in slender spikes (Fig. VI–1). The base of each node is surrounded by a two-pointed membranous sheath with short hairs, formed by the scale-like reduced leaves at the tip of the lower joint. The flowers are arranged in alternating pairs along the terminal twigs. Details are depicted in Haines (1951). The perigone is five-parted, at the time of seed ripening expanding into greenish, pale yellow, pink or purple fruitwings (Fig. VI–1). When the fruits are ripe, the shrub has a solid, whitish or purplish appearance from a distance (Fig. VI–2).

H. salicornicum is a chamaephyte with a seasonal reduction of the transpiring body. The succulent, green, assimilating cortex dries out, becomes pale yellow to whithish, splits and is shed from the shoots. Zohary (1962) therefore grouped *Haloxylon* under the 'bark splitting plants'. It was found that the whole outer part of fruitbearing branches is often shed, falling with the fruits. During the second half of December a layer of seeds and small dry branches was often seen under the shrubs.

Full-grown *Haloxylon* plants in Iraq occasionally reach a height of about one

Figure VI–2. Stand of well developed shrubs of *Haloxylon salicornicum* mixed with some *Artemisia herba-alba* and *Astragalus spinosus*. The compact, whitish appearance of the *Haloxylon* shrubs is due to the just developed fruitwings. (October 1971, Rutba – Nukhaib track)

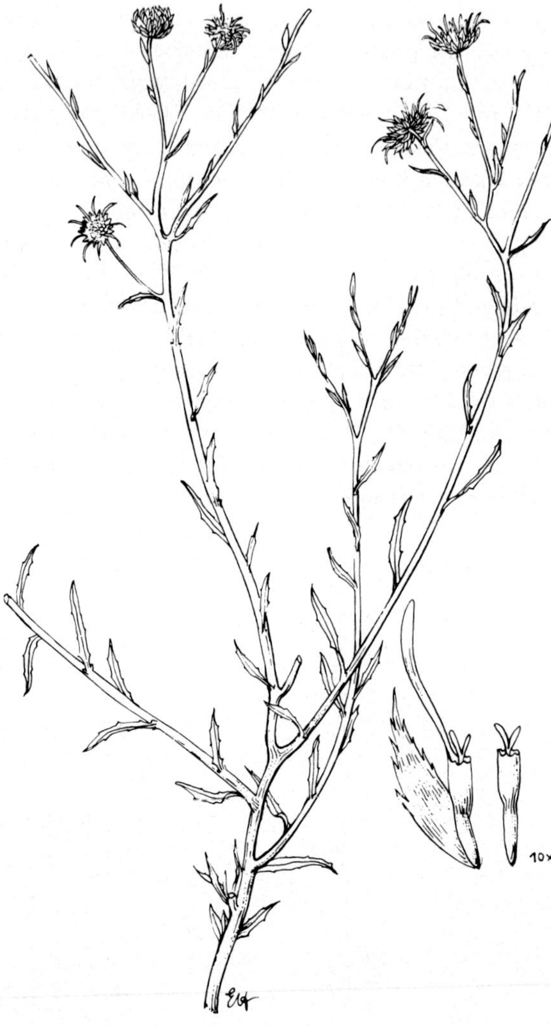

Figure VI–3. *Rhanterium epapposum* Oliv. (2/3)

metre, but the normal height range is 50–75 cm. Soil accumulation at the base of the plant, forming hummocks, can add another 50 cm to the height. Plant diameters in such situations can be two to three metres but normally, with no or only small hummocks, the diameter does not exceed one and a half metre. The individual plants of *H. salicornicum* grow often widely spaced, with a crown cover of 1–15%. In depressions and wadibeds crown covers of 20–25% have been measured locally. Densities for adult plants hardly exceed 0.5 plants.m $^{-2}$.

Outside Iraq, the species has been occasionally studied, especially in comparison with performance of other desert plants, e.g. Haines (1951), Hammouda & Bakr (1969) and Zohary (1973). A preliminary account of the species as a forage producer in Iraq was given in Thalen (1974 b), a summary of the stands and autecological data collected during the period 1969–1973 by Kaul and Thalen in UNESCO (1975).

The taxonomical place of the species with respect to related Chenopodiaceae is a controversial issue. In Täckholm (1974) the species is given under *Hammada elegans*, in Zohary (1973) under *Hammada salicornica*. The last name will probably be shared in the section on Chenopodiaceae for the Flora of Iraq, yet to be covered (Townsend pers. comm.).

VI.B.2. *Rhanterium epapposum* Oliv.

Rhanterium epapposum is a fine branched dwarfshrub of the family Compositae. The shrubs of this species are much branched from the base and 'numerous fine faggoty branchlets together form a spherical bush' (cf. Vesey Fitzgerald 1957b; Fig. III–2). The slender branches are during the dry part of the year whitish -canescent due to a bloom of very fine hairs. The leaves, remote, linear or linear oblong, small (maximum about 0.5 x 3.0 cm), develop at the old branches and from the young shoots, a few weeks after the first autumn rains have fallen (Fig. VI–3). They are shed during the dry season. The flowerheads (0.7–1.5 cm) are solitary and terminal. Full grown shrubs in the stands examined reached a hight of 50–70 cm. Rechinger (1964) gives 20–50 cm as height and Vesey Fitzgerald (1957b) for Saudi Arabia: '. . . growing to a height of about 1 m'. As mentioned for *Haloxylon salicornicum* in the previous section, it may well be the difference in estimating the height without and with the hummock formed under the shrubs. *R. epapposum* is highly drought resistant. It completes its growth-cycle in a few months during and shortly after the rainfall season. In terms of the Raunkiaer Life Forms the species is a tall woody chamaephyte.

R. epapposum shrub vegetation types always show a rather wide spacing between the individual full grown plants (Fig. VI–4). Stands with a crown cover of 40–50% were, therefore, considered excellent. Most common were covers between five and ten percent but very open stands with a cover of less than one percent were also seen. Vesey Fitzgerald (1957b) mentioned for Saudi Arabia spacings of 'several meters apart', '1 meter apart' and 'evenly spaced at 2x2 m'. The observation of Guest (1966) of a stand near Um Quasr with nearly 90% (!) ground coverage in broad sandy depressions, alternating with almost barren ridges of coarse gravel with 5% cover, may be biased due to oblique examination. Little has been published sofar on this species. In Thalen (1974a) an appraisal was given as a range species in Iraq and Kernick (1966) evaluated this aspect for Kuwait where *R. epapposum* is perhaps the most common shrub species. Vesey

Figure VI–4. Good stand of *Rhanterium epapposum* in the Southern Desert showing the spacing of the shrubs.

Fitzgerald (1957b) discussed the '*Rhanterium* – steppe' for Saudi Arabia and Zohary (1973) for the entire Middle East.

The species was first described by D. Oliver as given in Hooker's Icones Plantarum Part III, no. 1367 (Hooker 1881) and is very similar to *Rhanterium suavolens*, a species covering vast stretches in the arid lands of North Africa (Quézel 1965).

VI.B.3. *Artemisia herba-alba* Asso.

Artemisia herba-alba is a somewhat grey woolly (canescent) shrublet of the family compositae. The many stems, branching from the base, end in an oblong broad panicle with rigid branches, spreading rectangularly (Fig. VI–5). The leaves are small and dissected. Different leaf types may develop: 'Root-leaves and leaves of sterile branches petioled, ovate-orbicular in outline, bipinnate partite into oblong to oblong-linear lobes, those of flowering branches much smaller, few-lobed and clustered' (Rechinger 1964). The flower heads are numerousm sessile, blong shaped and small; 7 mm long, 4 mm wide and 2–44 flowered, according to Rechinger (1964).

A. herba-alba is a chamaephyte; the living parts of the aerial shoot are reduced in the hot months. It has been observed that the bigger leaves, developed during winter and spring, are shed during summer and replaced by smaller summer leaves (Orshan & Zand 1962). The species is polymorphic. Zohary (1973) mentiones the existence of a number of ecotypes and refers to Russian work to split *A. herba-alba* into microspecies.

Artemisia shrubs in Iraq usually had a height of 30–60 cm with exceptions upto 75 cm. For the crown diameters these figures are 40–80 cm and 100 cm respectively. A wide variety of stands was seen, the difference in appearance most likely due to biotic and terrain factors. Local densities of three full grown plants per square metre were found and crown covers of 50–70%. On the other hand deteriorated stands, by overgrazing or ploughing, were quite common and also stands with a very sparse cover (less than one per cent) under apparently unfavourable conditions. An average good stand is shown in Fig. VI–6.

Of the three key species discussed in this chapter, on *A. herba-alba* most has been reported, especially in studies dealing with autecological and production ecological aspects (e.g. Orshan & Zand 1962; van der Veen 1967; Kappen *et al.* 1972; Rodin *et al.* 1972; le Houérou 1969, 1975; Zohary 1973). A preliminary ecological appraisal of *A. herba-alba* vegetation was published by Kaul & Al-Mufti (1974).

VI.C. Distribution of the species

VI.C.1. *General distribution*

Haloxylon salicornicum is found from Afghanistan and Pakistan (Baluchistan) in the east to Libya in the west, which means that southern Iraq and the northern part of the Arabian peninsula form the central part of its geographical distribution. Hooker (1885) gives in the Flora of British India for its distribution Baluchistan and Afghanistan. The occurence of *H. salicornicum* as one of the common shrubs in the chenopodiaceous communities of the extreme semi deserts of Afghanistan has been described by Freitag (1971 a,b). The species is well known from the arid zone of Iran (Zohary 1973), but is absent in Turkey (Davis 1975) and finds its northernmost distribution in this part of the Middle East in Iraq (see VI.C.2). *H. salicornicum* is one of the common species of the hard gravel

Figure VI–5. *Artemisia herba-alba* Asso. (2/3).

Figure VI–6. Well developed stand of *Artemisia herba-alba* in the Western Desert.

and sandy gravelly areas in Kuwait (Dickson 1955; Kernick 1966) and the species was described by Vesey Fitzgerald (1957 b) for a variety of environmental conditions in central and eastern Arabia and also for the area towards the Red Sea coast (Vesey Fitzgerald 1957 a). The southern boundary of its distribution in the Arabian peninsula is not clearly defined. In the Kew Herbarium specimens are kept from Dhahran, Bahrein, Abu Dhabi, Muscat and Oman at the east coast, and from areas near Jeddah at the west coast. Vesey Fitzgerald (1955) in his description of the vegetation of the coastal area south of Jeddah does not mention the species. A line roughly from Muscat ((23°N,58°E) to Jeddah (21°29'N,39°16'E) seems therefore a first approximation, confirmed by Batanouny (pers. comm.) who has travelled extensively in the area. To the west *H. salicornicum* is known from Jordan (Poore & Robertson 1963) and Israel (Zohary 1973), and common in Egypt, into the Libyan Desert, west of the Nile (Batanouny 1974; Täckholm 1974). Probably the species extends into Libya, but there *H. schmittiana* is the western vicariad according to Zohary (1973). In North Africa various names have been used to indicate the members of a group of very similar or the same species, e.g. Täckholm (1974) for Egypt, and the Preliminary checklist of the Libyan Flora by Keith for Libya, and for Algeria Quézel & Santa (1962) give in total some six different names for *H. salicornicum* or apparently closely related species.

Rhanterium epapposum is far more limited in its distribution than *Haloxylon salicornicum*. The species is found in central and northern Arabia, Kuwait, southern Iraq and the southwestern part of Iran. The occurrence of the species in

Iran seems very limited. Some specimens were found in the Herbarium of the Faculty of Pharmacy (Tehran University) of the extreme southwest coastal area. Across the Arabian Gulf *R. epapposum* has been found on the island of Bahrein (R. Good in: Dickson 1955). It is the most common shrub species in Kuwait according to Dickson (1955) and Kernick (1966). For Iraq the species is confined to the southeastern part of the country (see VI.C.2). In northeastern Arabia the species covers extensive areas as described in Vesey Fitzgerald (1957b):

> '*Rhanterium* is the characteristic and most abundant perennial over wide tracts of north central Arabia. Steppe vegetation dominated by this plant extends from about 30°N to about 25°N. It is never a common plant west of 40°E, but eastwards it extends to within a short distance of the Persian Gulf coast in Kuwait and the Hasa Province.'

Batanouny (pers. comm.) gives as southern boundary of the species in the Arabian peninsula a line from Abu Dhabi to the west just south of Riyadh and Anaiza to Al Jauf and from there north into Iraq near Judaidat Arar (see also VI.C.2).

Artemisia herba-alba has the widest general distribution of the three species, ranging from southern France (Pyrenees) and Spain and the Atlantic coast through North Africa and the Middle East to Turkey in the north and in the east and northeast into Pakistan, Afghanistan and the USSR. It is a polymorphic species, but Zohary (1973) after having compared many populations in the field from Iran to Egypt saw no justification for a further subdivision as given by some Russian authors. However, in view of this aspect, a clear picture, delineating its distribution can not be given without a detailed study of populations in the field and Herbarium material. *A. herba-alba* covers vast areas in Iran. Mobayen & Tregubov (1970) mapped 756 817 km^2 out of the 1 628 782 km^2 of the entire country (46.5%) under steppe vegetation with *Artemisia herba-alba* and *Astragalus* spp. as the leading species in a large number of associations. Together with *Haloxylon salicornicum, A. herba-alba* is probably the most common species in the desert rangelands of Iraq. Davis (1975) gives for Turkey: 'probably restricted to the Mesopotamian area, where its range may have been much reduced by the spread of cultivation'. Being common in Syria (Davis 1975), Jordan (Poore & Robertson 1963), Israel and northern Arabia (Vesey Fitzgerald 1957b; Zohary 1973), Egypt (Batanouny 1974, Täckholm 1974) and northern Africa (le Houérou 1969; Quézel 1965), *A. herba-alba* is perhaps the most common species of North Africa and the Middle East.

VI.C.2. *Distribution in Iraq, relation to climate*

The distribution of *Haloxylon salicornicum, Rhanterium epapposum* and *Artemisia herba-alba*, each as a dominant of a group of range cover types has been shown in the map of Al-Ani *et al.* (1974) for the Southern and Western Deserts. This very small scale map only shows the regions where each of these species was

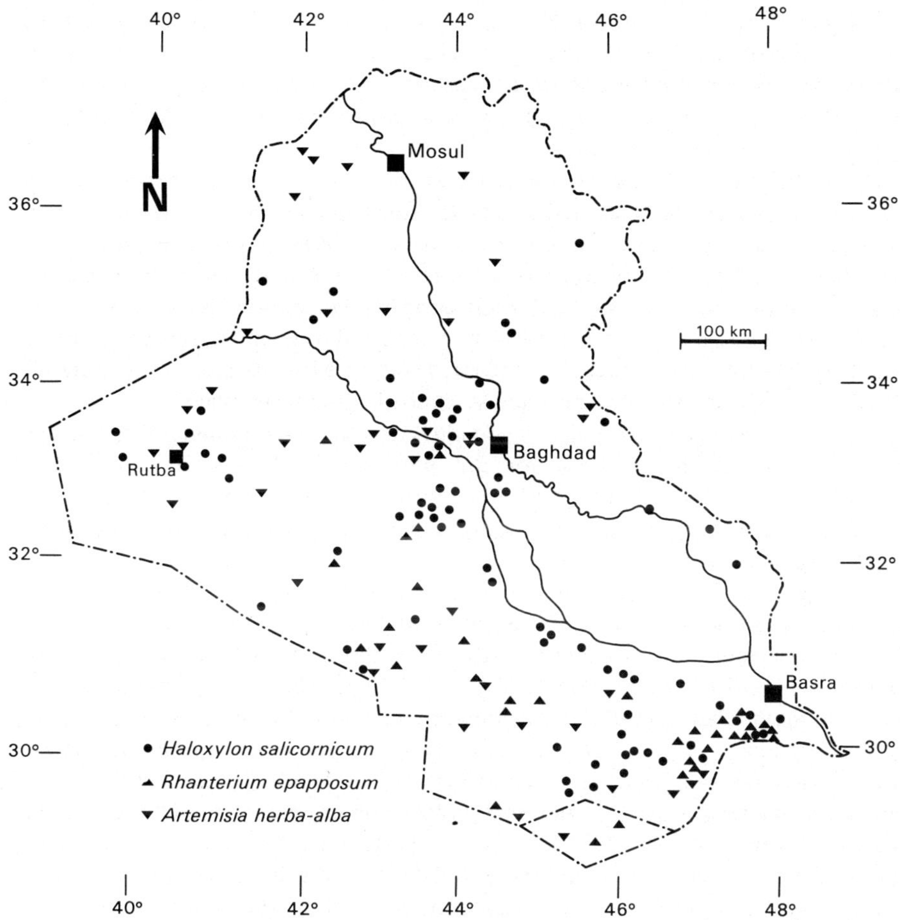

Figure VI–7. Distribution of *Haloxylon salicornicum, Rhanterium epapposum* and *Artemisia herba-alba* in Iraq. (Based on specimens kept in Herbaria in Iraq and, for *H. salicornicum*, at the Kew Herbarium).

found as the major dominant. The distribution of the species is more accurately depicted in Fig. VI–7, where the collection sites are shown for the specimens kept at the Herbaria in Abu Ghraib (National Herbarium and Herbarium of the College of Agriculture), Baghdad (Herbarium of the College of Science) and, only for *H. salicornicum*, the Herbarium at Kew.

Fig. VI-7. shows that *H. salicornicum* is found throughout Iraq, roughly south of 35°N, except in the Lower Mesopotamian Plain. The three major concentrations of collection sites are in the Lower Jezira (between Ramadi, Falluja, and the present Lake Tharthar), west of Kerbala, and in the southeastern part of the country between Nasiriya, the Neutral Zone and Basra. This illustrates perhaps more of the accessibility of the study area and collectors preferences for certain tracks than of the distribution of the species. The area between 41°–43°E and 31°–33°N shows only two collection sites (Nukhaib and Judaidat Arar) but is to a great extent *Haloxylon*-country.

The distribution of *Rhanterium epapposum* as shown in the same figure is confined to a belt of some 150 km wide from Nukhaib in the northwest and Safwan in the southeast. No observations have been recorded north of the Baghdad – Rutba road, the northermost specimen being collected near Habbaniya Lake, southeast of Ramadi. Little collecting has apparently been done in the Neutral Zone and the area some 100 km to the northwest, north and northeast of it, where *R. epapposum* is most abundant, together with the area between Basra, Busaiya and Khedr-el-Mai where intensive collecting has taken place.

For the third species, *Artemisia herba-alba*, only a limited number of specimens was examined in the Herbaria for this purpose (Fig. VI–7). Its distribution pattern in Iraq is less interesting while it finds apparently (see VI.C.1) no major distribution boundary in the country. Specimens of the species have been collected from throughout the country.

Considering the variation in soil conditions (and thereby opportunities for the species to establish themselves) in Iraq and especially in the study area, the master factors determining the distribution patterns as shown must be climatic. A comparison of the distribution patterns in Fig. VI–7 with the climatic data presented in IV.B. shows that *Rhanterium epapposum* in Iraq is by and large confined to an area with an average annual rainfall of 100 mm or less per year, in the hottest part of the country. *Haloxylon salicornicum* is found under these conditions too, but more extensively under somewhat milder conditions and an average annual rainfall of approximately 200–250 mm per year. *Artemisia herba-alba* occurs in the southeastern part of the country in isolated patches, but is the most common plant of the three in the western and northern part of the study area with an average annual rainfall of at least 100 mm. Such data can at best give a rough indication of the climatic conditions under which the species are found. Microclimatic conditions (runoff, shelter) can be very different and in places a climate may, therefore, be found that is conducive for the growth of the species in areas where the species normally would not occur. From the distribution pattern as discussed, the preliminary general conclusion can be drawn that *Artemisia herba-alba, Haloxylon salicornicum* and *Rhanterium epapposum,* in this order seem increasingly better adapted to extreme arid conditions.

VI.C.3. *Edaphic conditions*

In chapter V an indication has been given of the edaphic conditions under which the three key species were found in Iraq (Figs. V–5 and V–6). For *Haloxylon*

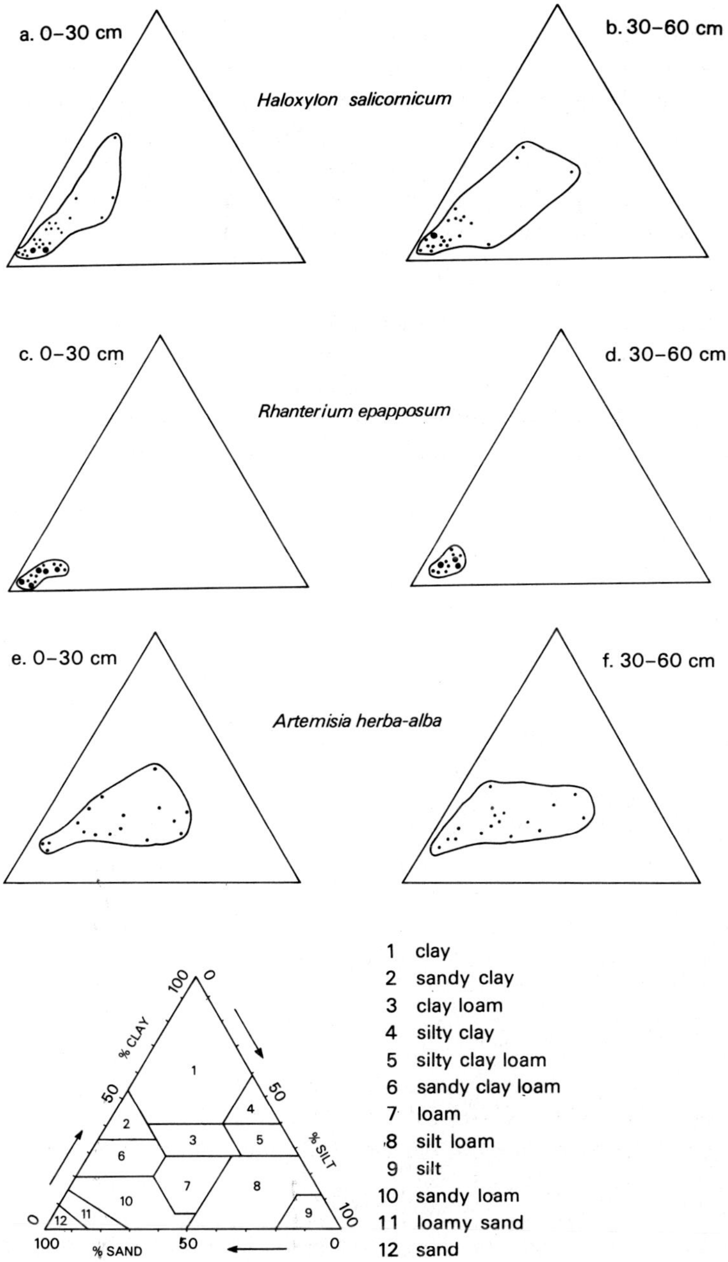

Figure VI–8. Texture of the soils from two depths (0–30 and 30–60 cm) under vegetation types where *Haloxylon salicornicum* (a, b), *Rhanterium epapposum* (c, d), and *Artemisia herba-alba* (c, f), respectively, were the dominant species. Large dots indicate more than one observation.

salicornicum these were generally the coarser textured soils, often with a compact top layer or a gravelly topsoil and hard gypsiferous or calcareous layers, at different depths. The textural class of the soils under stands where *H. salicornicum* was the dominant was sand, loamy sand or sandy loam, but in a few cases the species was found on finer textured soils such as sandy clay loam, clay loam and even clay (Fig. VI–8 a, b). High salinity levels were never observed, but moderate salinity did occur. On a few occasions the species was encountered on low dunes, e.g. near Tel Lachm between Nasiriya and Busaiya in the Southern Desert (see Dougrameji & Kaul 1971) and near Traifawi in the Jezira.

The indications of edaphic conditions under *H. salicornicum* from the surrounding countries are more or less similar to the above observations for Iraq. For Kuwait 'hard gravel' areas in the north and southwest and 'sandy gravelly' areas in the west are mentioned (Kernick 1966). In central and eastern Arabia, Vesey Fitzgerald (1957b) found the species amongst others in shallow runnels, on alluvial pans, and in shallow silty pans, in the last habitat together with *Artemisia herba-alba* and *Zilla spinosa*. Interesting is the following observation on the valley salt-bush associations of Wadi Hamdh (25°30'N, 39°E):

> 'Lower down, where the soil is gritty, dwarf shrubs of *Haloxylon salicornicum* are evenly spaced, forming a pure stand. In the flat bottoms which are liable to flooding, the soil is a fine silt. This is very slippery and muddy when wet, but becomes baked hard with a flaking surface which breaks up into a fine dust when dry. Here several species of halophytes grow on hummocks. *Seidlitzia rosmarinus* is very abundant and *Suaeda vermiculata* is frequent but the *Haloxylon* is completely absent'.

The soiltype last described here by Vesey Fitzgerald is found in Iraq amongst other places in Ar Rahab (see IV.C.2.f.), where *Seidlitzia rosmarinus* and *Suaeda* spp. with other halophytes are growing. Although a common species in nearby areas, *H. salicornicum* is here also absent. Vesey Fitzgerald (1957b) also describes a situation where *H. salicornicum* was apparently optimally developed:

> '. . . relatively shallow bed of sand overlying a salty pan. Here the *Haloxylon* flourishes to the exclusion of all other perennial dwarfshrubs or tussock grasses; it averages 1 metre high and the shrublets are well grown and evenly spaced.'

In Jordan the species has been found on 'sandy slopes or dunes and other areas of blown sand', on 'steep screes and coarse block fans on granite', 'gravel fans and gradual granite covered slopes in the sandstone and granite desert region', 'gravel covered plains usually derived from soft limestone and marls' and 'sandy wadis' (cf.Poore & Robertson 1963). In view of the variety of environmental conditions recorded for *H. salicornicum* areas in Iraq as well as in the neighbouring countries, the following statement by Guest (1966), though correct, is an oversimplification:

> 'We agree with Long (1956) that the community is mainly found on rather sandy loam soils with underlying calcareous matter; casual excavation at one or

two places has revealed a thin layer of finely-broken calcareous rock under the sandy top soil, with solid gypsum below at a depth of 25–50 cm'.

The underlying calcareous or gypsum layers can also directly be seen. The spiny-tailed lizard (*Uromastix aegyptius*) is often found in areas with a vegetation of *H. salicornicum*. Near their burrows conspicuous heaps of excavated (whitish) pieces of gypsum can be found, e.g. between Kerbala and Ukhaidir and near Khedr-el-Mai.

Rhanterium epapposum is in Iraq the species of the not too shallow sandy areas in the southeast. The textural classes recorded for the soils the species grow on are sand and loamy sand with a few cases of sandy loam (Fig. VI–8 c, d). There is an overlap with the soiltextures of *Haloxylon salicornicum* and the combined occurrence of the species has been shown in the previous chapter. A representative profile of the soils under *R. epapposum* has been given in table IV–10. Sometimes the sandy soils are overlain by gravel, giving the impression that *R. epapposum* grows in stony soils or rocky areas. *R. epapposum* has also been found in Iraq on sandy soils overlying deep silty or clayey layers. In these cases the species can be recorded in unusual combinations with others, e.g. with *Artemisia herba-alba, Zilla spinosa, Achillea fragrantissima* etc.

Kernick (1966) wrote that in Kuwait the species is most abundant on 'wind-blown silty sands' and 'gravels'. The following quotations from Vesey Fitzgerald (1957b) all point in the same direction for the *Rhanterium* substratum in Saudi Arabia:

– (in the vicinity of Jauf) 'on a plain of cretaceous limestone which is much overblown with sand. The rock is white, close-grained and wind-etched and the exposed surface is cracked so as to form a rough pavement, or broken loose to form cobbles.'

– (within the sands of the Great Nafud) 'in places the gravel is overlaid with ribs of wind-blown sand, which arrest and absorb sheet flow: in such places *Rhanterium* is again characteristic.'

– (the Shama district) '. . . a vast limestone pavement more or less overlain with red sand, and here *Rhanterium* grows under optimum conditions . . .'

– (eastern edge of the Dahama between 27°–26°N) 'characterised by ridges of red sand overlying a rough limestone pavement with huge potholes' and 'Although absent from the high dynamic dunes and from the gravel plain, *Rhanterium* is abundant over areas of shallow, level or undulating sand overlying the gravel throughout the Dahama'.

The same author gives a number of more or less similar statements of *Rhanterium* growing on sand overlying limestone pavement or pebbly plains. The information from southern Iraq, Saudi Arbia and Kuwait leads to the conclusion that *R. epapposum* is found on sandy soils, usually overlying stony or gravelly layers, mainly of limestone. The nature of the terrain is such that windblown sand deposits alternate with stony areas (see Fig. I-1). *Rhanterium* is found especially on such sandbodies, but disappears when they become real high sanddunes. In the latter case it was observed that the species was replaced by *Calligonum comosum* and ephemeral species.

Compared to the two species discussed above, *Artemisia herba-alba* occurs on the finer textured soils (Fig. VI–8 e, f), although there is a considerable overlap with the texture of the soils found under *Haloxylon salicornicum*. The textural classes for *A. herba-alba* range from sandy loam and sandy clay loam to soils with a higher silt content, loam and clay loam. In the Western and Southern Deserts the species is found on the silty flats and in depressions and wadis, receiving runoff water from the surrounding higher land. Zohary (1973) formulated the phytogeographical aspect of the occurrence of the species in this part of the Middle East as follows:

> '*A. herba-alba* penetrates deeply into Saharo-Arabian territories only where the deficiency in atmospheric moisture is compensated for by favourable edaphic conditions, so that its occurrence in these territories does not affect its value as an indicator of Irano-Turanian conditions.'

In Iraq the species was never found under (even slightly) saline conditions. Kaul & Al-Mufti (1974) found *A. herba-alba* in the Jezira less restricted in habitat and distribution than in the Southern and Western Deserts. In the Jezira they recorded the species for deep, grey steppe soils, without stones or with a few pebbles only on top, a soil similar to the one described by Poore & Robertson (1963) for *A. herba-alba* in Jordan.

The differences in texture of the soils where *A. herba-alba* grows best, compared

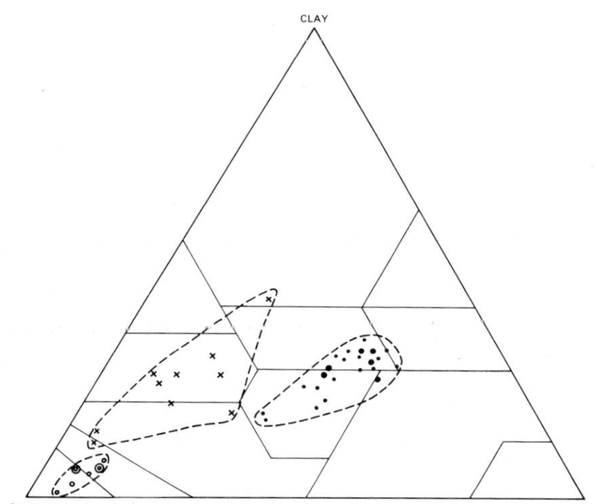

Figure VI–9. Texture of soils (0–30 cm) under virtually pure, well developed stands of *Artemisia herba-alba* near Rutba (dots), *Haloxylon salicornicum* near Khedr-el-Mai (circles), and three mixed stands of these species, near Nukhaib and along the Rutba-Nukhaib track (crosses); large dots and circles indicate more than one observation.

to those where *Haloxylon salicornicum* is very well developed, is illustrated in Fig. VI–9. The soils at the few locations where the species were found together in mixed stands hold an intermediate position (Fig. VI–9).

VI.D. Phenology of the species

The main phenological characteristics of the three species are shown in Fig. VI–10. The figure is based on field observations, herbarium material

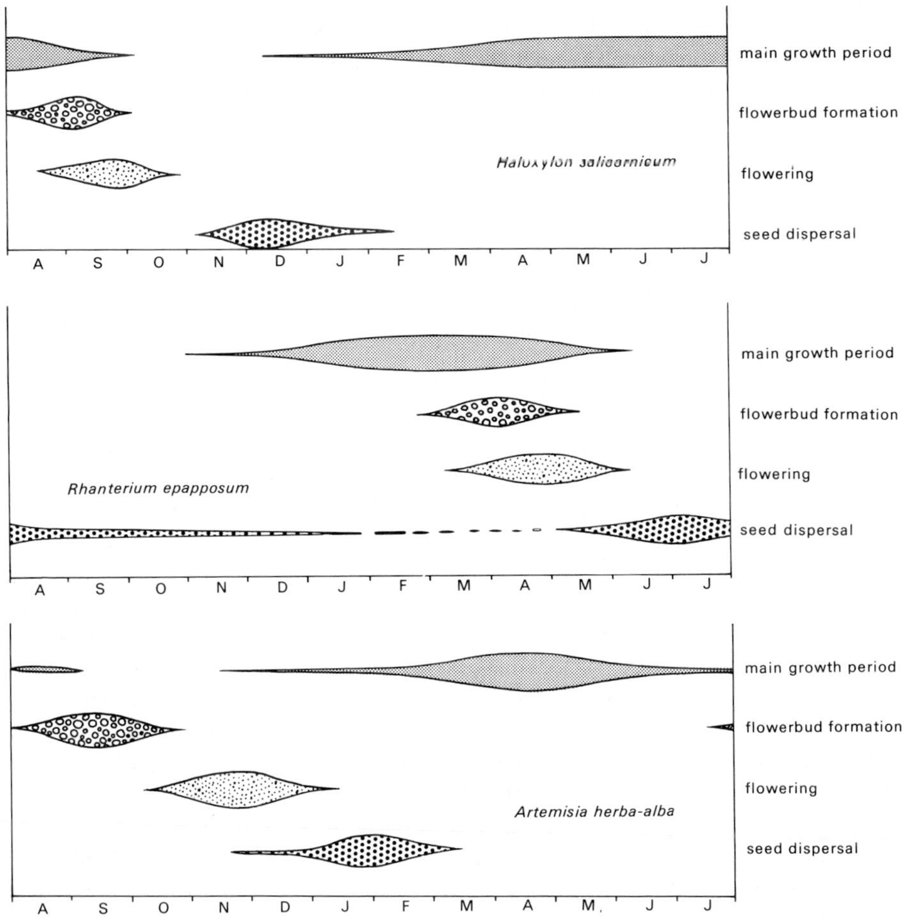

Figure VI–10. Phenology of *Haloxylon salicornicum*, *Rhanterium epapposum* and *Artemisia herba-alba.*

(National Herbarium and Herbarium of the College of Agriculture, Abu-Ghraib; Herbarium of the Royal Botanic Gardens, Kew), and on data in Haines (1951), Dickson (1955), and Kaul & Al-Mufti (1974).

There are marked differences between the species. All three of them start a new growth period after the autumn rains, but this is not a quick response. The first new shoots on *R. epapposum* were observed in November (1972). Dickson (1955) mentions for this species:

> 'It sheds its leaves each summer, and shoots out six weeks after the first autumn rain; the small yellow flowers appear in April giving the desert a golden sheen.'

A. herba-alba forms new leaves in late autumn and winter, but the main growth period is spring. The larger leaves grown during this period may be replaced later by smaller summer leaves. New green buds on the dry, pale yellowish shoots of *H. salicornicum* can sometimes already be seen in December, but it is usually not until March or April that they have developed to green shoots, thereby giving the shrubs a new colour. *R. epapposum* completes its growth cycle in a few months. At the end of the new shoots flowerbuds develop in March and April. Full flowering is in the second half of April. From May onwards the species withers, gradually shedding its dry leaves and the ripened seeds in the flowerheads. There is an overlap in phases. In the same bush flowerbud formation may still occur while the first dry flowerheads have already fallen. During the hot, dry summermonths, *R. epapposum* looks completely dry and is leafless. A 'waxy bloom' covers the vital parts and

> 'A thread of living tissue survives within the fibrous rootstock during prolonged dry periods, thus allowing the plant to resume growth when rains at last fall' (Vesey Fitzgerald 1957b).

A. herba-alba and *H. salicornicum* continue their growth during the hot rainless summer, although in the first species at a reduced rate. Fastest growth in *H. salicornicum* is in the period April-July. In early August many small lateral shoots develop from the erect young succulent twigs, on which the flowerbuds develop. Flowering, ripening and seed dispersal take place in September-December. In early September the first open flowers can be found. They are inconspicuous, arranged in alternating pairs along the terminal shoots (Fig. VI–1). In October, for a very short period the ground under the shrubs can be yellow from the fallen anthers. The lobes of the perianth develop into fruitwings with a yellowish green to sometimes pinkish colour. This gives the shrubs by the end of the month and in November their dense, solid outward appearance (Fig. VI–2). The dry fruits with wings break off from the shoots in December and are blown over the ground. Sometimes whole shoots with the fruits are shed. Bark splitting and shedding was seen in summer, autumn and early winter at the base of the main shoots. For a short period in the cold winter months *H. salicornicum* looks dry, greyish and almost lifeless. At a closer look the green buds of the new shoots at the nodes van be discerned. The phenological stages of *A. herba-alba* run rather parallel to those of *H. salicornicum* but are more extended. Development of new leaves is earlier, seed dispersal later (Fig. VI–10).

VI.E. Reproduction and development of 'Haloxylon salicornicum'

VI.E.1. *Seed production and dispersal*

The first ripe fruits of *H. salicornicum* are shed in early November, most fall in December and early January, but often some can still be found on the bush as late as February. Haines (1951) gave an illustrated description of the aerial parts of the species including details on the flower and development of the fruit. Clor *et al.* (1974a) described the morphology of the germinating seed. After flowering the lobes of the perianth expand to give overlapping fruitwings. The fruits break off from between the bracts and once fallen may be carried over large distances by the wind. They are easily blown over the surface and can be found in large quantities in wind sheltered places; in small depressions, soil cracks, between plants and stones, etc. Runoff water could be a carrier too. Haines (1951) mentioned the distribution by ants in the desert near Cairo. The fruits do not easily stick in hairs or fur and zoochory by larger animals, including livestock must be of little or no importance. Small rodents may collect the seeds in their burrows (see IV.F.2.). Large amounts of seed are produced. An idea about quantities was obtained from measuring, weighing and sample counting on 19 shrubs of which as far as possible all fruits were collected. The shrubs were randomly chosen and sampled on 19 and 20 December 1972. An average of 185 (±74) fruits per dm^3 crown volume was found. Small plants produced more than large plants per unit crown volume. The five smallest bushes, with a crown volume of 0.082 (±0.022)m^3, had a production of 244 (±92) fruits per dm^3 the five largest ones, with a crown volume of 0.387 (± 0.075)m^3, a production of 135 (±34) fruits per dm^3. Extrapolating these data for larger areas means that a well developed stand of *H. salicornicum* with a crown cover of ten per cent produces some 40 million fruits per hectare annually. A minor percentage of the seeds of these fruits will germinate, develop into a seedling and finally and adult plant. Heavy seed predation was seen. Birds and other vertebrates take a toll, but by far the largest amount is eaten by lower organisms (table VI-1). Some insect damage was observed to seed on the plant, but heavier is the attack after shedding.

From a mixed sample the average fruit weight was determined. For 500 unhusked fruits, weighed in 10 subsamples of 50, a weight of 132.6 (±7.4) mg per 50 fruits was found; an average of 2.65 mg per fruit. After removal of the fruitwings, the average weight of husked fruits was 1.13 mg. These were fruits collected in December. Observations on husked fruits, collected at different dates, later learned that the weight of the husked fruits decreases with time.

Table VI–1. Insect damage to seeds of *Haloxylon salicornicum.*

Collection			Examination after storage in unsealed bags at room temperature				
						Seeds damaged	
Date	Location	Remarks	Date of examination	Number of fruits examined	Seeds undamaged (%)	Due to insects* (%)	Other** (%)
9 Nov. 1971	Ukhaidir	Fruits difficult to remove from the shoots, tips of shoots with fruit collected	17 Jan. 1972	500	43	47	10
27 Nov. 1971	'Mia-sittien'	Fruits dry; easily stripped from the shoots; no collection from the ground	30 Jan. 1972	500	28	67	5
8 Dec. 1971	Ukhaidir	Most seeds already shed; collected from the ground and from the plants	18 Jan. 1972	100	2	93	5
12 Jan. 1972	Ukhaidir	Most seeds shed; collection mostly from the ground	17 Jan. 1972	500	20	71	9

* The term 'insects' is used to indicate that small invertebrate animals, most likely of different species, caused the damage. Most of these were insects.
** Includes amongst others fruits without properly developed seeds.

VI.E.2. *Germination*

Information on factors affecting germination of *Haloxylon salicornicum* was reported in Al-Ani et al. (1971c) and Clor *et al.* (1973, 1974a, 1976) for Iraq, and in Hammouda & Bakr (1969) for Egypt. Additional work was done in particular for aspects related to management practices e.g. seed harvest, storage, reseeding, etc. Germination tests were carried out in petri dishes with a double layer of filter paper moistened with about 5 ml distilled water. Usually four replicates of 25 seeds per dish were examined till no new germination was observed. Unless mentioned otherwise all tests were done with husked seeds, either at constant temperature under continuous light in an incubator, or at room temperature (about 20°C) under the in the laboratory prevailing light conditions.

VI.E.2.a. Morphology of the seed and germination rate

The seeds of *H. salicornicum* are rather uniform in shape. The spire-like embryo is surrounded by a membranous seed coat. The compactly coiled green cotyledons form the inner one loop and a half and the hypocotyl radicle portion forms another loop and a half. This seed has been formed within the thin transparent ovary wall. It rests in a sort of urn formed by the fruitwings and is partly covered by small inward protuberances developed from the perianth lobes.

Germination is epigeal. Upon imbibition of water, the embryo uncoils and stretches, easily rupturing the thin seed coat. Clor *et al.* (1974a) showed the characteristic behaviour of the elongating embryo axis. They described it as twisted, bent, coiled and curved in various patterns. It was found that the radicles in the early phase grow not only downward (when seeds are germinated on vertical beaker walls with wet filterpaper or on a floating pad) but also upward and in various other directions. Three to four days after germination, however, the rootlet grew down. Germination rate depends on temperature and seed viability, but can be extremely high (Table VI–2). At temperatures of 30°C–40°C some husked seeds germinated within one hour after water imbibition had started. Criterium for germination was the protruding of the radicle to a length of at least 2 mm. At these temperatures, seeds collected later in the season attained maximum germination in 24-48 hours.

VI.E.2.b. Seed viability and longevity

Seeds stored for 15 months at room temperature failed to germinate at temperatures of 5, 15, 25 and 35°C. Seeds stored at very low temperature, 2–4°C, maintained their full germinability when tested at 5 and 15°C for the whole testperiod, lasting to 18 months after collection. When tested at 30° a slight decrease in germinability was seen. After a cold storage period, seed exposed to 30°C during two months had completely lost their germination capacity. When kept at 15°, germination at 5, 15 and 30°C had reduced to 40, 34 and 2%, respectively. These results do fit in the conclusions reached by others. Hammouda & Bakr (1969) found that seed viability in *H. salicornicum* gradually decreased, reaching zero in just over a year. Clor *et al.* (1973, 1976) observed a 50% decrease in germination capacity over the March (collection) – December (testing) period for seeds stored at room temperature, against full maintenance of germination capacity for seeds stored at 5°C. From these findings it may be concluded that in the field, where the seeds will normally be exposed to high temperatures for several months during summer, seed longevity will be less than one year. Considering the time of seed distribution and the rainfall period, *H. salicornicum* seeds should germinate within a few months after being shed.

VI.E.2.c. Husked and unhusked seeds

An effect of removal of the perianth covering the seeds was reported by Al-Ani *et al.* (1971 c). They found 51% germination of husked seeds against 14% for unhusked ones, Sankary & Barbour (1972c) testing this aspect in the closely related species *Haloxylon articulatum*, conclude that 'the fruit appears to lack inhibitors since the seeds germinate readily without fruitwing removal or leaching'. Their data show a somewhat increased germination after fruitwing removal. Important from a practical point of view is their observation of the two to six times higher establishment percentage after 20 and 40 days of unhusked seeds compared to husked ones. They conclude that this may be due to such factors as protection of the embryo, reduction of desiccation hazards and provision of a sort of mechanical resistance, aiding the radicle in breaking through the soil.

VI.E.2.d. Optimal temperature and time of seed harvesting

Hammouda & Bakr (1969) gave 15°C as the optimal temperature for the germination of *H. salicornicum*. Al-Ani *et al.* (1971c) found the lower range of temperatures 5–15°C best, but did not state age or storage conditions of the seeds. Clor *et al.* (1974a) refined the picture. For highly viable seeds they found no difference when tested at 6, 13, 20 and 27°C; percentage germination was 96, 96, 98 and 96, respectively. Aged seeds, with a reduced viability germinated best at the lower temperatures. Hammouda & Bakr (1969) in their discussion on temperature already noted that the optimal temperature of germination could shift with time from harvesting, and that this would be connected with ripening of the seeds.

This was tested. Unripe fruits, harvested from the shrubs in the first half of

Table VI–2. Effect of different constant temperatures on germination (%) of *Haloxylon salicornicum*. Seeds collected 20 December 1972, stored four months at room temperature and 10 months at 2–4°C; 100 seeds per treatment.

	time (hrs)													
	1	2	3	4	8	12	16	20	24	48	72	96	120	144
5	0	0	0	0	0	3	5	6	8	26	63	85	93	97
10	0	0	0	0	6	9	14	25	43	96	97	98	98	98
15	0	1	5	8	17	41	76	92	97	99	99	99	99	99
20	0	2	6	10	30	59	85	92	93	98	98	98	98	98
25	0	2	6	13	36	60	80	87	94	98	98	98	98	98
30	1	3	15	21	46	66	80	89	92	100	100	100	100	100
temp. 35	1	7	12	23	55	77	84	92	92	94	94	94	94	94
(°C) 40	2	8	16	24	58	72	78	84	84	84	84	84	84	84

November, germinated best at low temperature (10°C). In a 9 November 1971 sample germination was 76% at 10°C and only 25% at 20°C with the 15°C test in between; for a 13 November 1972 sample 91% at 10°C, 79% at 20°C and 55% at 30°C was found. For seeds collected later in the season, in December, January and March, germination at the tested temperatures ranged from 85% to 100% and 15–20°C was found as the optimum. A final test with December seeds, which had been stored for four months at room temperature, and others stored for 10 months at 2–4°C, proved that germination was equally well at all temperatures from 5–35°C and only slightly reduced at 40°C. At a constant temperature of 5°C full germination was reached after six days, at the temperatures 15–35°C in two days, and at 40°C within one day (table VI–2).
The experimental facilities did not allow testing of the effect of alternating temperatures as compared to constant temperatures. However, such tests carried out for other chenopodiaceae of the Middle East (*Salsola vermiculata* var. *villosa* (Del.) Moq., and *Haloxylon articulatum*) and of California (*Atriplex polycarpa*) did show an indifferentness of these species, closely related to *Haloxylon salicornicum*, to alternating temperatures in comparison to the weighted constant temperatures (Sankary & Barbour 1972, a, b, c).

VI.E.2.e. Light conditions

Hammouda & Bakr (1969) found a higher germination of *H. salicornicum* in continuous dark (32% after 11 days), than in continuous light (8% in 11 days). No indication of storage conditions, seed age etc. is given. Rate of germination and final percentage both are low. Al-Ani *et al.* (1971c) tested *H. salicornicum* germination under continuous light, intermittent light and continuous dark and found no significant differences. Al-Mufti (pers. comm.) observed a slightly better germination in continuous dark. Sankary & Barbour (1972c) found for *H. articulatum* no influence of light conditions.

VI.E.2.f. Seed size

Separation between small seeds (<1 mm diameter) and large seeds (1.5–2 mm) was achieved by sieving and visual selection. Small seeds showed about 60% of the germination capacity of large seeds, even when seeds collected from the same shrub were compared.

Clor *et al.* (1976) found that small seeds lost their viability faster than larger ones. Seeds stored in a refrigator, collected in March and tested by them in July and December showed a difference of only about 5% germinability between small and large (germination over 90%), while for seeds kept at room temperature in December large ones showed about 60% and small ones less than 20% germination.

VI.E.2.g. Hydration and dehydration

The influence of different artificial rainfall levels on germination was tested by Hammouda & Bakr (1969) with seeds on loamy and sandy soils. *H. salicornicum* showed no germination at 5 mm precipitation, 10% at 10 mm and 16% at 20 mm. A daily addition of 1 mm slightly increased germination. No details on seed viability are given. Interesting results were obtained by Clor *et al.* (1974a) with repeated hydration and dehydration of seeds for various periods. After allowing imbibition, seeds were placed on dry filterpaper and left on the laboratory bench

Table VI–3. Effect of repeated hydration for different periods and dehydration of seeds of *Haloxylon salicornicum* on germination (derived from Clor *et al.* 1974).

Hydration, first imbibition (hours)	Dehydration, first drying (days)	Hydration, second imbibition (hours)	Dehydration, second drying (days)	Germination at room temperature (%)
0	7			97
½	7			90
1	7			98
2	7			95
4	7			94
8	7			95
0	7	2	10	98
½	7	2	10	70
1	7	2	10	90
2	7	2	10	94
4	7	2	10	90

till the germination tests. The capacity of *H. salicornicum* to germinate at a second or third 'shower' bringing sufficient moisture, after a first relatively short hydration period, was clearly shown (table VI–3).

VI.E.2.h. Seed harvest methods

For a range reseeding programme large numbers of highly viable seeds are required. For *H. salicornicum* they could be either stripped from the shrubs, or collected from the ground. In November relatively few seeds will have fallen and harvesting from the shrub will be the only method. At low temperature these seeds could germinate well (VI.E.2.d.). Later in the season it might be better to collect the already fallen seeds. A test between 30 January 1973 plant-collected and ground-collected seeds showed a highly reduced viability of the first group (25–47% against 97–100% germination). In considering the practical application of these results, however, the possibility of an increased insect damage for the ground seeds should also be considered.

VI.E.3. *Seedling development and establishment*

VI.E.3.a. Development during the first days after germination

Depending on viability of the seeds and temperature, the development of the young seedling can be very fast. Ripe seeds, collected in December and March produced seedlings with a length of about 50 mm in five days after germination at 20 and 30°C (table VI–4). At 30°C a length of 40 mm was already reached after two days. These were total lengths, measured from the tip of the cotyledons to the tip of the radicle or rootlet. The figures are mainly due to a rapid extension of the radicle: 5 mm in 12 hours and 10 mm in 24 hours has been observed (Clor *et al.* 1974a). Further observations with marked positions of cotyledonary poles and radicle tips showed growth rates of up to 5 mm in five hours and 15 mm in 24 hours. Hypocotyl extension is after three or four days followed by the development of a fine rootlet. No extension of the cotyledonary poles in these first days was seen (Clor *et al.* 1974a).

Seedling development at low temperatures and of seeds collected early (first half of November) was markedly reduced. Best development was found at a constant temperature of 20°C, although little difference with 30°C was found during the first days.

Very young seedlings of *H. salicornicum* with a radicle extension less than 10 mm, have shown the capacity to resume growth after being kept one week completely dry at room temperature. Seedlings of 48 hours old with a length exceeding 15 mm failed to recover. The stage of radicle extension is apparently crucial.

Table VI–4. Seedling development in *Haloxylon salicornicum* the first days after germination, as affected by seed collection date and temperature; date of experiment 13–18 March 1972; 100 seeds per treatment (mean value and standard deviation).

	Length of the 12 longest seedlings (mm)					
Temperature (°C)	Seeds collected 13 November 1972		Seeds collected 19 December 1972		Seeds collected 7 March 1972	
	after 2 days	after 5 days	after 2 days	after 5 days	after 2 days	after 5 days
10		13.3±2.0	12.8±1.1	20.2±1.8	9.8±1.3	16.2±1.6
20	20.8±1.3	36.3±2.6	26.5±1.2	51.3±3.1	28.8±2.0	49.2±5.2
30	22.1±2.9	32.1±3.3	40.6±4.0	49.7±4.3	40.5±4.7	48.3±3.5

VI.E.3.b. Root and shoot development

Seedlings in all stages of development were examined in the field and five convenient classes were distinguished (table VI–5, Fig. VI–11). The widely differing developmental stages observed at the same time in the same area may be due to differences in germination time, microclimatic conditions and local availability of water in the (sub) soil. The smaller plants (stage 1) were mostly dry and dead. The slender rootlet develops into a deep taproot (Fig. VI–12, table VI–6). Laterals were usually found at depths greater than 20–30 cm. Shoot development is slow in the beginning. Sankary & Barbour (1972c) showed that low temperatures may be the limiting factor for early development in *Haloxylon articulatum*. In parallel to their findings, from field observations and some preliminary growth experiments, stage 1 of Fig. VI–11a depicts seedlings of about one to two months old when growing under normal spring conditions. Stage 5 (Fig. VI–11e) is reached after five to six months by very well developing plants. Under sufficient available moisture in the subsoil active growth continues during the hot summer. Within a year a height and diameter of 50 cm can be reached. The few exceptionally well developed plants of such size sometimes produce a few flowers already in their first year, but flowering is normally seen first in the second growing season. Normal height at the time when new rain falls is in the 5–20 cm range, diameters are then between 10 cm and 20 cm. Smaller and larger plants are also commonly found. Occasional excavations of seedlings revealed a taproot reaching depths of one metre and more. In the second and third year very long shoots may develop (Fig. VI–13 a, b). Third year's plants can often not be distinguished from older plants than by yearring counts.

Figure VI–11. Seedlings of *Haloxylon salicornicum* in different stages of development, collected 26 July 1972, along the Rutba – Ramadi road, about 14 km from Rutba. Representatives of five distinguished classes are shown; details in table VI–5; scale is in cm and mm.

a) stage 1

b) stage 2

c) stage 3

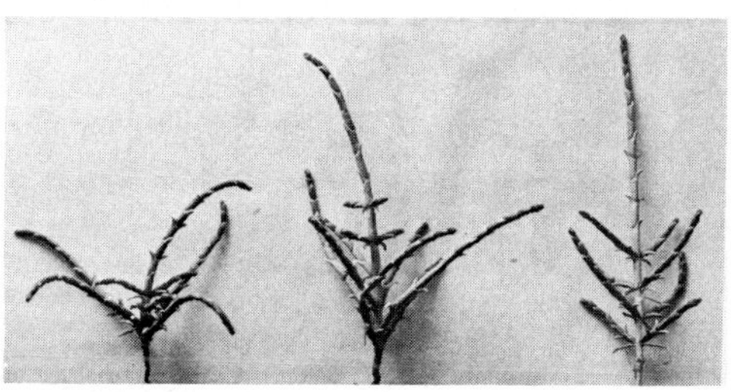

d) stage 4

e) stage 5

Table VI–5. Quantitative characteristics of seedlings of *Haloxylon salicornicum* in various stages of development; measured 26 July 1972 along the Rutba-Ramadi road, about 14 km from Rutba (25–50 seedlings per class, mean value and standard deviation)

Seedling stage	Dry weight aerial parts (g)	Length of the longest shoot (cm)	Length of the 2nd longest shoot (cm)	Number of 'branches'	Lateral spread (cm)
1 Only one shoot, or one shoot with laterals just developing (Fig. VI–11a)	0.06±0.02	2.1±0.7	0.4±0.3	2.0±1.0	1.2±0.3
2. The first lateral shoots developed (Fig. VI–11b)	0.17±0.10	3.3±0.8	1.2±0.4	4.4±1.2	1.9±0.5
3. Several lateral shoots, or at least one well developed (Fig. VI–11c)	0.30±0.10	5.3±1.4	3.0±1.0	6.3±1.8	4.1±1.5
4. Main shoot and laterals rapidly extending (Fig. VI–11d)	0.67±0.21	8.6±1.5	5.8±1.4	8.5±2.4	7.9±2.0
5. Lateral shoots either again branching, or approaching the main shoot in length (Fig. VI-11e)	1.50±0.40	14.6±3.4	11.4±3.5	21.7±11.9	13.8±4.2

Table VI–6. Rootlength of 15 excavated seedlings of *Haloxylon salicornicum* of different height.

		Shoot			Root	
Observation date and location	Soil conditions	Height (cm)	Diameters (cm) D_1	D_2	Length (cm)	Remarks
25 July 1972	Shallow gravelly	1	–	–	11	complete
Rutba–Nukhaib	soil with large	1	–	–	>13	broken
19 km after leaving	stones (10–15 cm	2	–	–	> 8	broken
the Rutba-Ramadi	deep), overlying	3	–	–	>14	broken
road at 'km 55'	hard stony layer where roots go down in cracks*	4	–	–	>23	broken
28 June 1972		2	1.0	0.5	18	complete
		2	1.5	0.5	> 9	broken
About 3 km east of Nukhaib, near Wadi Ubayidh	Deep sandy gravel with pebbles and some thin silty	2	1.5	1.0	>13	broken
	layers in the	5	2.5	2.0	33	complete
	profile.	5	4.0	2.5	24	complete
		5	4.5	2.5	36	complete
		6	1.5	2.5	23	complete
		6	3.5	2.5	>39	broken
		6	4.0	2.5	28	complete
		9	4.0	2.5	46	complete

* See Fig. VI-12.

Figure VI–12. Seedling with partly excavated root of *Haloxylon salicornicum.* The shoot is about 6 cm. The root penetrates some 25 cm through gravel layers and disappears in cracks of the underlying hard rock. Moisture may accumulate in such cracks and provide enough water for the dry season.

A

B

Figure VI–13. Two stages in the development of *Haloxylon salicornicum* as encountered in the field in July 1972: A. Second year plant, sprouted from a grazed seedling; B. Plant of at least three years old, showing excellent growth of current year with longest sprouts of about 50 cm.

VI.F. Reproduction and development of 'Rhanterium epapposum'

VI.F.1. *Seed production and dispersal*

Flowering of *Rhanterium epapposum* is usually at its peak by the end of April. Soon afterwards the first flowerheads are being shed and from the end of May onwards the ground under and nearby the shrubs is covered by these heads (Fig. VI–14). The topsoil at this time of the year is dry and no rain will fall in the three or four months to come.

Under such conditions germination will not commençe before the autumn rains. No literature record of the germination process in this species was found. Vesey Fitzgerald (1957b) is his description of the *Rhanterium* steppe' in Saudi Arabia writes:

> 'Regeneration from seed has not been observed although this must be the normal method of propagation. In all probability seedlings are only able to establish themselves during seasons of exceptionally good rainfall'.

From field observations and laboratory experiments an insight was gained on the regenetation from seed in this species. Establishing and established *Rhanterium* seedlings were found in late winter and in spring of all three years of observation: 1971, 1972 and 1973. According to police-officers posted in Samah and Busaiya

Figure VI–14. Dropped flowerheads of *Rhanterium epapposum* as found under and nearby the parent shrubs. Note number of heads in proper germination position (scale in inches and centimeters).

in and near the areas where these observations were made, only 1972 could be regarded an exceptionally good rainfall year.

Rhanterium seeds germinate while remaining in the flowerhead. There is no distribution of individual achenes. The cotyledons emerge from between the withered flowers in the heads. Rootlets develop at the base where the flowerhead broke from the stem (Fig. VI–15). Usually more than one seed germinates at the same time and small groups of seedlings establish at the same time. They are in the beginning still kept together in the flowerhead, which breaks at a later stage when the shootstems grow thicker. The seed dispersal mechanism of *Rhanterium epapposum* is, therefore, one of the distribution of flowerheads.

In order to obtain an idea about the amount of heads produced, counts were done on shrubs of different sizes, *viz.* 14 shrubs, ranging in height from 32 cm to 76 cm and in crown cover from 0.11 m^2 to 1.40 m^2. No significant difference was found between the amount of 'seeds' produced by small plants and large ones per unit crown volume. Five plants with a crown volume of less than 0.05 m^3

Figure VI–15. Excavated seedlings with roots in heads of *Rhanterium epapposum* a few days after germination (scale in mm).

(0.036±0.008) produced an average of 2.8 (±0.9) flowerheads per dm^3 crown volume. For five plants with a crown volume between 0.05 m^3 and 0.20 m^3 (0.138±0.039) this was 3.0 (±1.1), and for the remaining four plants with crown volumes between 0.20 m^3 and 0.75 m^3 (0.459±0.188) a number of 2.1 (±1.4) per dm^3 crown volume was found. An overall average of 2.68 (±1.09) flowerheads per dm^3 crown volume was calculated. A stand of *Rhanterium* with a crown cover of about 25% may have a crown volume of 100 dm^3 per square metre. Extrapolating the above findings, such a stand would produce about two and half to three million flowerheads per hectare annually. Counts on individual flowerheads showed that the amount of seeds per head was highly variable. The largest number of seedlings found developing from one head was seven.

Shedding of the seeds in May was already mentioned. The dropping of the heads, however, is a gradual process which continues in the months to follow. Some heads will remain on the shrub for over a year and the phenomenon of heads of two different years at the same shrub is not at all an uncommon one. It was also observed in an herbarium specimen (no. 2022, Herbarium Iraquense, College of of Agriculture, Abu Ghraib), collected 21 April 1962. The dropped heads are blown over the ground surface or transported by water after exceptionally heavy showers. Most heads remain under or very near the parent plants, due to their shape, and are partially or completely buried under the sand, awaiting sufficient soil moisture for germination.

A count of ten random strips of 10 x 50 cm in a good stand under or near shrubs and a similar count in between the shrubs, produced amounts of 1554 (±528) and 442 (±141) heads per square metre, respectively.

Seed dispersal over great distances may be zoochore. Dry *Rhanterium* heads stuck in the wool of grazing sheep and in the hair of camels' heads and legs were observed. Dickson (1955) makes a remark that the dry flowerheads, seeds, of the 'Arfaj'-bush are considered good grazing for sheep and in late summer are readily eaten by them (see Fig. VII–5). Other livestock such as goats and donkeys and wild mammals such as the common hare (*Lepus europaeus*) could be an instrument of dispersal. Agnew & Flux (1970) examined 369 hares in Kenya and found that 160 had a total of 810 disseminules of 17 plant species in their fur.

VI.F.2. *Germination*

VI.F.2.a. Position of the flowerheads

It was assumed that the mechanism of germination in *Rhanterium*, seedlings emerging from the upper side of the flowerhead and roots from the underside, would require a proper position of the heads on the ground to guarantee germination. Field observations proved this. In all examined seedlings where the flowerhead was still present, the head was 'upright', never upside down and only in a few

Table VI–7. Outcome of four counts on *Rhanterium epapposum* flowerheads with respect to their 'proper' germination position; three counts in the field, one laboratory test.

Description of the sample / experiment.	Heads in 'proper germination' position.	
	Number	%
Ten sets of ten heads at random sites, near Samah, 24 May 1972	71	71
All heads in ten strips of 10 x 50 cm, randomly placed, under or near plants, near Samah, 27 January 1973	556	72
All heads in ten strips of 10 x 50 cm, randomly placed, in between plants, near Samah, 27 January 1973	149	67
25 heads, each dropped 20 times from about 50 cm height on a smooth surface in a test experiment	330	66

cases slightly turned sideward, positions as shown in Fig. VI–17. For a high germination chance the flowerhead should, therefore, be in upright position on the ground. To investigate this aspect several counts were made, in the field as well as on heads dropped artificially in the laboratory (table VI–7). An overall proportion of two thirds or more were found in the proper germination pesition.

VI.F.2.b. Temperature and germination rate

Technical difficulties beyond control restricted the laboratory work on germination of *Rhanterium*. Results of a germination experiment with 100 heads each at three constant temperatures are shown in table VI–8. The figures indicate that *Rhanterium* germinates best at low temperatures, but that germination at higher temperatures is also possible. These results were obtained with seeds kept for over a year at room temperature. Mean monthly temperatures for November-March, when the species normally germinates, range for the *Rhanterium*-area in Iraq roughly from 7°C (Rutba 7.0°C, January) to 20°C (Basra 19.3°C, November).
At 30°C the first seedlings were visible three days after beginning the experiment; at 20°C this was five days, at 10°C it was 12 days. At 20°C and 30°C new seedlings emerged during a period of 14 days after the first were seen. At 10°C this period was twice as long.

The more than four times higher germination rate of seeds collected from the ground compared to those picked from the shrub in the same area, might be an

Table VI–8. Results of germination tests with *Rhanterium epapposum* heads at three different temperatures. Material collected January 1973, tested April 1974, after storage at room temperature.

Sample	Observation	Results of four replicates (a, b, c, d of 25 heads each) at three different temperatures (°C)														
		10°				total	20°				total	30°				total
		a	b	c	d	(%)	a	b	c	d	(%)	a	b	c	d	(%)
Heads collected from the plants	Heads germinated	2	2	0	1	5	1	0	1	1	3	0	0	0	0	0
	Total number of seedlings	2	3	0	1	–	1	0	1	1	–	0	0	0	0	–
Heads shed, collected from the ground	Heads germinated	6	6	3	8	23	2	2	4	5	13	2	1	0	3	6
	Total number of seedlings	10	9	3	10	–	2	2	4	5	–	2	1	0	3	–

indication that a certain ripening is required and maturity reached when the flowerheads are dropped.

VI.F.2.c. Longevity

Rainfall in the *Rhanterium* distribution area is scarce and unpredictable. A long survival of the seeds would be highly favourable for the species. Flowerheads collected near Samah from the ground at 26 January 1973 were stored dry at room temperature and 100 heads tested in June 1977 at room temperature. A germination of 60% was found, proving the exceptional longevity of the species unde;dry conditions.

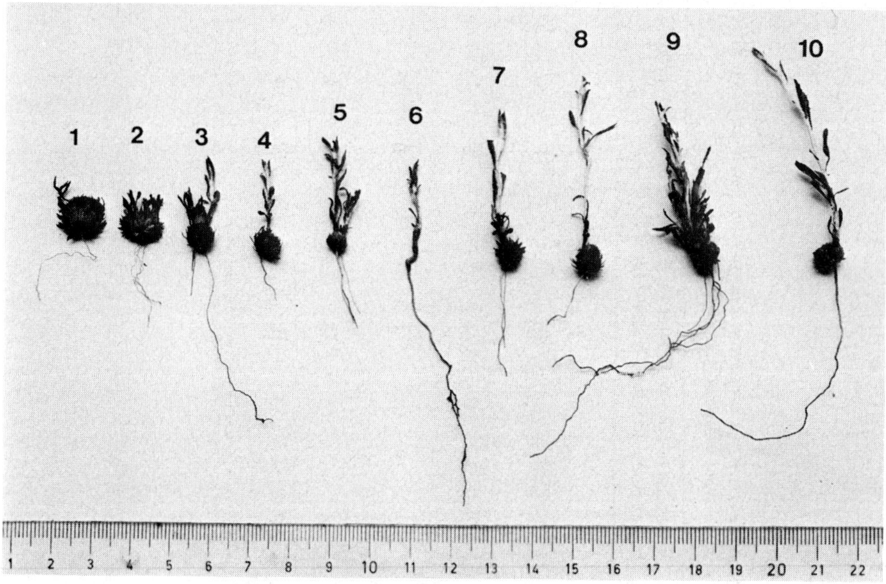

Figure VI–16. Excavated seedlings of *R. epapposum* in different stages of development (scale in cm and mm). 1 and 2 seedlings of a few days, 3 – 10 older seedlings of a few weeks to a few months. 3, 5, 7 and 9 more than one seedling developed per head, 4, 6, 8 and 10 with only one seedling (6 head removed).

VI.F.3. *Seedling development and establishment*

Insufficient data on *Rhanterium* grown under greenhouse conditions were available to establish distinct seedling growth stages. Field observations in stands of Rhanterium showed the following.

Figure VI–17. Positions of *Rhanterium epapposum* flowerheads on the ground. Left three heads germination impossible, right proper germination position.

On 19 December 1971, 20 heads with seedlings of *Rhanterium epapposum* were collected, including their fine roots as far as possible, about two kilometres north of Samah. The number of seedlings per head was 3.9 (±1.6) length of the longest shoot 11.9 (±1.3)cm and length of the longest root 30.4 (±8.0)cm. Number of seedlings per head, their (combined) diameters and height were determined for 50 *Rhanterium* seedlings collected in random transects across a stand near Joë Liban (20 km southeast of Busaiya) at 29 January 1973 for an idea about the average size of *Rhanterium* seedlings at the end of the first growing season. It was found that in exceptional cases as much as six or seven seedlings may develop from the same flowerhead. In the December 1971 sample this was also observed, but 92% of the January 1973 sample and 70% of the December 1971 sample had only 1–4 seedlings per head.
A height of 10–12 cm and a crown cover up to 100 cm^2 were found to be the most normal size for established seedlings after the first growing season in that area in that year.

In the same area an attempt was made to excavate the rootsystem of some of

Table VI–9. Results of attempts to excavate rootsystems of established *Rhanterium epapposum* seedlings.

Seedlings height (cm)	Length of longest excavated root (cm)	Remarks
11	91 (+)	Root broken
10	38 (+)	Root broken
12	54 (+)	Root broken
15	45 (+)	Root broken
16	104	Complete, only some more lateral spread; very hard soil layer at about 100 cm depth
15	78	Almost complete, branching just above a hard layer at about 55 cm depth

Figure VI–18. Four seedlings of *Rhanterium epapposum* equally well developing from the same head (scale in cm).

the seedlings. Excavation was hampered by the occurrence of very hard compact layers with gypsum at different depths. The tiny roots disappeared in small cracks, broke easily and could not be followed to any great depth (table VI–9). A rootlength of five to ten times the shoot at the end of one growth season may well be normal.

Few signs of competition between seedlings developing from the same head were observed. In most cases germination was followed by normal seedling development, whether one, two or more seedlings per head were found (Fig. VI–16 and Fig. VI–18). Detailed experimental work will be necessary for definite proof of this aspect of competition. That more than one seedling can finally develop to an adult plant was proved by excavating some older shrubs. An exceptional case is shown in Fig. VI–19. One bush could be separated into five parts, highly intertwined at the base, but nowhere connected. After separation the height and crown measures of the individual plants were (in cm): 69–83/54, 67–73/68, 70–52/48, 75–100/73 and 58–87/82.

Figure VI–19. Excavated *Rhanterium epapposum* shrub, consisting of five separate intertwined plants, most likely due to germination from the same flowerhead: upper left: in situ during excavation, upper right: partly separated, below: spread out after complete separation.

VI.G. Reproduction and development of 'Artemisia herba-alba'

VI.G.1. *Seed production and dispersal*

Fruiting of *A. herba-alba* takes place mainly in December and January. The fruit is small, dry and single seeded; an achene. It has an average length of 1.3 mm, a width of 0.6 mm and an obovate, slightly compressed shape. The pericarp serves as a seed coat, while the true seed coat is not well developed (Clor *et al.* 1974b). These achenes, below for practical reasons referred to as 'seeds', have a dark brown to black colour and about 6000 seeds weigh one gram (Kaul & Al-Mufti 1974). After shedding they are easily windblown over large distances. Animals may also play a role in distribution, but no data ara available. Large amounts of seed are produced and seeds can build up to a thick layer under the older plants.

VI.G.2. *Germination*

VI.G.2.a. The process

In contact with water, a transparent gelatinous envelope soon develops around the seed. When germination proceeds, the small, somewhat thick, pointed radicle visually first emerges from the narrow end of the achene. This radicle is clearly separated from the tiny cotyledons by a narrower part in the embryo axis. Details are shown in Clor *et al.* (1974b). These authors also investigated the role of the gelatinous substance and found that its main function may bc one of water retention, supplying the germinating seed and young seedling with moisture for several days without additional water being required.

VI.G.2.b. Afterripening

Increasing germinability with time during the first year in *A. herba-alba* has been observed several times (Sankary, referred to in Sankary & Barbour 1972a; Al-Mufti pers. comm.; Clor *et al.* 1974b). The last authors found a gradual germination increase from just over 20% in February to about 80% in December, followed by a decrease to 40% in July of the next year in seeds kept at room temperature. This afterripening was not affected when storage was at low temperature (6°C) in sealed bags. Germinability, however, was reduced to one and two per cent when seeds were kept at low temperature in unsealed bags.

VI.G.2.c. Temperature

Optimal temperatures for seeds at their maximum germinability are in the range of 10°–23°C. Less viable seeds showed best germination at 10°–15°C. No germination has been found with tests at 5°C and 6°C (Al-Ani et al. 1971c; Clor et al. 1974b).

VI.G.2.d. Hydration and dehydration

Results of experiments on (repeated) dehydration of hydrated seeds reported in Clor *et al.* (1974b) are summarized in table VI–10. It can be seen that seeds of *A. herba-alba* maintain full germinability after hydration for ½–24 hours followed by a week dehydration and also after two such cycles with hydration times of half an hour and one hour.

VI.G.3. *Seedling development and establishment*

In *Artemisia herba-alba* the hypocotyl and primary root can be clearly separately discerned in the embryo. When germinating, the radicle usually protrudes first, but in some cases the cotyledons may emerge earlier. The radicle of the emerged embryo soon develops into a rootlet with extensive root hairs and a total seedling length of about 1.5 cm can be reached in 48 hours under favourable conditions as was seen in laboratory experiments. As observed in *Haloxylon salicornicum*, the rapid extension is mainly due to the fast development of the rootlet. Plants grown in the greenhouse developed within three weeks small roots, sometimes already

Table VI–10. Effect of repeated hydration (for different periods) and dehydration of seeds of *Artemisia herba-alba* on germination (derived from Clor *et al.* 1974b).

Hydration first imbibition (hours)	Dehydration first drying (days)	Hydration second imbibition (hours)	Dehydration second drying (days)	Germination room temperature (%)
0	7			74
½	7			74
1	7			69
3	7			76
6	7			74
12	7			70
24	7			72
0	7	3	7	74
½	7	3	7	81
1	7	3	7	82

Figure VI–20. Seedlings of *Artemisia herba-alba* in different stages of development collected at the end of July 1972 near Rutba. Representatives of five distinguished classes are shown; details in table VI–11, scale is in cm and mm.

branched, with a length of over 10 cm while the height of the shoot was barely 1 cm. The development of *Artemisia herba-alba* seedlings is shown in Fig. VI–20 and table VI-11. Together with the extension of the rootlet some more leafpairs gradually develop and after about four weeks a shoot starts growing fast, sometimes together with two or three smaller shoots. By that time the rootdepth is at least half a metre. From the main shoot sidebranches soon develop. A new plant is then established, but only a few of the once germinated seeds reach this stage.

It has been observed that seeds sown at the soil surface had a higher germination than those sown at 1, 2 and 3 cm depth (Al-Jibouri, quoted in Kaul & Al-Mufti 1974). Seeds accumulate in small soil hollows and under the parent plants. After the rains such a hollow may also contain some freshly washed seeds in fine textured soil and accumulated moisture. Germination was especially seen in such favourable micro-habitats. For the rather fine textured soils in which *Artemisia herba-alba* grows (Figs. VI–8 and VI–9) surface sealing does occur that can be broken by the trampling of livestock. The hoofprints may create more favourable germination conditions. This is supported by the observation reported in Kaul & Al-Mufti (1974), who found a regeneration of 2.30 seedlings per square metre in a moderately grazed area against 0.56 in a comparable protected situation after counting 44 randomly laid out one square metre quadrats.

Many just germinated seedlings die after a prolonged drought period. Dead *Artemisia* seedlings with a few well developed leafpairs, completely dried out,

Table VI–11. Seedling stages of *Artemisia herba-alba* with their height and average root depth as determined under field conditions (mean value and standard deviation), with an indication of the number of days after germination required to reach the different stages, as worked out under greenhouse conditions. See Fig. VI–20 a-e; based on Kaul & Al-Mufti (1974) and Al-Jibouri (pers. comm.).

Seedling stage	Length of the longest shoot (cm)	Depth of the root (cm)	Age (days after germination)
1. One-pair-leaf stage (Fig. VI–20a)	0.9±1.0	4.8±0.8	10
2. Two-pair-leaf stage (Fig. VI-20b)	1.7±0.4	14.3±4.5	23
3. Six-pair-leaf stage without development of a main shoot (Fig. VI–20c)	2.4±0.5	45.6*	69
4. Nine-pair-leaf stage with distinct development of a main shoot (Fig. VI–20d)	5.0±1.1	54.5*	75
5. Main shoot well developed with side branches (Fig. VI–20e)	about 11	no data	about 120**

* Roots seen to continue deeper in the soil.
** Field estimate on the basis of rainfall and observation date.

were sometimes seen in great numbers during the summer months. Repeated counting of 23 quadrats of 50 x 50 cm laid out along a transect across an unprotected, moderately grazed wadi showed a decrease from an average 15.6 per quadrat in June, through 8.7 in October to 4.5 in December. Most of the seedlings occured in clusters. Still more would die, perhaps in competition for moisture with root-systems of already established plants or as a result of repeated grazing, because in the stands measured in the field, densities of full grown *Artemisia* plants were usually only 1–4 plants/m^2 (table VII–21).

Kaul & Al-Mufti (1974) found that only those seedlings which had nine pairs of

Figure VI–21. Old plant of *Artemisia herba-alba* with new leaves and sprouts developing at the base in early spring after a good rainfall period.

leaves with a distinct main shoot by the end of June could survive the rainless conditions of the following months until the next autumn rainfall. After these rains new leaves and sprouts will develop at the base and about one and a half year after germination a young plant as depicted in Fig. VI–5 may have developed.

Contrary to *Haloxylon salicornicum* and *Rhanterium epapposum*, where usually a few shoots only will finally develop as main branches, in *Artemisia herba-alba* older plants are primarely recognised by a larger number of shoots of about equal length developed from a common base close to the ground (Fig. VI–21). Thick woody parts, about two centimetres, can only be found at this base and the species has, therefore, contrary to *Haloxylon salicornicum* and *Rhanterium epapposum* a very low fuel value.

VI.H. Adaptive characteristics, comparison and discussion

The three species under discussion are probably the most common shrub species of the desert shrub rangelands of Iraq. What enables them to survive under the harsh environmental conditions? Are there any marked differences between them that could be related to the differences in habitat and distribution as pointed out under VI.C.? Answers to these questions may be found when considering the following three aspects, crucial for the survival of a plant species in an area:
(i) the reproduction mechanism, in the present case the production, distribution and germination of seed;
(ii) establishment, the development of the seedling to an adult plant;
(iii) maintenance, the adaptations of the adult plants enabling survival.

(i) Reproduction. It was calculated that an average stand of *Haloxylon salicornicum* would produce some 40 million seeds per hectare annually (VI.E.1.) and a stand of *Rhanterium epapposum* 2.5–3 million flowerheads with a number of seeds each (VI.F.1.) while for *Artemisia herba-alba* this figure is even much higher. The number of plants per hectare is never more than 40 000 for *A. herba-alba* and much less for the other two species. Assuming a lifetime of ten years for the shrubs (a very low estimate; Kernick (1966) found 15–20 year old plants of *H. salicornicum*, and *R. epapposum* in Kuwait) means that every year only a few thousand, perhaps only a few hundred shrubs per hectare need replacement. Several million seeds are being produced for this purpose. In other words only one seed in 10 000 to 100 000 needs to develop into an adult plant to ensure the species' continued occurrence in the same density.

For germination soil moisture in the topsoil is required and the best chances for rain are between November and April. *H. salicornicum* and *A. herba-alba* both shed all their seed in the middle of this period, December to February. In case of *H. salicornicum* germination must take place within a few weeks. Seed longevity is less than one year (VI.E.2.b.) and there is a high seed predation. The mechanism should, therefore, be such that seeds have a good chance to germinate as soon as they have fallen. Germination of almost 100% was found for the range of temperatures which is normal in the November–April period. The optimum temperature of 15–20°C (VI.E.2.d.) concides with the mean air temperature (Fig. IV–1) and topsoil temperatures (table IV–2) in the area in February-April, the time for germination. Moreover, the observation that 'early' seeds germinate better at relatively low temperatures and 'late' seeds at higher temperatures indicates an additional adaptation of the species. It has been shown that seed of *H. salicornnicum* germinated under light as well as dark conditions (VI.E.2.e.) perhaps with a somewhat better germination in the dark. The species can, thereby, start germination after a night shower, while the light condition for buried seeds will not prevent them from germinating. The hydration – dehydration experiments (VI.E.2.g., table VI–3) provide evidence that the species can very well survive one or more light showers and continue germination and establishment at heavier rainfall not too long afterwards. A final aspect of the germination of *H. salicornicum* germinated under light as well as dark conditions (VI.E.2.e.) higher temperature range. Within an hour after the rainfall the first seedlings emerge from the seed and over 90% germination was seen within a day at temperatures ranging from 15 to 35°C (table VI-2).

The seed (or flowerhead) distribution of *Rhanterium epapposum*, being spread over a long period, nearly the whole year (Fig. VI–10, VI.F.1.), may at first glance seem a waste of seed, while relatively few will meet moisture conditions conducive for germination. However, of the three species *R. epapposum* is the most confined to arid conditions and even occurs under extremely arid conditions. Rainfall here is not only low, but also highly unpredictable. Its seed distribution being spread in time, *R. epapposum* ensures that at any moment of rainfall at least some newly dropped seed is on the ground. The seeds in the flowerhead have an extreme longevity under warm, dry conditions (VI.F.2.c.), a factor which can also contribute to germination under unpredictable rainfall conditions. The distribution and germination mechanism of the seed remaining in the flowerhead seems most appropriate for an arid, sandy environment. The sandy soil allows the heads to come in proper germination position, often not too far away from the parent plant because of the roughness of the surface, and bc (partly) buried and fixed in this position. On soils with a higher silt content often surface crusts develop on which the flowerheads are blown away over large distances and not easily buried. On a gravelly or stony topsoil on the other hand seeds are trapped in great quantities in between the stones and the proportion found in proper germination position on a soil that would allow root penetration is very limited compared to sandy soils. The hydration-dehydration effect on germination may well be solved in *R. epapposum* by the mechanism that at the first rain only one or a few seeds germinate and produce seedlings which die when a dry period follows. During continual showers new seeds may germinate, allowing at least some to achieve successful establishment. Vesey Fitzgerald (1957b) has described a similar mechanism for *Neurada procumbens*. The observation that at prolonged moisture supply to the seed of *R. epapposum* new seedlings still emerge after some days, points to this direction. In one experiment with 75 flowerheads, after three days six heads had together eight seedlings. Three days later 25 heads had 32 seedlings, about the same ratio (1.3), another four days later 27 heads 43 seedlings, a ratio of 1.6. Seedlings of different size in the same head were common (Fig. VI–16).

Of the three species *Artemisia herba-alba* produces the largest amounts of seeds, but is also the one with an afterripening (VI.G.2.b.), together perhaps a mechanism of spreading the possibilities for germination over more than one rainfall season. Most interesting is the development of a transparent gelatinous envelop around the achene within several minutes of contact with water. This envelop can help the germinating seed to adhere to the substratum. Its role in a continued moisture supply to the developing seedling during the first few days after germination was shown by Clor *et al.* (1974b). The ability of the seedlings to withstand periods of dehydration was also shown (table VI-10).

The above data show that all three species have their specific propagation mechanism, well adapted to the prevailing conditions. In *H. salicornicum* and *A. herba-alba* a certain soil moisture, concentrated in the December – March period is a necessity, as otherwise an entire annual seed production will be lost. *R. epapposum* can produce some seedlings after any moderate shower and hence seems best equipped for extremely arid conditions.

(ii) Establishment. For survival and growth of the seedling continued access to moisture is crucial, although for *H. salicornicum* and *A. herba-alba* it was shown that temporary desiccation is not necessarily fatal if it does not last too long. In the *H. salicornicum* habitat of sandy or gravelly plains the infiltration rate can be very high and a few hours after rainfall the upper soil layers may be already relatively dry. The fast germination of *H. salicornicum* (table VI–2) and the rapid growth of the rootlet (table VI–4, table VI–6, Fig. VI–12) are, therefore, important properties for seedling survival. Branching of the rootlet in the upper soil layers was not found and shoot development in the first weeks is very slow. All efforts seem to be concentrated on developing soonest a deep (tap)root. In *R. epapposum* a somewhat similar development was found but less extreme than in *H. salicornicum*. Shoot development is faster and the rootlet develops laterals early, which may be an advantage in a sandy soil. In one germination experiment in a sandy soil at room temperature a week after seedling emergence from the head, the average shoot length was 1.1 cm (range 0.6–1.7 cm) and the average root lenght 7.7 cm, (range 5.5–10.5 cm). Tha data on *A. herba-alba* indicate massive germination in the suitable microhabitats, followed by the death of a large number of seedlings, possibly under competition for soil moisture. However, in this species too a very fast root development has been observed (table VI–11).

In spite of the large amounts of seed produced, the well adapted germination mechanisms and the ability of rapid root development after germination, relatively few seeds succeed in developing to an adult plant. Quantitative data are lacking but many seedlings seem to perish during the dry hot summer months following the rains that cause the germination. Extensive areas with dead, dry seedlings of *H. salicornicum* were seen and the dying of *A. herba-alba* seedlings in great numbers has already been mentioned. Dry, dead seedlings of *R. epapposum* were commonly found between the parent plants but never in large numbers together, which may be due to the adaptation of spreading the risk of failure. Haines (1951) observed in Egypt for *H. schweinfurthii (=H. salicornicum)* that in most years no seed germinated, but in an exceptionally wet season large numbers of seedlings appeared, some near the parent plants and some far away, and writes:

> 'When, as in 1950, a wet winter is followed by a hot summer, there is a heavy mortality amongst these seedlings, evidenced by the great number of dead plants, still retaining the remains of their leaves, though survivors of the same generation have shed them long ago.'

For rangeland reseeding programmes knowledge on this aspect is needed. Only fieldwork over a number of years with different amounts of rainfall can provide the necessary information.

(iii) Maintenance. Shrubs in arid lands have developed a wide range of physiological and structural adaptations in response to water scarcity and annual and seasonal variation in water supply. Of the three species under discussion only some physiological aspects of *Artemisia herba-alba* have been investigated. Orshan (1972) reported a water output in late summer (calculated as per cent of the maximum spring value) of 2.4%, a low value compared to Mediterranean plants with values of 11–19% and the desert shrubs *Anabasis articulata* (7.4%) and *Haloxylon articulatum* (9.5%). During the dry season Kappen *et al.* (1972) measured for the species a negative hydrostatic pressure in the xylem down to – 163 bars and an extreme of osmotic potential in the leaves of – 92 bars. All three species under discussion reduce their transpiring body. *A. herba-alba* gradually sheds the larger winter and spring leaves during summer and replaces them by smaller summer leaves. *Haloxylon salicornicum* sheds the transpiring cortex with part of the previous year's growth (see Orshan & Zand 1962; for a review Orshan 1972). *Rhanterium epapposum* is most extreme in its adaptation to the dry season. All leaves are being shed and the plant can enter a sort of dormant stage, referred to before in VI.D., in which it has a completely dry appearance. Orshan & Zand (1962) suggested the term 'partial drought evasion' for this process whereby an increasing part of the body of the transpiring plant is shed with the increase of water shortage during the long and dry summer, while the plant as a whole retains its activity. This term is certainly appropriate in the case of *R. epapposum*.

Another, structural, adaptation of the species are the rootsystems. Occasionally, attempts were made to investigate the underground parts, be it with primitive techniques and water, for a careful removal of soil particles, not in sufficient amounts available. The rootsysteem found for *H. salicornicum* in the Khedr-el-Mai exclosure (see X.B.) was most difficult to dig out. It answers the description given by Noy-Meir (1973) for a certain category of desert plants:

> 'Another specialization is induced by soil heterogeneity, in particular stoniness. In a rocky or stony soil, moisture is very unevenly distributed in the profile, being concentrated in soil pockets and fissures, at soil-rock interfaces, and under stones. This favors plants with flexible "exploring" roots, capable of penetrating cracks, following tortuous paths, and expanding whenever a favourable pocket is found.'

The main root penetrates very deep. In one case a diameter of over two centimetres was still found at a depth of about one and a half metres. Extensive development of laterals in the upper soil layers was not observed in this species. Excavation of three large plants of *A. herba-alba* in the Rutba exclosure (see X.C.) showed a system of one or two thin, but deep taproots, going deeper than one metre and three types of laterals (see Kaul & Al-Mufti 1974), *viz.* (i) numerous thin horizontal laterals of a few decimetres in the upper 10–15 cm (these roots may develop after the rains and die again during early summer), (ii) a few thick perennial laterals at 25–50 cm depth extending horizontally but also turning downward, and (iii) a few deeper laterals, also turning almost vertically downward at some distance from the taproot. Excavation of *R. epapposum* showed a branching related to certain soil layers. In the upper 10 cm of sand and sandy gravel there were only a few fine horizontal laterals, but in a more compact sandy soil, perhaps with better moisture holding capacity at 10–30 cm, there was an extensive network of laterals which was traced to some 50 cm from the main root but which still sprawling horizontally. One main root growing vertically had become twisted in the more stony subsoil and had grown deeper out of reach, more or less similar to the taproots seen of *H. salicornicum*.

In conclusion, the above data although in many ways fragmentary, clearly show that *H. salicornicum*, *R. epapposum* and *A. herba-alba* have a number of adaptive characteristics that enable them to make use of the limited amout of water becoming available at irregular intervals. Certain adaptations are common to all three, e.g. an extensive rootsystem with taproots, high germination rate and fast development of the rootlet after germination, seasonal body reduction to reduce transpiration losses, etc. Other seem specific adaptations related to the climatic and edaphic conditions prevailing in the areas where the species are optimally developed, e.g. the lateral roots in the topsoil of *A. herba-alba*, especially useful in the finer textured soils where desert rains do not always penetrate deep; the almost dormant stage of *R. epapposum* enabling the species to withstand extreme, prolonged drought and the seed distribution of this species, well adapted to sandy soils, equipping *R. epapposum* best of all three to sandy, extremely arid and arid conditions.

Species survival in the desert shrub rangelands of Iraq is not only a matter of being adapted to the limitations set by the abiotic environment. Biotic factors, particularly man and his livestock, have an increasing impact, as will be shown in the following chapters.

PART THREE

Utilization

CHAPTER VII

GRAZING

VII. GRAZING

VII.A. Introduction

Utilization of the desert shrub rangelands has been going on for millennia, but has perhaps never been so intensive and destructive as in the present century, especially during the last few decades. In IV. E., IV. G., and IV. H. the opening-up of the desert areas was discussed and in IV. F.4. the increase in livestock population between 1938 and 1965 was shown. Harvesting of the relatively low amount of plant material produced, with the different types of livestock, has been man's main use of the desert shrub rangelands from olden times. The present chapter deals with this, from an economical point of view most important (see chapter XII), utilization type. First some aspects related to forage supply to the grazing* animal are discussed, including selectivity (VII.B.), chemical composition (VII.C.), moisture content ot the vegetation (VII.D.) and forage production (VII.E.), while in the last section (VII.F.) demand and supply of forage are compared when working out the grazing pressure and the grazing capacity of the area as a whole.

VII.B. Selectivity

All animals select their food. In the grazing process the animal makes a choice for plant species, individual plants and parts of individual plants. Selection varies for the different animal species and may even differ for individual animals. It changes in time with changing environmental conditions. In considering the utilization of a vegetation cover for forage and the impact of feeding on the vegetation, this process of selection plays the key role. *Preference* is the term used in relation to the animal reactions, *palatability* refers to the plant attributes which determine their acceptability by grazing animals. In selectivity, the two concepts are mixed and inseparable. Heady (1975) points out that 'forage selectivity results from a highly complex interaction among three sets of variables operating over time: the plants being eaten, the animals doing the grazing, and the environment of both'. The complex nature of the set of interacting variables is probably the main reason that research on selectivity has started rather late. Comparison between grazed and ungrazed plots, direct observations on grazing animals, ranking of species from observations on experimental choise situations and retrieval of eaten material from esophageal fistulated animals, provide useful data nowadays. These data however, are only of value for the particular situation investigated. Extrapolation of the results to other areas with a different set of environmental conditions,

* In the desert shrub rangelands of Iraq where most species do not attain a height over 50 cm and in many species no distinct leaves develop (Chenopodiaceae), the terms 'grazing' and 'browsing' can be equally well applied. For convenience in the present account the first term is generally used.

Figure VII–1. Camel feeding on the spiny *Astragalus spinosus*.

different plant species combinations and different animals, is virtually impossible. No work on food selection in the natural rangelands and on plant attributes affecting palatability had been carried out in Iraq prior to the late sixties. An important point in the present study was to establish a certain ranking of species or species groups with respect to preference shown by grazing animals with sheep and camels as the important animal species. This ranking could provide clues to explain the relative increase or decrease of certain plant species and thereby processes in the observed vegetation cover of the desert areas. No experiments with animals could be carried out and the following lines of data collection on preference were, therefore, adopted: (i) observations on feeding animals and on grazed plants in the field, (ii) questioning graziers and Bedouins herding flocks, and (iii) literature sources.

Direct observations in the field showed that in the desert rangelands of Iraq virtually all plant species are being grazed at one (seasonal) stage of development or another. Species ignored by all livestock types over all seasons were hardly found. Even such spiny species as *Astragalus spinosus, Zilla spinosa* and *Zizyphus nummularia* were seen or found to be eaten by camels and goats (Fig. VII–1). Bitter tasting species, that are considered poisonous *e.g. Peganum harmala* and *Citrullus colocynthis*, were seen to be eaten at a certain stage by sheep. Of the first species the dried leaves and stems, of the latter the fruits only. When ephemeral species were available, alle livestock preferred these, sheep and goats

Figure VII–2. Flock of sheep grazing in a vast stand of *Haloxylon salicornicum* with some plants of *Ephedra alata*. Virtually all animals eat from the *Ephedra*, neglecting for the time being the *Haloxylon* (foreground and background). Animal at the right foreground eats the rosette leaves of *Artemisia scoparia* (Faidhat-al-Jeez, February 1972).

Table VII–1. Preference of sheep for *Artemisia herba-alba*, grasses and forbs (a.o. *Poa bulbosa, Carex stenophylla, Malva* spp., *Plantago* spp.), recorded for four ewes individually on 12 March 1969 in the Rutba fenced area. Thanks are due to the late S.Y. Serkahia who organised the observations, and kindly allowed publication (pers. comm. Serkahia 1970).

	Morning grazing period		Afternoon grazing period		Total one day grazing	
	A. herba-alba	grasses+forbs	*A. herba-alba*	grasses+forbs	*A. herba-alba*	grasses+forbs
total number of bites for four sheep (thousands)	1.7	18.8	3.0	20.5	4.7	39.3
(%)	8	92	13	87	11	89

Table VII–2. Relative preference of sheep and camels for the most common perennial woody species in the Nukhaib-area (information from Bedouins).

Preference (decreasing)	Sheep		Camels	
	Species	Local name	Species	Local name
1	*Artemisia herba-alba*	SHEEH	*Salsola rigida**	RUTHA
2	*Achillea fragrantissima*	GAISUM	*Anabasis setifera***	SHAHRAN
3	*Haloxylon salicornicum*	RIMITH	*Haloxylon salicornicum*	RIMITH
4	*Anabasis setifera***	SHAHRAN	*Artemisia herba-alba*	SHEEH
5			*Achillea fragrantissima*	GAISUM

* Mentioned with the remark that the species had virtually disappeared.
** Only locally in reasonable supply

highly selectively taking leaves or the top parts of certain species, camels pulling out complete plants often including the shallow roots. However, it was noticed in various situations that shrub species were occasionally selected as well. This is clearly shown in some data collected by Serkahia and co-workers. Four observers each followed a sheep over a full day's grazing period in the spring of 1969 in an area where at that time plenty of grasses and forbs were found between the *Artemisia herba-alba* plants. On an average 11 bites of the shrub were taken against 89 of grasses and forbs combined (table VII–1). A number of cases were examined and it was recorded where livestock had the choice between two or more shrub species. They would usually eat from all species, sheep and goats being selective in the parts of each plant taken (Fig. VII–5) and all livestock, including camels and donkeys, having a preference for a particular species. A few examples may be given. In February, sheep were seen to eat all green parts of the few *Ephedra alata* plants, growing in a vast stand of *Haloxylon salicornicum*, just putting on its first new growth. There was a clear preference for *Ephedra* and whenever *Haloxylon* was taken (Fig. VII–2), this was confined to the outer parts of the top branches. Goats in a similar stand also showed a preference for *Ephedra* but less pronounced and taking whole branches of *Haloxylon salicornicum*. Camels roaming the area were rather indifferently taking complete branches of both, *Haloxylon* and *Ephedra*. They were, in addition, eating from the freshly sprouting but very spiny *Astragalus spinosus* plants, chewing away complete branches, including the spines (Fig. VII–1). Goats would carefully select the fleshy outer part of the branches and nibble the leaves from between the spines.
The combination of such data gave an overall impression of the preferences of sheep and camels especially with respect to the commonest woody perennials. Discussions with Bedouins in some cases confirmed these data. An example is given in table VII–2. In such discussions, in establishing an order of preference for a particular area for a particular season, hardly any difficulties were faced. However, when taking into account other seasons and species from other areas the issue often became too confusing for a reliable preference ranking. Moreover, Bedouins were apt to forget species which had already largely disappeared from an area as a result of overgrazing.

Salsola rigida (local name RUTHA) is such an example. When brought to the Bedouins attention, this species was qualified by them as the best grazing for camels. They remarked that the species had disappeared and was, therefore, not useful any more. Evidence of this process was found and at one location a stand was quantified (Fig. VII–3 and Fig. VII–4). It was said that the area had been covered by a dense RUTHA-vegetation until the mid-fifties. In 1973 only a few living *Salsola* plants were found and no specimen of these higher than 10 cm. A succession to a mixed *Haloxylon salicornicum – Artemisia herba-alba* community with the first species as a dominant was seen and can be read from Fig. VII–3.

Literature data ons preference and palability were very scarce. No systematic coverage of the preference of livestock for the (especially woody perennial)

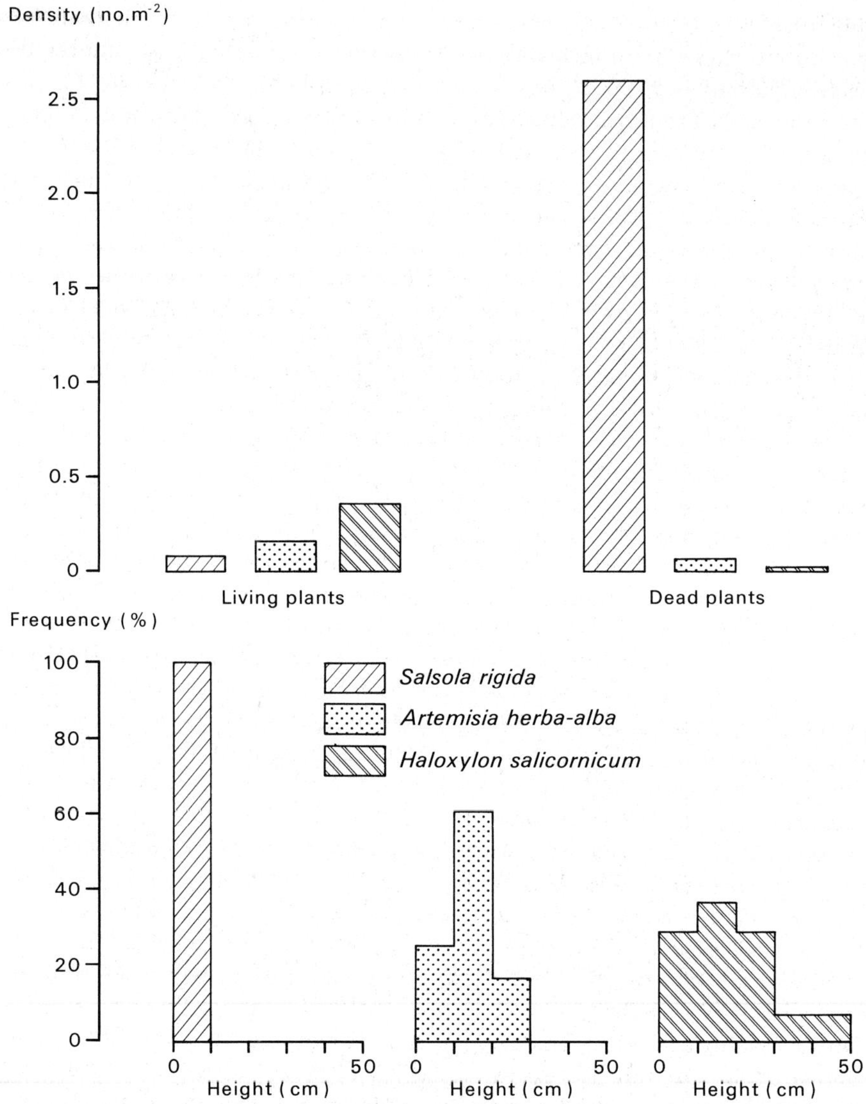

Figure VII–3. Density and frequency distribution of height for the three shrub species encountered in a degenerating 'RUTHA' (=*Salsola rigida*) community, about 13 km from Nukhaib, east of Wadi Tibel. (Based on data of 50 quadrats of 4 m^2)

species has been published for Iraq and basic information is still lacking. Some data were found in (mostly unpublished) reports. For easy reference the information is compiled in table VII–3.

Figure VII–4. Above: General view of a degenerated stand of *Salsola rigida* with young plants of *Haloxylon salicornicum* and *Artemisia herba-alba*.
Below: Close-up of maltreated (cut and subsequently grazed) specimen of *Salsola rigida* (right) and well developing *Haloxylon salicornicum* plant (left).

Based on the above information some general conclusions on selectivity can be given:
Grasses and forbs are generally preferred as bulk forage over the woody shrubs by all types of livestock in the Iraqi desert. However, even during the (short) period that these species groups are plentiful, all livestock, but especially camels and goats do eat green material of shrub species too.

Table VII–3. Indication of forage value given by selected authors for the most common woody perennials and some often mentioned perennial grasses.

Shrub species		Guest (1933)	Gillett (1948c)	Serkahia (1956)	Kernick (1966)
Name and family	Local name				
Achillea fragrantissima Compositae	GAISUM	*Achillea* spp. Some spp. eaten by camels and other animals and provide useful grazing in the desert.	*Achillea* spp. Listed amongst chief shrublets of grazing value in the north and mentioned as 'to have been named as most valuable'.	Poor quality grazing shrub, but useful in summer.	–
Artemisia herba-alba Compositae	SHEEH	Grazed by camels.	Listed amongst chief shrublets of grazing value in the north. More useful than SALMAS.	Winter and spring grazing plant. Grazing value: poor.	–
Artemisia scoparia Compositae	SALMAS	Said to be grazed by sheep and camels.	Listed amongst chief shrublets of grazing value in the north. (*cf. A. herba-alba*).	Eaten when other shrubs are over-grazed or absent	–
Rhanterium epapposum Compositae	ARFAJ	An important grazing plant for camels and also, when young, for sheep.	Listed as chief shrublet (of the Compositae) of grazing value in the	Used as fodder and a very valuable grazing shrub.	Camels, goats and sheep.
Halogeton alopecuroïdes Chenopodiaceae		–	–	Listed as grazing plant.	Camels; especially eaten when dry and yellow in winter.
Anabasis setifera Chenopodiaceae	SHAHRAN	–	Mentioned as useful Chenopodiaceae species.	Listed as grazing plant.	Camels, good salt grazing.
Haloxylon ammodendron Chenopodiaceae	GHADA	–	Mentioned under Chenopodiaceae without further [illegible]	Useful shrub for camels.	–

Haloxylon articulatum Chenopodiaceae	NETUL	–	Listed amongst chief Chenopodiaceae in the northern deserts without further comment.	Good summer grazing shrub.	–
Haloxylon salicornicum Chenopodiaceae	RIMITH	A valuable grazing plant for camels since there is often little else for them to eat.	Listed amongst chief Chenopodiaceae in the south and quoted as disappeared from Jezira according to Bedouins.	Good grazing plant for camels.	Camels; valuable salt grazing. Sheep and goats graze when dry and yellow in winter.
Noea mucronata Chenopodiaceae	SIRR	Grazed by camels.	–	Listed as grazing plant.	–
Salsola rigida Chenopodiaceae	RUTHA	Grazed in summer and autumn by camels.	Mentioned as 'to have been named as most valuable' and quoted as disappeared from Jezira according to Bedouins.	A first class grazing plant. Over-grazing is so close and heavy that no seeding branches are left in late autumn. A spring and summer grazing plant.	–
Salsola vermiculata Chenopodiaceae	(RUTHA?)	–	–	Grazing value good to fair.	
Astragalus spinosus Leguminosae	CHEDAD	–	–	Because of its thorns it is eaten only lightly by camels.	Camels eat it after the thorns are burnt off.
Calligonum comosum Polygonaceae	ARTA	–	–	Listed as grazing plant.	Camels.
Ephedra alata Gnetaceae	ALENDA	Valuable grazing plant for camels.	Valuable for camels only.	Listed as grazing plant.	–
Teucrium polium Labiatae	JA'ADA	A useful grazing plant.	–	A useful grazing plant.	Camel, sheep and goats.

Table VII–3. (Continued)

Perennial grasses		Guest (1933) and Guest in Bor (1968)	Gillett (1948c)	Serkahia (1948)	Kernick (1966)
Name	Local name				
Stipagrostis plumosa	NISSI	Grazed by camels, sheep and other livestock and is considered by the nomed herds and shepherds to be an excellent spring forage grass . . . a most important pasture plant in the desert.	Provides a certain amount of grazing in the southern deserts shortly after the first rains.	A useful grazing plant in the early growth.	Camels, sheep, goats and horses.
Aeluropus lagopoides	ICHRISH	Grazed by stock.	–	–	Sheep and goats.
Lasiurus hirsutus	(SKHABAR, THAMMAM)	Excellent desert fodder grass, grazed by sheep, camels and other animals.	–	Listed as grazing plant.	Sheep and goats.
Panicum turgidum	THAMMAM	Much prized fodder for camels, as well as horses, donkeys and other animals.	–	Listed as grazing plant.	Horses and sheep.
Poa bulbosa (and *P. sinaica*)	QUBA	Invaluable, if rather poor and scanty, pasture for sheep. Provides spring grazing for livestock.	Give welcome pastu-rage for sheep at difficult time of the year (directly after the first rains).	Poor feeding plant but valuable owing to its abundance.	–

There is a definite preference of the different kinds of livestock for particular woody species. This changes with the season. All observations indicate that, of the most common species in the Iraqi deserts, *Artemisia herba-alba* is generally preferred by sheep and *Haloxylon salicornicum* by camels. Some, once perhaps widespread and common, species, earlier preferred to these, may have disappeared already due to selective overgrazing.

Species which are considered unpalatable in other areas (and even considered as such by some graziers in Iraq) because of their morphology or taste, have been observed to be eaten in Iraq, apparently due to the lack of alternatives. *Astragalus spinosus* and *Zilla spinosa* (spines) and *Achillea* spp., fruits of *Citrullus colocynthis* and dried plants of *Peganum harmala* (bitter taste) are examples.

An order of overall relative preference of sheep and camels for the most common woody perennials and other species groups in the desert rangelands of Iraq is presented in table VII–4. This table can only show a general trend, derived from the various data sources. It should be mentioned (again, but now based on Iraqi data) that an animal does not normally limit itself to one species, but consumes a mixture and has in addition an individual food selection behaviour, which changes over the **season.** depending on the variation of food available. This makes generalization difficult and unreliable.

Table VII–4. Relative preference of sheep and camels for the most common shrub species and other species groups in the desert areas of Iraq. Compiled from information given by different groups of Bedouins, contacted during fieldtrips over the years 1970–1973, personal observations in the field over the same period and from selected references.

Qualitative preference class	Sheep	Camels
Always preferred but only occasionally encountered	Grasses* (with few exceptions) and (mostly annual) forbs*	(Mostly annual) forbs* and certain grasses*; *Salsola rigida*
Highly preferred	*Artemisia herba-alba* (both dry and green)	*Haloxylon salicornicum* (both dry and green) *Ephedra alata* (both dry and green)
Moderately preferred	*Artemisia scoparia* (green) *Ephedra alata* (green)	*Artemisia herba-alba* (dry and green) *Astragalus spinosus* (green)
Usually not preferred, but nevertheless eaten	*Artemisia scoparia* (dry) *Haloxylon salicornicum* (dry) *Astragalus spinosus* (green)	*Artemisia scoparia* (green)
Never preferred, only eaten as last choice	*Achillea fragrantissima* ** *Haloxylon articulatum*	*Achillea fragrantissima* *Haloxylon articulatum* (dry)

* A listing of species would be of little value. Virtually all species are eaten when accessible to the animal. Important genera include: *Trigonella, Medicago, Trifolium* (NAFAL, QURT), *Malva* (KHUBBAZ), *Plantago* (ZIBAD, RIBLAH), *Matricaria* (HUNKRAIS, BAIBUN), *Stipa* (SAMA'AH), *Helianthemum* (JARAID), (cf. Gillett 1948c; Kernick 1966; Bor 1968).

** Only the young green leaves and shoots.

VII.C. Chemical composition of the forage

VII.C.1. *Introduction*

Before 1970, little attention had been given in Iraq to the investigation of the nutritive value of the various kinds of animal feeds through chemical analysis. In Labib *et al.* (1968) some data are presented on all sorts of feedstuff, but no native range species are included. In a later publication of the same research division a considerably extended amount of data was given including more data on cultivated forage crops and also the first data on native range species (Khawaja *et al.* undated). Chemical analysis as a basis for nutritive value evaluation of native range species was initiated at the Institute for Applied Research on Natural Resources late 1969 and a large number of samples was analysed in the next three years. Some results have been published in technical reports of the Institute (Al-Ani *et al* 1971a, b; Al-Ani & Jawad 1973, 1974). A summarized account of this information, extended with new data is given below. The proximate-analysis was used for the chemical analysis of the plant material.

Although subject to criticism, proximate-analysis is a widely used technique, which groups the chemically similar components of foods into six categories, indicated between parentheses in the following outline of the chemical organization of foods (after Moen 1973):

I. Water
II. Dry matter
- A. Organic substances
  - 1. Nitrogenous compounds (Crude Protein)
    - a. True protein
    - b. Nonprotein nitrogenous materials
  - 2. Non-nitrogenous substances
    - a. Carbohydrates
      - (1) Soluble carbohydrates (Nitrogen-Free Extract)
      - (2) Insoluble carbohydrates (Crude Fibre)
    - b. Fats (Ether Extract)
- B. Inorganic substances (Ash)
  - 1. Salts
  - 2. Mineral matter

Water as a component of the forage is separately discussed for the Iraq data in VII.D. All results of the chemical analysis are presented as percentages of dry matter.

Crude Protein is calculated by multiplying the total nitrogen by the factor 6.25, a factor which may in certain cases be disputed (Crampton & Harris 1969), but which is generally used as a valid approximation. Protein content in feeds is an extremely important component although its value has been overemphasized. Forages high in protein are usually low in Crude Fibre and highly digestible. The digested and absorbed protein is used for energy and required for repair and new growth of animal tissue, including such animal products as milk, wool and hair. The protein requirement in feed of ruminants is still a matter of discussion. As a rough indication the following figures of Preston (1972) may be quoted: 6–8% for maintaining ruminants, 10–13% for growing ruminants and 12–16% for lactating ruminants, all in Crude Protein percentages. Figures given elsewhere were found to be comparable or lower (Blair Rains 1970; Cook 1972; Stoddart *et al.* 1975). Actual utilization is usually first assessed by determining the protein digestibility, but no such data are available for Iraq. Preston (1972) has shown the similarity between five different sources giving the direct relation between Crude Protein and Digestible Crude Protein. Keeping this direct relation in mind, the earlier quoted figures of Preston may well be used for a first rough assessment, using Crude Protein figures.

Crude Fibre is the residue, insoluble after boiling with dilute alkali and dilute acid. Although resistant to this treatment, this component does not represent an indigestible portion of the feed, as originally supposed. Its main component is cellulose, which is broken down by the rumen microorganisms. In ruminants as much as 50% to 90% of the Crude Fibre in plants can be digested (Crampton & Harris 1969), but the digestibility of Crude Fibre in dormant or dead plant material may be considerably less than 50% (Moen 1973). As a forage component it needs to be considered in calculating the energy supply while its physical role in giving the forage its effective bulk is important too.

Ether Extract, sometimes referred to as Crude Fat, consists of a number of different components and its food value is, depending on their proportions, highly variable. Crampton & Harris (1969) show coefficients of digestibility of this component for various feedstuffs, ranging from 36–100% in cattle. Main value of this component is its energy supply, often (oversimplified) taken as 9 kcal of metabolizable energy per gram Ether Extract, about twice the amount of energy derived from carbohydrates and protein.

Ash is the inorganic residue after heating a sample, usually to about 600°C. Its component parts in plants can be many and as such it is a poor index of inorganic nutrients. In the data for Iraq some minerals (Calcium, Phosphorus, Potassium, Magnesium and Sodium) were, therefore, separately determined. Ash content can be high as a result of adhering soil particles. Herbage generally containes sufficient amounts of minerals to meet the animal's requirements, except Phosphorus. Stoddart *et al.* (1975), quoting National Research Council figures, recommend for Calcium 0.18–0.28% and 0.22–0.30% for beef cattle and sheep, respectively, depending on body weight and lactation. For Phosphorus these figures are 0.16–0.22 and 0.17–0.22%. Blair Rains (1970) gives the following tentative levels, expressed in percentage dry matter:

	Deficient	Uncertain	Theoretically adequate for production
Phosphorus	<0.10	0.10–0.20	0.14
Calcium	<0.15	0.15–0.30	0.20
Potassium	–	–	0.34
Magnesium	–	–	0.10
Sodium	<0.01	–	0.02

Nitrogen-Free Extract is not separately analysed in the proximate-analysis but determined by subtraction of the other components from the total sample and therefore subject to relatively large errors. It is made up of starches and sugars, some hemicelluloses and lignin. This part of the feeds represents a non-specific source of energy. Its components differ again in digestibility and in the yield of net energy. Crampton & Harris (1969) give an average of 64% for forages or roughages and of 90% for grains. Stoddart *et al.* (1975) give a range of 53–78% for eight browse species in the USA for mule deer.

The proximate analysis gives, as the name indicates, an approximation of the chemical composition of plants and 'It is an index of nutritive value only because the fractions that it isolates are correlated with some of the properties of feeds that have nutritional significance' (Crampton & Harris 1969). A chemical analysis, without an assessment of how the animal digests the feed, is not necessarily a correct index for nutritive value. As Stoddart *et al.* (1975) put it: 'A mixture of feathers, sawdust, and charcoal can be formed that, on the basis of chemical analyses, would appear to be a satisfactory animal food'. New techniques have been developed the past ten years to evaluate the nutritive value of forages by chemical analysis. An example is the method of van Soest (1975, and earlier publications), who separates cell wall and cell contents and fractions these into chemical groups using designated chemical reagents, a technique discussed with other techniques in the review of Barnes (1973). In spite of the shortcomings of the proximate-analysis applied for the present study, a first general picture of the nutritive value of the desert shrub vegetation of Iraq may be obtained from the data presented below.

VII.C.2. *Differences within and between selected shrub species*

Although chemical analysis of the total plant may provide a rough indication of the nutritive value of a species, selective grazing can cause considerable differences in the nutrients obtained by the animal from that species. Table VII–5 shows the results of chemical analysis of different parts of *Rhanterium epapposum* plants, collected at late flowering stage in May 1971. Selective picking of leaves and flowers will give the animal a diet containing two to three times the amount of Crude Protein, about half the amount of Crude Fibre, and about twice the amount of minerals, compared to a diet obtained by eating whole branches. Such selective picking was observed for sheep and especially goats (Fig. VII–5). Camels on the other hand were seen to be not so selective, eating the shrubs gradually from the outer side inwards, not only from the top, but also from the sides. Keeping this in mind, from five shrub species samples were collected from the outher side inwards, by folding the branches together and wrap them in paper, cutting at five centimetre intervals. For each segment a chemical analysis was carried out. Fig. VII–6 shows the increase in Crude Protein from base to top and the decrease of Crude Fibre in this direction. In both the *Haloxylon* species the decrease in Crude Fibre from base to top in most distinct, while in *Rhanterium epapposum* there is hardly any difference. This may be due to the date of collection. *Rhanterium* is already very dry by the end of May and the Crude Fibre of even the small branches is very high by that time (table VII–5). Moreover, leaves are usually rather equally distributed throughout the shrub, while in *Haloxylon* most young shoots are concen-

Table VII–5. Chemical composition of different parts of *Rhanterium epapposum* plants, collected 31 May 1971 (per cent dry matter).

Plant part analysed	Crude Protein	Crude Fibre	Ether Extract	Ash	Na	K	Ca	Mg
Thick branches	4.3	44.8	1.5	8.8	0.2	0.8	1.0	0.4
Thin branches	7.9	35.3	1.9	14.8	0.1	1.8	1.7	0.4
Leaves	18.8	12.3	2.9	23.7	0.2	3.8	3.3	0.8
Flowers	15.5	32.0	1.0	17.6	0.1	2.6	1.8	0.6

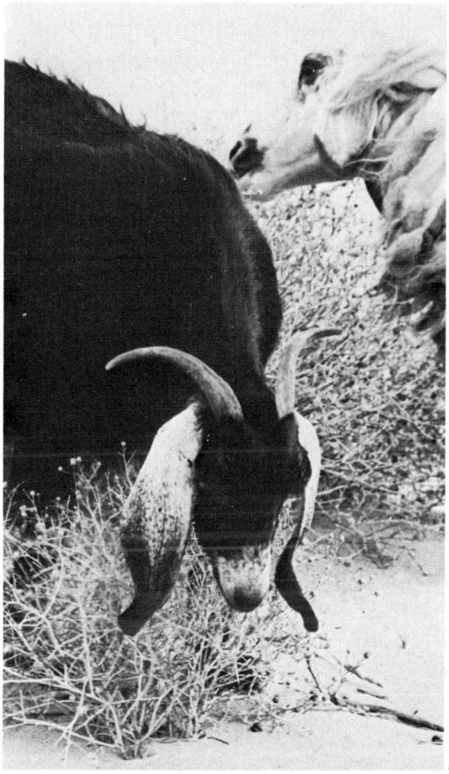

Figure VII–5. Goat and sheep, nibbling selectively flowers, leaves and young shoots from plants of *Rhanterium epapposum*. (Between Um Quasr and Safwan, April 1973)

trated at the end of the older branches. Large differences in Crude Protein were found.*Artemisia herba-alba* and *Achillea fragrantissima* showed a range from 2.2% and 3.7% at the base to 10.3% and 11.6% at the top, respectively. Values in *Haloxylon salicornicum* ranged from 7.2% to 9.4%, in *Rhanterium epapposum* from 4.4% to 5.7%, in both the species without a clear distribution pattern. Most remarkable is the high Crude Protein content of *Haloxylon articulatum*, from 10.1% at the base to 28.1%(!) at the top. Khawaja *et al.* (undated) found a value of 21.0% for the whole plant, the highest value for any of the range species analysed. The distribution of the determined minerals is given in table VII–6.

Ash content increases from base to top in all five species, most significantly in *Haloxylon articulatum* (4.3% to 16.4%), and was highest in *Haloxylon salicornicum* (14.9% at the base and 36.4% at the top). Na-content in this species ranged from 0.5% to 2.3%, while in all other species it was lower than 0.5% in all segments analysed. K-content increased from about 0.5% to 1.5% in *Achillea fragrantissima*, *Haloxylon salicornicum* and *Rhanterium epapposum* and up to

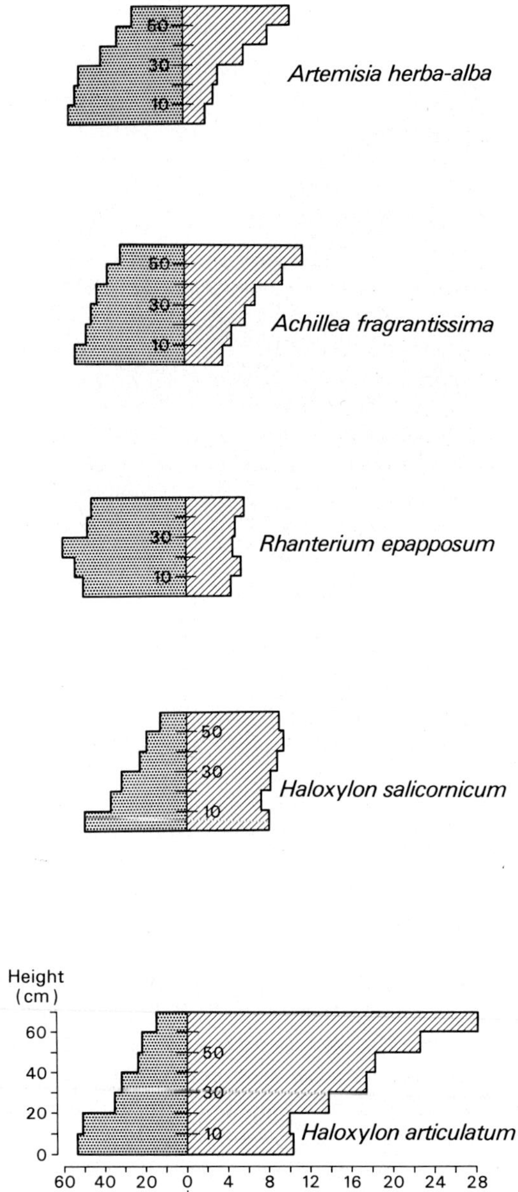

Figure VII–6. Height distribution of Crude Fibre and Crude Protein contents in five important shrub species from samples collected in May and June 1971.

Table VII–6. Distribution of selected mineral constituents in five important shrub species, expressed in percentages of dry matter (partly calculated from Al-Ani & Jawad 1973).

Species and date of collection		height (cm) base 0–10	10–20	20–30	30–40	40–50	50–60	top 60–70
	Na	0.2	03	0.4	0.4	0.3	0.4	–
	K	0.5	0.7	0.8	1.1	1.4	1.6	–
Achillea fragrantissima	Ca	0.7	0.9	0.8	0.7	1.4	1.9	–
(29 June 1971)	Mg	0.2	0.4	0.3	0.3	0.5	0.4	–
	Ash	8.2	9.3	8.0	9.3	9.6	10.6	–
	Na	0.1	0.1	0.1	0.2	0.2	0.1	–
	K	1.5	1.4	1.5	1.9	2.2	2.4	–
Artemisia herba-alba	Ca	1.0	1.3	1.5	1.6	1.6	1.4	–
(28 June 1971)	Mg	0.6	0.6	0.4	0.6	0.6	0.5	–
	Ash	3.2	9.1	7.5	8.2	9.8	9.7	–
	Na	0.2	0.2	0.3	0.3	0.3	0.3	0.2
	K	0.5	0.8	1.7	1.8	2.3	2.6	2.7
Haloxylon articulatum	Ca	1.0	1.0	1.8	2.2	1.8	1.8	1.4
(28 June 1971)	Mg	0.5	0.5	1.6	1.4	1.4	1.4	1.2
	Ash	4.3	7.3	13.4	16.4	16.1	16.4	15.8
	Na	0.5	1.0	1.3	1.6	2.2	2.3	–
	K	0.7	1.0	1.1	1.0	1.2	1.2	–
Haloxylon salicornicum	Ca	3.5	5.6	6.2	7.9	6.3	10.3	–
(12 May 1971)	Mg	0.7	1.6	2.0	2.1	2.0	2.6	–
	Ash	14.9	25.3	29.3	37.0	36.4	–	–
	Na	0.1	0.1	0.1	0.1	0.1	–	–
	K	0.4	0.7	0.9	1.4	1.4	–	–
Rhanterium epapposum	Ca	1.9	2.0	1.7	1.7	2.0	–	–
(25 May 1971)	Mg	0.3	0.2	0.3	0.3	0.4	–	–
	Ash	8.9	8.7	11.9	12.7	11.8	–	–

2.4% and 2.7% in the top parts of *Artemisia herba-alba* and *Haloxylon articulatum*, respectively. Ca-content is very high in *Haloxylon salicornicum* ranging from 3.5% to 10.3% from base to top. *Haloxylon salicornicum* is also highest in Mg-content.

Literature data on the chemical composition of the most common shrub species of the Iraqi desert shrub rangelands have been compiled in table VII–7. Crude Protein contents of over 20% are no exception for selected parts of some species, e.g. leaves or young shoots of *Anvillea garcini, Artemisia herba-alba, Calligonum comosum, Haloxylon articulatum* and *H. salicornicum, Nitraria retusa* and even *Rhanterium epapposum* in its very early growth stages. Crude Fibre in these cases is low, under 20% and in most species even under 10%. Ether Extract reaches over 5% in *Artemisia herba-alba* and leaves of *Zygophyllum coccineum* and is in all species much lower in the stems compared to the leaves with ranges of 0.2–3.2%

Table VII–7. Chemical composition of some common shrub species of the desert rangelands of Iraq. Data compiled from the available literature. Figures express percentages of dry matter. A dash indicates no data available.

Species	Plant part analysed	Date of collection	Source*	Crude Protein	Crude Fibre	Ether Extract	Ash	K	Ca	Mg	P	Na
Achillea fragrantissima	leaves (l)	21 April 1970	1	18.13	12.10	4.6	19.06	2.97	3.0	0.80	–	0.46
Achillea fragrantissima	stems (s)	21 April 1970	1	5.42	47.30	3.2	8.34	1.14	1.60	0.50	–	0.21
Achillea fragrantissima	–	–	3	12.65	35.55	2.4	6.22	–	0.97	–	0.75	–
Anvillea garcini	leaves	26 March 1970	1	22.34	9.70	3.9	18.28	3.16	1.82	0.94	–	0.12
Anvillea garcini	stems	26 March 1970	1	9.52	29.00	1.5	11.67	2.49	1.27	0.39	–	0.19
Artemisia herba-alba	leaves	12 Nov. 1969	1	25.50	10.60	3.3	15.30	3.11	1.20	0.70	–	0.24
Artemisia herba-alba	stems	12 Nov. 1969	1	4.70	43.80	2.3	8.00	1.20	0.86	0.24	–	0.07
Artemisia herba-alba	–	–	3	11.60	26.14	5.1	9.59	–	1.24	–	0.55	–
Artemisia scoparia	–	–	3	13.79	19.38	3.6	8.74	–	1.07	–	0.80	–
Atriplex leucaclada	new green (l+s)	8 Sept. 1962	2	13.03	16.37	–	21.25	1.97	1.92	–	0.38	–
Atriplex leucaclada	yellow green (l+s)	20 Aug. 1962	2	7.35	20.05	–	16.33	2.27	1.51	–	0.15	–
Atriplex sp.	–	–	3	15.40	20.59	0.8	14.53	–	1.25	–	0.45	–
Calligonum comosum	leaves	April 1970	1	23.26	5.70	3.9	21.13	3.07	2.22	0.56	–	0.13
Calligonum comosum	stems	April 1970	1	10.72	28.30	2.3	16.28	3.74	1.67	0.48	–	0.37
Calligonum comosum	young (l+s)	6 July 1962	2	19.95	11.00	–	18.26	1.47	1.61	–	0.40	–
Calligonum comosum	mature (l+s)	1 Oct. 1963	2	3.76	38.47	–	6.35	1.33	0.36	–	0.32	–
Cornulaca sp.	leaves + stems	24 Febr. 1970	1	11.57	17.10	1.1	20.83	3.53	1.75	1.36	–	1.10
Ephedra alata	young shoots	Oct. 1969	1	14.82	32.00	2.9	12.00	3.18	1.11	0.34	–	0.06
Ephedra alata	older material	Oct. 1969	1	7.75	46.40	2.2	9.32	0.78	1.68	0.35	–	0.04
Haloxylon articulatum	old material	29 Febr. 1970	1	7.52	46.10	0.2	10.57	0.38	3.73	0.35	–	0.56
Haloxylon articulatum	young + old mat.	3 March 1970	1	23.71	11.03	1.3	13.98	2.62	1.90	1.03	–	0.17
Haloxylon articulatum	young + old mat.	12 Nov. 1969	1	13.03	9.90	2.6	15.80	1.32	2.15	2.18	–	0.55
Haloxylon articulatum	–	–	3	21.02	19.05	1.0	12.58	–	1.96	–	0.38	–
Haloxylon salicornicum	green shoots+stems	April 1969	1	11.44	14.76	2.8	27.74	1.64	2.48	2.43	–	1.64
Haloxylon salicornicum	green shoots+stems	26 Aug. 1962	2	24.06	11.27	–	17.07	3.01	1.64	–	0.49	–
Haloxylon salicornicum	dry shoots+stems	19 July 1961	2	12.51	20.46	–	24.21	0.02	3.92	–	0.01	–
Heliotropium ramosissimum	leaves	21 April 1970	1	11.68	12.00	1.9	17.26	1.47	7.50	0.90	–	0.30
Heliotropium ramosissimum	stems	21 April 1970	1	8.04	36.60	2.3	12.42	1.38	5.40	0.50	–	0.20
Nitraria retusa	new shoots+leaves	28 Febr. 1962	2	22.57	5.68	–	26.50	0.59	2.00	–	0.64	–
Nitraria retusa	mature shoots+leaves	31 July 1962	2	9.01	6.11	–	36.36	1.13	0.73	–	0.09	–
Rhanterium epapposum	green (l+s)	27 Jan. 1962	2	20.47	18.12	–	9.91	0.03	0.86	–	0.48	–
Rhanterium epapposum	leaves + stems	23 April 1970	1	5.47	44.20	3.5	9.96	1.19	2.20	0.24	–	0.23
Rhanterium epapposum	dry (l+s)	30 June 1962	2	9.80	41.10	–	10.27	1.76	0.91	–	0.01	–
Salsola rigida	–	–	3	15.44	20.63	2.4	13.91	–	1.44	–	0.45	–
Salsola rigida	leaves + stems	18 Nov. 1969	1	7.50	32.70	2.2	22.74	1.14	8.40	0.22	–	1.30
Teucrium polium	leaves	21 April 1970	1	9.80	14.50	3.3	25.34	0.78	8.00	0.80	–	0.13
Teucrium polium	stems	21 April 1970	1	3.10	50.30	3.0	7.54	0.50	2.60	0.27	–	0.07
Zygophyllum coccineum	green (l+s)	8 April 1962	2	11.20	15.48	–	13.86	0.86	2.70	–	0.08	–
Zygophyllum coccineum	dry (l+s)	20 Aug. 1962	2	7.26	14.50	–	9.71	0.63	2.24	–	0.06	–
Zygophyllum coccineum	leaves	20 Nov. 1969	1	6.74	3.60	5.3	32.74	0.23	4.64	4.68	–	0.23
Zygophyllum coccineum	stems	20 Nov. 1969	1	5.76	36.30	2.7	12.48	0.69	1.40	1.02	–	0.69

* 1 – Al-Ani *et al.* (1971a) 2 – Kernick (1966) 3 – Khawaja *et al.* (undated).

and 1.9–5.3%, respectively. Ash-content varies widely with high values, roughly from 24% to 36%, for the succulent species often found in a saline environment (*Haloxylon salicornicum, Nitraria retusa* and *Zygophyllum coccineum*), and low values, even under 10% in the stems of *Artemisia herba-alba, A. scoparia* and *Teucrium polium* and in young material of *Rhanterium epapposum.* Minerals content is adequate to high in almost all analysed samples compared to the requirement figures given before in VII.C.1.

In a number of cases the Phosphorus content was found to be inadequate in older, native plant material, notably of *Atriplex leucaclada, Haloxylon salicornicum, Nitraria retusa, Rhanterium epapposum* and *Zygophyllum coccineum.* It is likely that in such cases the grazing animal will compensate by foraging highly selectively.

VII.C.3. *Seasonal variation*

From table VII–7 it can be seen that samples from the same species, collected at different seasons, varied widely in chemical composition. This aspect was elaborated for *Rhanterium epapposum* and *Haloxylon salicornicum.*

Fig. VII–7 shows the increase in current season's growth for Crude Fibre content from 11.7% at the beginning of growth, to 30.0% later in the year in *Rhanterium epapposum.* For Crude Protein there was a decrease from 29.5% to 4.7% over the same period. In the older woody material little variation was found for both, Crude Fibre (high) and Crude Protein (low). The 33.2% Crude Fibre recorded in the November sample may well be due to the sampling. Some dry current season's growth may have been included while distinguishing between current season's growth and older material becomes increasingly difficult in November and December. Crude Protein values and Crude Fibre values for *Haloxylon salicornicum* for nine different dates are shown in Fig. VII–8. Crude Protein in the current season's growth drops from almost 17% at the end of March to about 7% in December when it is all dried up. Crude Fibre fluctuates between 10% and 20%, increasing in December to about 30%. In the older branches a Crude Protein content between 5% and 10% was found, with distinct highest values in June and August (between 9% and 10%) at the onset of flowering. Crude Fibre figures were high, from about 40% early in the year to over 50% in December. Al-Ani & Jawad (1974) present figures on the minerals content. No distinct trends can be derived from these. An intensified sampling programme might be required to bring these out. Data on *Artemisia herba-alba* from the same authors show a Crude Protein level between 18% and 23% for February, March and April (rapid leaf production) against 11% to 14% from June until December. It should, however, be pointed out that these figures pertain to green material separated subjectively from the older material on the basis of its palatability for sheep. The older material showed no distinct variation, the level ranging from about 3.5% to 6% Crude Protein.

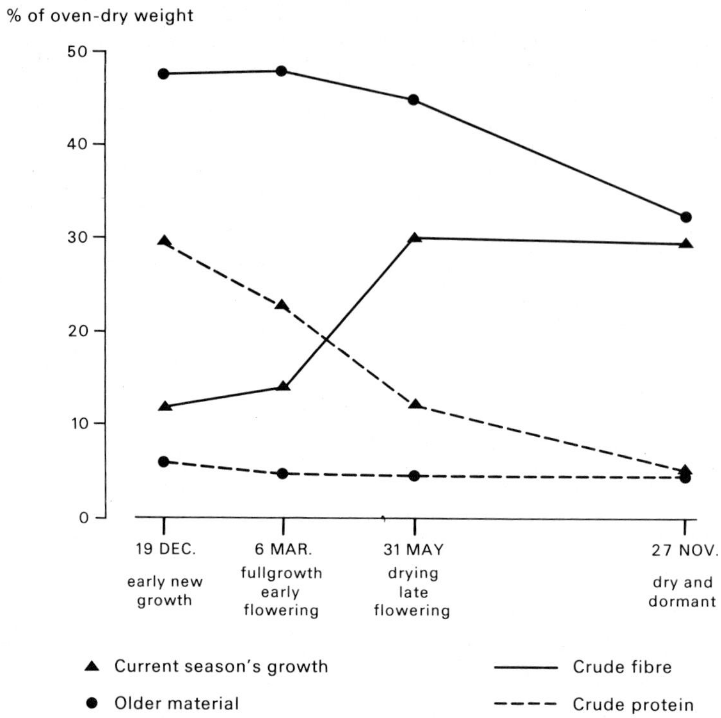

Figure VII–7. Seasonal variation in Crude Protein and Crude Fibre content in selected parts of *Rhanterium epapposum* plants (mixed samples of several plants).

The above data indicate the strong seasonal changes of forage quality in shrubs, especially as related to Protein and Fibre content. Similar changes take place in annual forage plants and even at faster rate. No data from Iraq are available on this point. The figures presented in table VII–8 were calculated from data for Kuwait and are representative for at least the southern part of Iraq too. In the annual species Crude Protein drops by a factor two and a half between January and May; Crude Fibre increases by about half in the same period.

Cook (1972) discussed nutritional content and stage of growth *comparatively* for three forage classes: shrubs, forbs and grasses. He reports that in Carotene content, Digestible Protein and Phosphorus shrubs show increasingly higher figures from the vegetative, through flowering and fruiting to the mature (pheno-

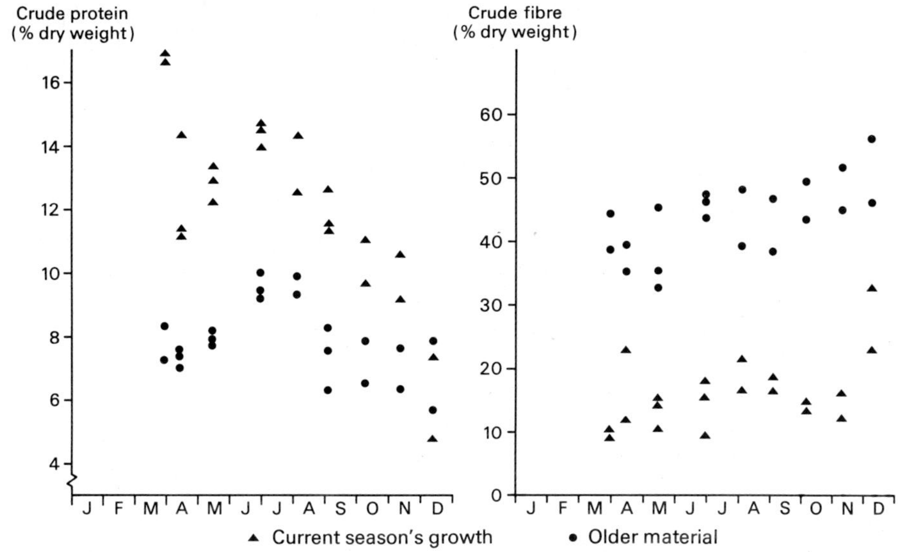

Figure VII–8. Changes in Crude Protein and Crude Fibre content over the season in *Haloxylon salicornicum*, shown on samples from nine dates of collection from the same site.

Table VII–8. Comparison of Crude Protein and Crude Fibre content (percentages of dry matter) of annual forage species, collected early and later in the growing season. Calculated from Kernick (1966) for data from Kuwait ($\bar{x}$ – mean, s – standard deviation).

Nature of the sample	Date of collection	Crude Protein		Crude Fibre	
		$\bar{x}$	s	$\bar{x}$	s
Green leaves and stems of annuals (16 species)	Dec. – March	23.28	4.19	15.07	5.26
Dry leaves, stems, and seeds of annuals (6 species)	April – May	10.19	3.37	22.77	5.94

logical) stage, compared to the other classes. Such data, as well as the data for Iraq and Kuwait discussed above, emphasize that due attention should be given to the forage quality of each forage class in relation to the season for which a forage assessment is attempted.

VII.C.4. *General assessment*

Crude Protein.

Of the 41 shrub samples of table VII–7 almost half of them meet the requirement of lactating ruminants and almost three quarters that of growing ruminants, while less than 15% fall short of the maintenance level requirement. The last category comprises all samples of older material, generally dry branches, only eaten as last choice. In the case of *Haloxylon salicornicum* it can be seen that Crude Protein level (even of the older material!) is sufficient for maintaining body weight all over the year and that in the new growth Crude Protein content is sufficient for growing ruminants until October, when it falls below 10% (table VII–7, Fig. VII–8). In the woody material of *Rhanterium epapposum* an amount between 4.3% and 7.9% was found (table VII–7, Fig. VII–7). The current season's growth of this species has an extremely high Crude Protein content dropping sharply when maturing to a level below ruminant maintenance requirement in fall and winter. In a third key shrub species of the Iraqi deserts, *Artemisia herba-alba*, Crude Protein in the green plant material was above 10% over all seasons and above 16% from January to April, sufficient for growing and lactating ruminants, respectively (Al-Ani & Jawad 1974). Dry material of *A. herba-alba* showed a Crude Protein level at the lower side of the maintenance requirement.

Minerals.
Comparison of the mineral levels in the samples with the requirements (given in VII.C.1.) shows that mineral levels of the shrub species are generally high and well above the requirement of cattle and sheep. An exception has to be made for Phosphorus in a few samples. Deficiencies in this element were observed in dry, old material of *Haloxylon salicornicum, Nitraria retusa, Rhanterium epapposum*, and in green and dry material of *Zygophyllum coccineum* (table VII–7).

Ether Extract, Crude Fibre (and Nitrogen-Free Extract).
As pointed out before these three groups of components mainly supply energy to the animal, but their contribution depends heavily on the digestibility. High values for Ether Extract were found in the leaves of *Achillea fragrantissima, Anvillea garcini, Artemisia herba-alba* and *A. scoparia, Calligonum comosum, Rhanterium epapposum* and *Zygophyllum coccineum*. These values may include an amount of non-digestible oils. In general the percentage Crude Fibre increases when Crude Protein, the simple Carbohydrates and Fats decrease. The latter three are usually related to a high palatability. High Crude Fibre contents were in almost all cases found in the older material of the shrubs analysed (table VII–7). This material has a low palatability because in the process of maturation lignification takes place which reduces digestibility. Shrubs are, therefore, generally not considered good sources of energy after they reach the fruiting stage. This is the only nutrient aspect for which shrubs are rated lower than grasses and forbs (Cook 1972).

VII.C.5. *Shrubs in comparison to other feeds*

Instead of comparing the nutrient content of shrubs with animal requirements,

Table VII–9. Chemical composition of selected samples of the natural vegetation and samples of cultivated fodder crops and other animal feeds (percentages of dry matter).

Nature of the material	Crude Protein	Crude Fibre	Ether Extract	Ash	Nitrogen-Free Extract
Cultivated fodder crops*					
Alfalfa hay	11.56	30.70	2.08	12.36	43.30
Barley hay	5.97	28.77	0.82	9.67	54.77
Sorghum plant	3.23	29.53	1.60	6.71	58.93
Other feeds*					
Date pulp (dried)	5.17	22.84	0.89	5.97	65.13
Sugarbeet pulp (dried)	8.45	23.41	0.48	5.17	62.49
Barley (grain)	9.92	6.58	2.02	3.77	77.71
Annual grasses and forbs**					
Leaves (7 species)	22.85	9.96	2.10	23.16	41.93
Stems (7 species)	11.30	28.05	1.10	15.94	43.61
Stems + leaves (11 species)	15.21	20.07	2.30	17.37	45.05
Shrubs**					
Leaves (12 species)	18.17	13.06	3.80	18.43	46.54
Stems (11 species)	7.41	37.38	2.30	11.13	41.78
Stems + leaves (16 species)	13.26	23.49	2.60	17.99	42.66

* Data calculated from Khawaja *et al.* (undated); mean values of a number of samples.

** Data from Al-Ani *et al.* (1971a)

shrubs as forage could also be compared with other types of food. From table VII–9 a comparison can be made with annual grasses and forbs, some cultivated fodder crops and other feeds. From this comparison and the data presented and discussed above the following conclusions can be drawn:

– Shrubs have a high Crude Protein content, especially in the leaves. The values are higher than in cultivated fodder crops and other foodstuffs. Annual forbs in their early growth stage only rank slightly higher.

— Crude Fibre content of shrubs is comparable to that of date – and sugarbeet pulp and lower than that of alfalfa – and barley hay or sorghum plants.

— Crude Fat content (Ether Extract) in shrubs is higher than in annual forage plants or (the selected) fodder crops and other feedstuffs.

— Total mineral content of shrubs (Ash) varies highly, but is usually far higher than in the cultivated fodder crops.

The above conclusions show the high nutritive quality of shrubs, particularly in their vegetative and flowering stage. Shrubs rank higher than forbs and grasses, except perhaps for the aspect of digestible energy.

It may finally be stressed that selective grazing may give the animal an even more nutritious menu than suggested by the figures from the chemical analyses. It has been stated that where selective grazing is possible, the composition of the animal's food may be 20% to 100% better than the results of herbage analyses suggest (Blair Rains 1970) Table VII–10 can illustrate this. A camel grazing in a

Table VII–10. Chemical composition, digestibility and nutritive value estimates for selected parts of *Haloxylon salicornicum* and *Rhanterium epapposum*.

Nature of the sample	Date of collection	Moisture (%)	Ether Extract (%)	Crude Fibre (%)	Crude Protein (%)	Ash (%)	Nitrogen-Free Extract (%)	D.M.D.*	N.V.I.**
Haloxylon salicornicum									
Older branches	Oct. 1972	4.73	0.83	28.68	7.10	5.96	52.70	10.2	15.6
Current growth	Oct. 1972	5.15	1.07	15.67	8.15	16.26	53.70	35.8	56.5
Seeds	Dec. 1972	2.91	2.70	10.20	14.34	24.23	45.62	46.9	74.3
Rhanterium epapposum									
Older branches	Nov. 1972	3.63	1.63	33.16	4.55	5.19	51.84	19.8	30.9
Thin, outher branches	Nov. 1972	8.61	1.80	29.49	4.72	4.53	55.85	23.9	51.9
Leaves	Jan. 1973	6.26	4.15	12.83	13.49	13.33	49.94	50.4	79.9

* D.M.D. – Dry Matter Disappearance, indicating the percentage of the original sample that has become soluble by an 'in vitro' digestibility analysis.

** N.V.I. – Nutritive Value Index, a forage nutritive value estimate calculated from the D.M.D. value.

mixed stand of *Haloxylon salicornicum* and *Rhanterium epapposum* has in the period between October and January, covered by the sampling, the choice between parts of shrubs which differ a factor five in estimated nutritive value. The higher nutritive value from fistula samples compared to clipped samples has definitely proved that the animal is able to take the best part of the supply whenever the supply is more than adequate.

VII.D. Moisture content of the forage

VII.D.1. *Significance of moisture content data*

Living plants always contain water. This water, under certain conditions, can be sufficient to cover an animal's total water requirement. Investigations on the water economy of the camel have shown that in winter the amount of water in the plants consumed by the animal may allow him to go for two months without drinking, while still having normally hydrated blood, tissues etc. (Schmidt-Nielsen 1956). Stoddart *et al.* (1975) quote an example of sheep being grazed in Montana with absolutely no watering for three summer months. In spring 1972, when a wealth of annuals was available for grazing after heavy rains in Iraq, Bedouins told that sheep had not been watered for six weeks. In these situations flocks are free to wander over the areas where such vegetation is found. It makes the movement independent of water supply from wells. This may, under certain conditions, be a decisive factor in management and shows the importance of information on moisture content.

Sharif & West (1968) studied forage moisture content and provided data for another important application: deriving estimates of dry weights from green forage by using empirical formulae. They reviewed studies on plant water content and variations in water content and showed in their situation (a mountain summer range) significant differences due to aspect, shade, time of the day, season and interaction of some of these factors. Consequently, improved, more detailed green weight to air-dry weight conversion factors were suggested. Establishment of such factors for the Iraqi situation would be useful.

VII.D.2. *Determination and expression of the moisture content*

Moisture content of plant tissues can be determined by the difference between the field fresh weight of a sample (FW) and the dry weight (DW) after heating the sample in an oven (for the present study 24 hours at about 70° C in a forced draft oven) to constant weight. The quantity thus found has to be related to some other parameter of the sample to allow comparison with other samples, required for an assessment of seasonal trends, comparison between species, etc. Difficulties in establishing a basis for the expression of the moisture content have been discussed in Kozlowski (1968). Moisture content values for the present study were determined not on particular tissues, but on the total aerial biomass of individual plants or stands and on samples of current season's growth. The analyses were carried out not for measuring water shortage as for most of the studies referred to in Kozlowski (1968), but for the amount of water available to animals and actually taken by animals when grazing and browsing. This aspect reduces the limitations of using total fresh or dry weight as a basis.

For reasons of practical application and also because more sophisticated determinations could not be made under the given circumstances, moisture content has been expressed as per cent of plant dry weight (Wd), calculated from

$$Wd = \frac{100\,(FW-DW)}{DW}$$

and as per cent of plant fresh weight (Wf), calculated from

$$Wf = \frac{100\,(FW-DW)}{FW}$$

Values have been derived from three types of data which are subsequently presented and discussed, viz. (i) seasonal measurements of moisture content in two shrub species, *Haloxylon salicornicum* and *Rhanterium epapposum*, (ii) moisture content of various mixed stands of shrubs, measured only once, and (iii) moisture content of the annuals component.

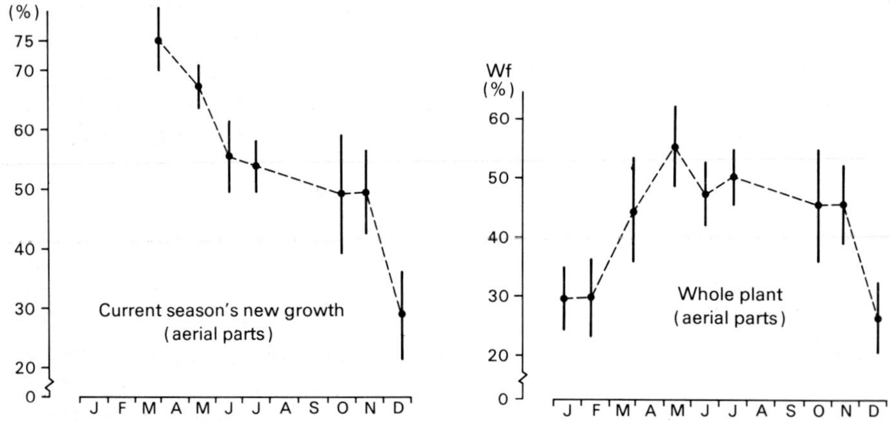

Figure VII–9. Moisture content of *Haloxylon salicornicum* at nine different dates in 1972. Means of 30 observations for each date are shown by the (connected) dots, lines show the standard deviation in observed values for each date.

VII.D.3. *Seasonal variation*

Moisture content determination for *Haloxylon salicornicum* shrubs was repeated nine times from January to December, 1972. Thirty plants were sampled at each date and fresh weight and dry weight determined for the whole plant as well as current season's new growth (Fig. VII–9). It may be recalled that in January and February no separation of old and new plant material was possible. Fig. VII–9, therefore, gives no figures for these months. In both the cases, for current season's growth and for the whole plant a clear trend can be seen. The new growth, early in the season, has a moisture content of over 75%, quickly dropping to just above 50% in early summer. A slow, gradual decrease continues during the summer months to values just below 50%, followed by a sudden drop to below 30% in December. These values, combined with the increase in new growth compared to older (woody) material, determine by and large the picture as shown for the whole plant. An increase from about 30% to over 50% from January to May, a gradual decrease to about 45% in November, followed by a drop to lower than 30% again in December. The standard deviations indicate the dispersion in the recorded values. It was observed that small plants and large plants differed in their moisture content over the seasons. This is elaborated in table VII–11, for the current year's growth and the older plant material. Larger samples would be required to test higher significance of the differences, but a first impression was gained. In the current season's growth no significant differences were seen. In the older plant material, however, in small plants, the differences over the seasons were more pronounced than in the larger ones. A difference of 25.7% for the small plants against 14.7% for the large ones was found between July and December. In March the moisture content of old woody material or small plants was found to be 10% higher than for large plants, in December 6% lower. This difference may be explained by a greater amount of tissue not reacting to moisture differences in the larger plants. The medium size plants showed in all seasons values in between those of the small and large plants as might be expected. Table VII–12 shows the results of observations on *Rhanterium epapposum*. The highest moisture content, 81.5%, was found in the new growth in December, when the plants had just begun to sprout. The lowest value was 21.8% for the older material of the larger plants. The highest moisture content for the whole plant was 48.2%, in March, when the plants are in full growth. Similar to *Haloxylon salicornicum*, little difference was observed for the moisture content of the current season's growth at a particular date, comparing small, medium and large plants. Moisture content of the old woody material in large plants dropped to almost half (39.2% to 21.8%) from March to May, in small plants by about one third (42.2% to 29.7%) over the same period.

A comparison of the *Rhanterium* data of table VII–12 and the *Haloxylon* data of Fig. VII–9 and table VII–11 shows that the observed moisture content values for the two species are in the same order of magnitude at their respective growth

Table VII–11. Percentage moisture in different parts of *Haloxylon salicornicum* plants in three weight classes, collected near Ukhaidir at four different dates over the growing season ($\bar{x}$ – mean, s – standard deviation, n – number of observations).

	Month of collection	Per cent moisture for plants of three weight classes on fresh weight basis (Wf)								
		Small plants FW < 100 g			Medium plants 100 g < FW < 400 g			Large plants FW > 400 gr.		
		$\bar{x}$	s	n	$\bar{x}$	s	n	$\bar{x}$	s	n
Current season's growth	March	75.8	5.9	13	75.5	5.1	11	71.0	5.1	6
	July	52.4	4.2	9	56.0	4.9	11	52.9	2.9	10
	October	44.5	7.2	6	53.2	3.7	12	51.0	9.0	12
	December	28.3	8.3	10	30.4	5.7	14	29.2	4.6	6
Older material	March	37.9	3.2	13	32.4	3.3	11	27.5	7.2	6
	July	41.6	8.0	9	38.1	6.6	11	36.5	9.0	10
	October	17.2	5.3	6	19.8	5.4	12	20.3	4.5	12
	December	15.9	4.7	10	19.8	2.9	14	21.8	1.7	6

stages. For *Rhanterium*, however, the growth period is much shorter. In March both the species had a moisture percentage for the new growth of about 75%. Two months later it was 50–55%. *Rhanterium* had then about completed its growth for that season becoming dryer over the rest of the year. *Haloxylon*, however, is in May in full growth and the moisture content of the plants as a whole of this species only drops steeply in November. In planning the grazing for certain areas these facts should be considered, especially in relation to availability of drinking water from wells.

Table VII–12. Percentage moisture in different parts of *Rhanterium epapposum* plants, in three weight classes, collected near Samah at three different dates over the growing season ($\overline{x}$ – mean, s – standard deviation, n – number of observations).

	Date of collection	Per cent moisture for plants of three weight classes on fresh weight basis (Wf): Small plants FW < 100 g			Medium plants 100 g < FW < 500 g			Large plants FW > 500 g			Wf for all plants		
		$\overline{x}$	s	n	$\overline{x}$	s	n	$\overline{x}$	s	n	$\overline{x}$	s	n
Total plant	18 December 1971	41.2	10.8	13	27.1	7.8	8	23.0	10.3	9	32.0	12.7	30
	6 March 1972	54.2	7.3	19	44.2	4.8	21	45.1	4.4	10	48.2	7.4	50
	22 May 1972	43.7	6.5	12	45.0	4.6	9	37.3	3.8	9	42.2	6.1	30
Current season's growth	18 December 1971	–*			–*			–*			81.5	2.1	**
	6 March 1972	77.1	3.7	19	75.3	2.7	21	75.3	2.9	10	76.0	3.2	50
	22 May 1972	47.3	4.4	12	45.0	4.8	9	49.4	7.0	9	49.0	5.4	30
Older material	18 December 1971	–*			–*			–*			–*		
	6 March 1972	42.2	6.7	19	35.8	3.9	21	39.2	4.9	10	38.9	5.9	50
	22 May 1972	29.7	10.2	12	26.9	8.9	9	21.8	5.4	9	26.4	8.9	30

* No correct estimate possible.

** Calculated from three mixed samples of a large number of plants.

Table VII–13. Moisture content of some stands of shrub vegetation, arranged from highest to lowest measured values. Percentages are the mean value of three mixed samples per stand of several hundred grams fresh weight each, taken randomly from the material harvested from 50 quadrats of 4 m^2. Range indicates the difference between the highest and the lowest value.

Sampling date	Cover dominant species*	Co-dominant species*	Aerial biomass dry ($kg.ha^{-1}$)	Moisture content: Per cent fresh (Wf) Avge.	Per cent fresh (Wf) Range	Per cent dry (Wd) Avge.	Per cent dry (Wd) Range
23 March 1973	*Peg. harm.*	*Lag. farc.*	145	70	2	239	22
21 Dec. 1972	*Zyg. cocc.*	*Corn.* sp.	266	58	4	140	27
24 March 1973	*Peg. harm.*	*Art. scop.*	228	56	11	129	26
23 March 1973	*Sals.* sp.	*Art. scop.*	58	55	8	125	40
18 Dec. 1972	*Hal. sal.*	*Sals.* sp.	96	54	2	116	8
17 Dec. 1972	*Zyg. cocc.*	*Hal. sal.*	237	53	10	114	47
2 June 1973	*Ach. frag.*	*Zill. spin.*	166	53	4	113	17
22 April 1973	*Cyp. congl.*	*Rhan. epp.*	486	51	4	105	15
2 June 1973	*Peg. harm.*	*Art. scop.*	102	49	5	96	20
18 Dec. 1972	*Sals.* sp.	*Ael. lag.*	995	49	7	95	25
19 Dec. 1972	*Hal. sal.*	*Sals.* sp.	379	46	10	85	37
22 May 1973	*Art. scop.*	*Art. h.-a.*	56	45	6	81	21
20 March 1973	*Peg. harm.*	*Art. scop.*	200	44	5	79	19
3 June 1973	*Astr. spin.*	*Hal. sal.*	276	40	1	68	3
16 Dec. 1972	*Eph. al.*	*Hal. sal.*	261	39	10	65	26
19 Dec. 1972	*Hal. sal.*	*Sals. sub.*	244	37	7	58	18
4 June 1973	*Hal. sal.*	*Art. scop.*	1312	37	3	60	3
23 March 1973	*Peg. harm.*	*Art. scop.*	77	36	15	58	33
28 Jan. 1973	*Rhan. epp.*	*Eph. al.*	768	35	2	54	4
29 Jan. 1973	*Hal. sal.*	*Eph. al.*	444	34	2	51	4
4 June 1973	*Hal. sal.*	*Art. h.-a.*	99	34	5	52	10
20 Dec. 1972	*Hal. sal.*	*Corn.* sp.	379	34	8	52	17
28 Jan. 1973	*Eph. al.*	*Hal. sal.*	1156	32	14	47	28
25 Jan. 1973	*Eph. al.*	*Hel. lip.*	74	31	4	46	6
31 May 1973	*Eph. al.*	*Hal. sal.*	114	29	7	41	14
29 Jan. 1973	*Hel. lip.*	*Eph. al.*	66	29	9	41	22
6 June 1973	*Art. h.-a.*	*Hal. sal.*	406	28	10	39	19
17 Dec. 1972	*Rhan. epp.*	*Hal. sal.*	949	27	13	39	24
26 Jan. 1973	*Rhan. epp.*	*Hel. lip.*	608	27	2	38	5
29 Jan. 1973	*Rhan. epp.*	*Hal. sal.*	490	26	6	35	11
22 March 1973	*Art. h.-a.*	*Art. scop.*	165	26	1	36	3
5 June 1973	*Hal. sal.*	*Hal. art.*	130	24	8	32	15
28 May 1973	*Hal. sal.*	*Astr. spin.*	785	24	3	32	6
26 Jan. 1973	*Hel. lip.*	*Art. h.-a.*	99	24	3	31	5
25 Jan. 1973	*Rhan. epp.*	–	1274	24	5	32	4
24 Jan. 1973	*Sals.* sp.	*Stip. pl.*	99	24	17	32	28
24 April 1973	*Stip. pl.*	*Halo. al.*	196	23	1	31	1
21 March 1973	*Hal. sal.*	*Stip. pl.*	329	22	1	28	1
22 March 1973	*Art. scop.*	*Hal. sal.*	141	19	3	23	6
21 March 1973	*Art. h.-a.*	*Noe. mucr.*	390	19	7	23	10
22 May 1973	*Art. h.-a.*	*Sals.* sp.	198	17	7	20	8
24 April 1973	*Hel. lip.*	*Stip. pl.*	273	17	5	21	7
24 April 1973	*Halo. al.*	*Hal. sal.*	1093	13	5	16	6
25 April 1973	*Hal. sal.*	*Hel. lip.*	1069	12	4	10	5
21 April 1973	*Art. h.-a.*	–	43	9	4	10	5

* Species abbreviations: *Ach. frag.*= *Achillea fragrantissima*, *Ael. lag.*= *Aeluropus lagopoïdes*, *Art. h.-a.*= *Artemisia herba-alba*, *Art. scop.*= *Artemisia scoparia*, *Astr. spin.*= *Astragalus spinosus*, *Corn. sp.*= *Cornulaca* sp., *Cyp. congl.*= *Cyperus conglomeratus*, *Eph. al.*= *Ephedra alata*, *Hal. art.*= *Haloxylon articulatum*, *Hal. sal.*= *Haloxylon salicornicum*, *Halo. al.* = *Halogeton alopecuroïdes*, *Hel. lip.* = *Helianthemum lippii*, *Lag. farc.* = *Lagonychium farctum*, *Noe. mucr.* = *Noea mucronata*, *Peg. harm.* = *Peganum harmala*, *Rhan. epp.*= *Rhanterium epapposum*, *Sals.* sp. = *Salsola* sp., *Sals. sub.*= *Salsola subaphylla*, *Stip. pl.* = *Stipagrostis plumosa*, *Zill. spin.* = *Zilla spinosa*, *Zyg. cocc.*= *Zygophyllum coccineum*.

VII.D.4. *Moisture content of shrub vegetation and annuals*

When sampling selected stands of the shrub vegetation for establishing of range cover types, as explained in chapter V, fresh weight to dry weight conversion factors were determined by collecting three representative mixed samples of several hundred grams each and taking fresh weight and dry weight. Table VII–13 shows the variation in the results. The highest Wf-values observed are between 50 and 70%, measured in stands with mostly succulent chenopodiaceae as dominant and co-dominant species. The lowest values are less than 20%, measured in March, April and (in one case) May in stands with mostly *Artemisia* spp. as dominants.

The variation illustrated in table VII–13 means that a grazing animal may have a difference of half a litre of water per kilogram fresh plant material eaten when grazing in one stand compared to another. The seasonal aspect of water-intake with the forage is most pronounced in case annuals are available in good supply. In spring 1972 good growth of these drought evading species was seen and the moisture content aspect is shown in table VII–14.

Moisture content (Wf) was around 75% in early March. It was already mentioned that in spring camels, but also sheep and goats can rely completely on this water source for their daily need. This need is dependant on a great number of factors, such as air temperature, air-humidity, work done by the animal e.g. carrying loads, walking in loose sandy terrain or hard gravel, animal factors like breed, sex, age, condition of the animal, etc., and estimates have to be rough and generalized. Figures of four to five litres per day during the dry season have been given for goats and sheep (Beaudelaire 1972), but also much higher figures. Assuming that a sheep would need an intake of 4.5 litres water daily, the animal could cover this need with a forage intake of about 6 kg fresh plant material, representing less than 2 kg dry forage, a figure sometimes given as daily requirement. This explains the prolonged non-watering period in spring 1972 observed for sheep in Iraq.

VII.E. Forage production

VII.E.1. *Introduction*

Net aerial production in shrub vegetation can be determined in a number of ways. Considered as forage, animal performance is the most straightforward and best measure (Brown 1954). Measurement of aerial biomass differences ($B_2 - B_1$) over a given time interval ($t_2 - t_1$), taking into account losses due to shedding of flowers, seeds, leaves branches etc. (L) and consumer organisms (G), as expressed in the formula

$$Pn = (B_2 - B_1) + L + G$$

is another technique to measure the net aerial primary production (Pn) (Milner &

Table VII–14. Moisture content of some stands of annuals, sampled early March 1972 ($\bar{x}$–mean, s–standard deviation).

Description of the sample	Moisture content			
	Per cent of fresh weight (Wf)		Per cent of dry weight (Wd)	
	$\bar{x}$	s	$\bar{x}$	s
20 quadrats of 0.25 m^2 sampled 3 March 1972 inside the Khedr-el-Mai fenced area.*	73.9	7.1	308.5	104.5
18 quadrats of 0.25 m^2 sampled 3 March 1972 outside the Khedr-el-Mai fenced area.*	72.9	4.4	277.8	54.7
18 quadrats of 1.0 m^2 sampled 7 March 1972 in a *Rhanterium epapposum* stand near Samah**	76.9	6.0	357.5	108.5

* Contributing main annual species, ranked in estimated order of importance: the forbs *Plantago* spp. (*P. ovata, P. ciliata* and to a lesser extent *P. Boissieri*), *Savignya parviflora Matthiola oxyceras, Calendula aegyptiaca, Schimpera arabica, Picris babylonica, Hypecoum pendulum, Arnebia decumbens, Helianthemum salicifolium, H. ledifolium, Matricaria aurea, Erodium cicutarium, Trigonella stellata*, and locally highly contributing the grasses *Schismus barbatus, S. arabicus* and *Stipa capensis.*

** Excellent cover of a few species, viz. *Schimpera arabica, Plantago Boissieri, P. ciliata, Matthiola oxyceras, Savignya parviflora* and the grass *Schismus barbatus.*

Elfyn Hughes 1968). Both methods require a continuous control of the study site,which was lacking in the situation in the study area in Iraq.

An other technique, the 'direct method' was, therefore, applied (Gimingham & Miller 1968). Individual shrubs were separated by hand into current season's growth (recognizable as the young green shoots, later in the season with flowers and seeds) and older material. The fresh and dry weight of the current season's growth was determined and related to the total shrub weight, expressing it as a percentage. To this purpose ungrazed shrubs were selected. Losses due to death or shedding do not play a role till at the end of the growing season, when seeds are also dropped. It should be mentioned that the technique is biased with respect to the total aerial production, while the material laid down in the woody stems is not included. In an assessment of production as related to forage supply this aspect can be neglected while this woody material is usually not eaten by the larger herbivores.

In the following, results are given of this 'direct method' of measuring forage production in the key species treated before in chapter VI.

VII.E.2. *Seasonal aspects*

Using the 'direct method', forage production in *Rhanterium epapposum* was evaluated at three dates, December, March and May, being representative for the beginning, middle part and end of the growing season, respectively. Results are shown in table VII–15. At the end of the growing season almost three quarters of the fresh weight or two thirds of the dry weight of the shrubs is current season's growth and may be considered available forage. A high variation in the individual values per shrub was found, ranging from 60% early in the season to about 20% at the end of the growing season. Further analysis of the data showed an effect of shrub size. Small plants produced a significantly higher percentage of new growth than large plants. In May the difference between the two classes (plants with fresh weight less than 100 g compared to plants with fresh weight more than 750 g) was almost a factor two and at all three dates the difference was significant at the $p < 0.001$ level.

Contrary to *Rhanterium*, in *Artemisia herba-alba* the seasonal forage availability aspect is not very pronounced, at least for the second half of the year. This species has a period of rapid shoot and leaf development from January to May. Although some growth may continue afterwards, at the same time lignification and leaf-shedding takes place (Orshan & Diskin 1968). Consistent separation of old and current season's material is, therefore, impossible at that time of the year. In *Artemisia* a method of subjective separation of green forage from the more woody material was, therefore, adopted. Results are given in table VII–16. From April to June available forage more than doubled from about 15% to 42% of total aerial biomass, followed by a gradual decrease to about 29% in December. The starting point of putting on new growth differs considerably from one plant to the other. This explains the high value (72) of the coëfficient of variation in April, compared to values ranging only from 25 to 36 later in the year.

Table VII–15. Forage production of *Rhanterium epapposum* plants measured in a stand near Samah at the beginning (December), in the middle part (March) and at the end (May) of the growing season 1971/1972 ($\bar{x}$–mean, s–standard deviation, c.v.–coefficient of variation).

Date of observation and number of plants examined	Current season's growth as percentage of total plant weight							
	Fresh weight				Dry weight			
	$\bar{x}$	s	range	c.v.	$\bar{x}$	s	range	c.v.
18/19 December 1971 (13)	9.3	5.3	2.2–20.0	57	2.7	1.7	0.5–5.6	63
6/7 March 1972 (50)	24.8	9.7	12.0–55.6	39	11.9	6.0	4.8–32.4	50
22 May 1972 (30)	72.1	14.1	45.3–92.7	20	61.9	17.8	31.5–95.1	27

Table VII–16. Available forage from *Artemisia herba-alba* plants measured at eight dates in 1972 near Rutba in the Western Desert ($\overline{x}$–mean, s–standard deviation, c.v.–coefficient of variation).

Date of observation and number of plants examined	Available forage as percentage of total plant weight (dry weight basis)		
	$\overline{x}$	s	c.v.
13 April 1972 (28)	15.2	10.9	72
22 June 1972 (30)	41.9	15.1	36
23 July 1972 (30)	38.2	12.9	34
27 August 1972 (30)	33.5	10.2	30
27 September 1972 (30)	31.7	9.9	31
25 October 1972 (29)	34.3	8.7	25
19 November 1972 (30)	29.9	10.2	34
23 December 1972 (30)	28.8	9.0	31

The most intensive forage production sampling could be carried out on *Haloxylon salicornicum*. During two successive years, 1971 and 1972, from the onset of new growth in spring to the dried up stage in December almost monthly observations were taken on 30 shrubs harvested each date (Fig. VII–10). The production pattern is easily observed. There is an increase in available forage up to October, followed by a slight drop in November and December due to the shedding of seeds and outer branchlets. In the period of highest production

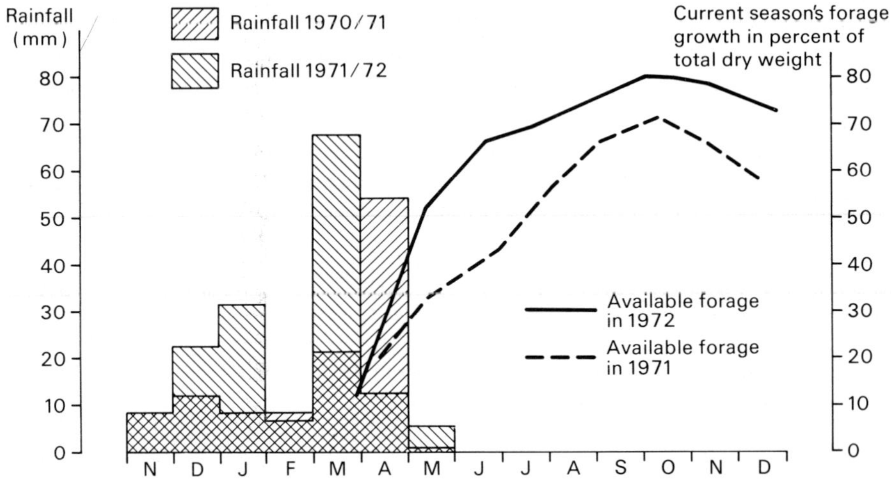

Figure VII–10. Available forage from *Haloxylon salicornicum* over the seasons during two years, 1971 and 1972, measured on 30 plants per date of observation near Ukhaidir. The rainfall data are mean values of the two nearest stations, Kerbala and Habbaniya.

achieved, 70 to 80% of the dry weight of the is accounted for by the current season's growth. For *Rhanterium* in May about 65% was found (table VII–15). Comparative figures for *Artemisia herba-alba* were given by Rodin *et al.* (1972), 80% for *Artemisia herba-alba* in Algeria. Such figures show the forage production capacity of arid shrub vegetation and at the same time stress the variation in the amount of available forage over the season.

VII.E.3. *Between–year variability*

Fig. VII–10 shows a clear difference between the amount of forage produced by *Haloxylon salicornicum* in 1971 compared to 1972. Production of shrub vegetation can be governed by a number of factors, including community type, soil fertility (especially availability of Phosphorus and Nitrogen) and climatic factors. In a comparison of the production of the same vegetation in a certain area between years, the amount of precipitation is the main factor (Noy-Meir 1973). Fig. VII–11 summarizes a comparison of three stands of *Haloxylon salicornicum* for production in 1971 and 1972. Cumulative rainfall data for the four nearest stations are also given. Production data represent the peak in October (see Fig. VII–10), rainfall is given for the preceding months. At all three sites, production in October 1972 was higher than in October 1971, by 14, 29 and 14% respectively. Rainfall differs by 20–30%. More important to shrubs than total rainfall is perhaps rainfall distribution. The most effective rainfall distribution for shrub growth is when larger amounts fall in a short period, preferably at a time of low evapotranspiration. Such rainfall will replenish the moisture reserve at deeper layers on which shrubs are dependent for a permanent water supply (Noy-Meir 1973). The situation in 1971/72 is in this respect also much more favourable in 1970/71. The Kerbala station recorded *five days* with rainfall over eight millimetres from November to February 1971/72, compared to only *two days* in 1970/71. The production curves of Fig. VII–10 illustrate the response of the shrub vegetation. *Haloxylon salicornicum* put on far more new growth early in the 1971/72 season: over 53% in May 1972 was found against 33% in 1971, while good rainfall in April and May was observed in both years.

Kaul & Al-Mufti (1974) observed even greater differences for three stands of *Artemisia herba-alba* near Rutba for the same years: a peak aerial biomass of 1796, 531 and 264 $kg.ha^{-1}$ in 1971, compared to 3939, 1327 and 568 $kg.ha^{-1}$ in 1972, respectively. Between-year variability, therefore, can be very high in the desert shrub rangelands. A badly needed quantitative assessment of this aspect of forage production can only be obtained by monitoring the production over a sequence of years in combination with rainfall recording at the observation sites.

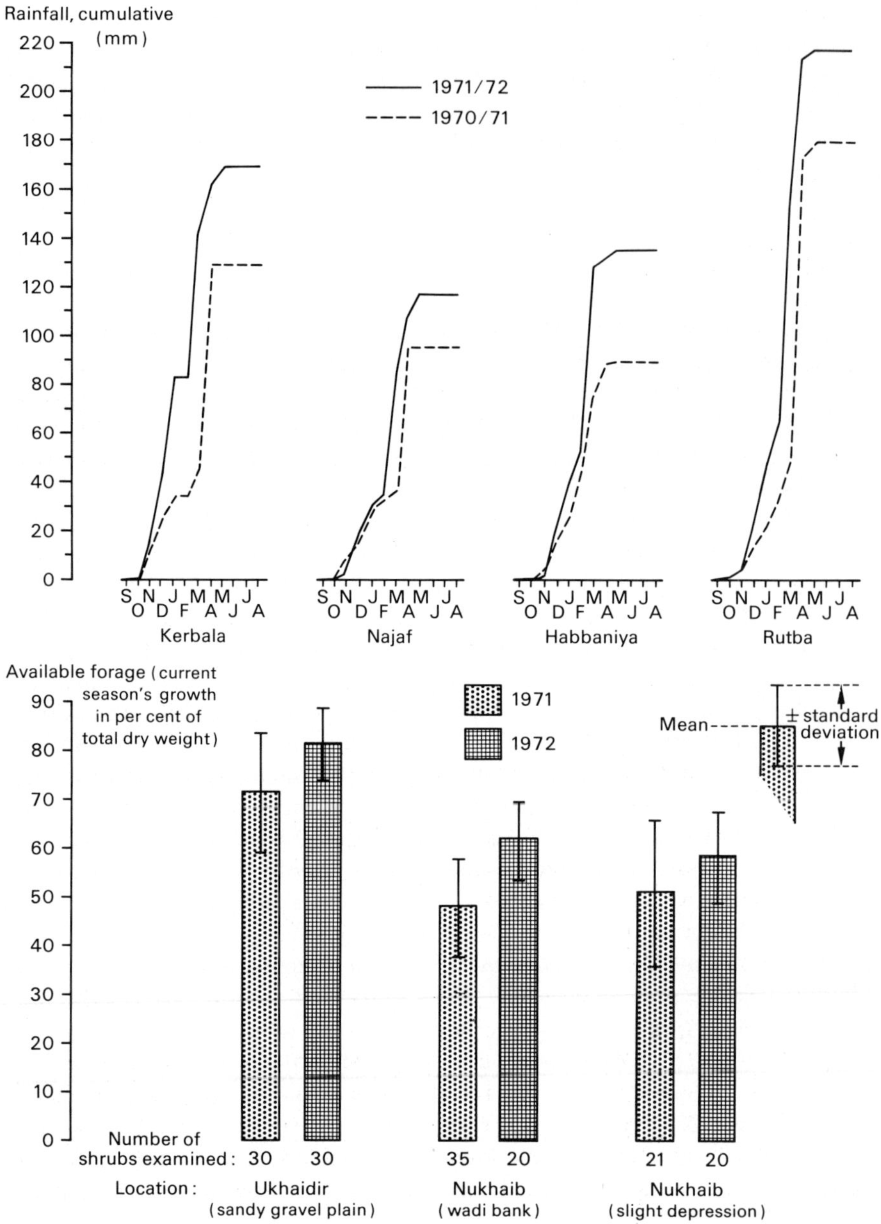

Figure VII–11. Forage production in three stands of *Haloxylon salicornicum* at peak aerial biomass in October for 1971 and 1972 and cumulative rainfall during the preceeding months in the four nearest stations.

VII.E.4. *Contribution of the annuals*

In considering the forage production of the desert rangelands the annuals' contribution is extremely important. Although no data have explicitly been collected on this aspect, some indication can be found in table VII–17 and seen from Fig. VII–12. They illustrate the tremendous variation, both in space and time. Annuals may produce over 2000 $g.m^{-1}$ fresh weight under favourable conditions. However, such figures should never be extrapolated to larger areas and are only of local significance. In the heavily overgrazed situation in Iraq (see VII.F.), a stage has been reached that in certain areas hardly any seed source seems to be left. Comparison of the data inside and outside the Khedr-el-Mai exclosure emphasizes this. Too early grazing, as often observed in the situation in Iraq, does not allow the plants to develop to flowering or seed setting stage. The effect becomes even more serious in areas where no perennials are present to protect at least part of the annuals against grazing.

Most literature on production in desert rangelands deals with one-time measurements, usually carried out at the time of supposed peak standing crop (le Houérou and other authors in Rodin (1972), Noy-Meir (1973). Orshan & Diskin (1968), however, reported on the 'annual march of the green plant material in a few typical desert stands', dominated by *Artemisia herba-alba, Zygophyllum dumosum, Anabasis articulata* and *Poterium spinosum*, from February 1963 to October 1964. The stands by and large exhibit the same 'march-pattern', an increase in dry green plant material from October to a peak in March/April (due

Table VII–17. Aerial biomass of some selected stands of annuals ($\bar{x}$-mean, s-standard deviation).

Location	Nature of the sample	Date of collection	Aerial biomass ($g.m^{-2}$)			
			Fresh		Dry	
			$\bar{x}$	s	$\bar{x}$	s
Khedr-el-Mai inside the exclosure	12 quadrats of 16 m^2	28 May 1971	–	–	35	29
Khedr-el-Mai inside the exclosure	40 quadrats of 0.25 m^2	3 March 1972	196	124	48	37
Khedr-el-Mai outside the exclosure	12 quadrats of 16 m^2	29 May 1971	–	–	1	1
Khedr-el-Mai outside the exclosure	40 quadrats of 0.25 m^2	3 March 1972	62	65	19	20
Near Samah in a shrub cover of *Rhanterium epapposum*	18 quadrats of 1 m^2	7 March 1972	503	356	104	66
Near Samah in a shrub cover of *Rhanterium epapposum* annuals extremely well developed	10 quadrats of 0.25 m^2	7 March 1972	2166	232	–	–

Figure VII–12. Importance of the seasonal contribution of the annuals' component, illustrated by three views of the same area between Khedr-el-Mai and Samah in the Southern Desert at respectively 17 December 1971, 4 March 1972 and 21 May 1972.

to the annuals' contribution), followed by a gradual drop. The results are of direct significance to the situation in Iraq. A better knowledge of the spatial and temporal aspects of forage production and thereby available forage is a prerequisite for grazing management in the desert shrub rangelands (see chapter XIII).

VII.F. Grazing pressure and grazing capacity – an overall assessment

VII.F.1. *Grazing pressure*

The total area of Iraq, without territorial waters, is 437 542 km^2 (Central Statistical Organization 1970). Out of this area approximately 241 000 km^2 is occupied by the Central and Lower Jezira and Western and Southern Deserts, some 55%. In table IV–14 the estimated total livestock numbers of Iraq were given for the census years; in table XII–5 the number of livestock estimated to be dependent on the forage supply of the desert rangelands is presented. With these figures a rough estimate can be made of the grazing pressure (table VII–18). The estimates

Table VII–18. Estimated grazing pressure for the whole of Iraq and for the desert rangelands, for the years for which census data or livestock estimates are available.

Year	Whole of Iraq*		Desert rangelands			
	animal unit.** ha^{-1}	ha. $a.u.^{-1}$	a.u. ha^{-1}	ha. $a.u.^{-1}$	sheep.ha^{-1}	ha.$sheep^{-1}$
1971	0.09	11.6	0.04	26.6	0.19	5.3
1965	0.13	7.8	0.08	13.2	0.38	2.6
1956	0.13	7.8	0.07	15.0	0.33	3.0
1950	0.10	10.1	0.06	17.0	0.29	3.4
1938	–	–	0.04	25.2	0.20	5.0

* Excluding 45 000 km^2 for actually cultivated area, inland water, build-up areas etc.
** a.u. = animal unit, for a definition see table IV–14.

for the whole of Iraq range from 7.8 ha per animal unit in 1956 and 1965 to 11.6 ha per animal unit in 1971. For the desert areas an estimated 2.6 ha per head of sheep in 1965 is the highest grazing pressure, against 5.3 ha per head of sheep in 1971. These figures, however, are of little significance unless they can be compared with forage production estimates.

VII.F.2. *Estimated total forage production*

For an estimate of total forage production the records continually kept during 3800 km reconnaissance survey over different tracks in the study area were re-examined and elaborated. The vegetation types, described in terms of dominant species, condition and use were grouped in four 'productivity' classes, including one class for bare or almost bare areas without any productivity of a visible perennial vegetation. The area under each of these classes was calculated from the proportion of the total 'intercept-distance' to the total distance traversed and the known total areas of the Central and Lower Jezira and Western and Southern Deserts. From table VII–19 it can be seen that almost 25% of the total desert area was found to be devoid of any visible perennial vegetation. Low productivity stands covered together 46%, moderate productivity stands almost 22% and high productivity stands only just over 7% of the total area. For each class a productivity figure under average rainfall was established. In view of its higher rainfall, the Jezira was treated separately. The productivity figures were based on the productivity figures measured for a number of stands in Iraq, mainly of the key shrub species (Thalen 1972, 1974 a, b; Kaul & Al-Mufti 1974), on measurements of aerial biomass and estimates of available forage in some 50 stands of shrub vegetation (Kaul & Thalen in UNESCO 1975), and on literature data from similar

Table VII–19. Estimated net aerial productivity in the desert shrub rangelands of Iraq for the perennial vegetation under average rainfall conditions in the early seventies.

Part of the desert rangelands	Vegetation productivity class	Net productivity of aerial dry matter ($kg.ha^{-1}. year^{-1}$)	Area (km^2)	Total net aerial dry matter produced ($kg.10^6. year^{-1}$)	
Central and	Completely or almost bare	0	9 070	0	
Lower Jezira	Low productivity stands	75	23 840	179	
Avge. annual	Moderate productivity stands	375	4 620	173	
rainfall	High productivity stands	750	1 470	110	
125–250 mm	All classes Jezira		39 000		462
Western Desert and	Completely or almost bare	0	50 750	0	
Southern Desert	Low productivity stands	50	87 130	436	
Avge. annual	Moderate productivity stands	250	47 990	1200	
rainfall	High productivity stands	500	16 130	807	
75–125 mm	All classes W. and S. Deserts		202 000		2443
Total desert rangelands	All classes		241 000		2905

vegetation types in comparable situations in South West Asia and North Africa, *viz.* Kernick (1966) for Kuwait, van der Veen (1967) for Syria, Floret (1971) for Tunesia, Floret & Pontanier (1974) for Tunisia, and examples of productivity estimates over larger areas, based on sample measurements, in the Chellala Region of Algeria, Central and Southern Tunisia, Eastern Morocco, and the Hodna Basin in Algeria, quoted in le Houérou (in press). Table VII–19 shows that an estimated total production of 2905 . 10^6 kg or almost three million tons was found as the total forage yearly produced by the desert shrub vegetation under average rainfall conditions in the early seventies. This means an overall average of approximately 120 $kg.ha^{-1}$ $year^{-1}$ of dry matter.

In view of the assumptions incorporated in this figure, a brief discussion on its reliability should be given. The figure was calculated from the formula

$$F_{tot} = \frac{A}{D} \sum_{i=1}^{n} d_i p_i$$

in which:

F_{tot} = total net aerial production of the perennial vegetation in one year ($kg.year^{-1}$)
A = total area of the desert rangelands (km^2)
D = total distance traversed (km)
d_i = distance covered in productivity class (i) (km)
p_i = net aerial productivity of the perennial vegetation of productivity class (i) ($kg.km^{-2}. year^{-1}$)

Three productivity classes were distinguished in the Jezira and three in the Western and Southern Deserts while in addition a class for bare areas was used where p = 0, which brings the total to n = 7 productivity classes.

A – the total area of the desert rangelands, was estimated to be 241 000 km^2. The estimate is the average value of repeated dot grid countings on three different maps of the country. Results were very close and the figure at highest includes a two or three per cent error.

D – the total distance traversed, was 3799 km subdivided in 26 one-day tracks, usually between 100 and 200 km, but with 12 and 300 as extremes. First the question arises whether these tracks may be considered representative transects. During the survey it was observed that tracks tend to follow the lower situated areas with the relatively better vegetation. They usually do not traverse stands but skirt them, rather than following the bare areas, which are often covered by stones and loose sand. If there is a bias on this point, it causes a too high total forage estimate. Another point is whether the 3800 km constitute a representative sample. Assuming a view of 750 metres from either side of the car, in the flat or undulating terrain a low estimate*, then 2.37% of the area was sampled and included in the descriptions. This figure should be adequate for the rough estimate here attempted.

d_i – The distance covered in each productivity class inevitably includes some subjective decisions. However, the high productivity stands were clearly indicated in the records, as well as the bare areas. In case of doubt for the low productivity or moderate productivity stands, the stands were put in the moderate productivity class. Bias from subjective decisions, therefore, results in a higher total production figure.

p_i – The net aerial productivity figures for the classes determine to a great extent the final total figure. The high productivity stands were given a two times higher productivity than the moderate ones. In some cases a productivity higher than the 500 or 750 $kg.ha.^{-1}.year^{-1}$ was found, but such cases were measured especially because they represented extremely good conditions, e.g. wadibeds or depressions after good rainfall (Kaul & Al-Mufti 1974, Thalen 1974 a, b) and are not representative for the average productivity of good stands. Kernick (1966) gives for Kuwait figures of 100–335 $kg.ha.^{-1}$ for stands of *Rhanterium epapposum* and *Haloxylon salicornicum*. Floret (1971) found for an *Artemisia herba-alba* vegetation in Tunisia after 300 mm rainfall 200 $kg.ha.^{-1}$ dry matter in the current season's growth and 400 $kg.ha.^{-1}$ in a stand of *Rhanterium suavolens*. For the *aerial biomass* in the desert rangelands of Iraq an average of 472 $kg.ha.^{-1}$ was found for 9 stands where *Artemisia herba-alba* was dominant (range 43–2228), an average of 435 $kg.ha.^{-1}$ for 26 stands where *Haloxylon salicornicum* was dominant (range 73–1313), and an average of 552 $kg.ha.^{-1}$ for six stands where *Rhanterium epapposum* was dominant (range 74–1274) (Kaul & Thalen in UNESCO 1975). Although not direct productivity measurements, these figures show that the productivity figures adopted may be at the higher side, certainly for the poor and moderate productivity classes. These classes occupy by far the largest area but for them relatively little productivity reference data were available.

From the above discussion it may be clear that the figure of 2905 . 10^6 kg net aerial production per year, although unlikely to be an accurate estimate, certainly must be of the right magnitude and, if biased, at the higher side.

VII.F.3. *Supply and demand of forage*

How does the above calculated supply compare with the demand? The 906 000 animal units given in table XII–5 as the estimated number of livestock dependent on the forage supply of the desert rangelands in 1971 represent a forage demand

* See Heady (1972) for a similar survey in Saudi Arabia.

of the equivalent of 4530 000 head of sheep. Per head of sheep this means a forage supply from the perennial vegetation of 641 kg dry matter per year or 1.76 kg per day, assuming that all dry matter produced is available to the animals. Figures on forage requirement in terms of dry matter without an indication of the forage quality vary in the literature. Stoddart *et al.* (1975) give figures of the National Research Council, ranging from 1.2 kg per day for a body weight of 45 kg during early gestation to 2.5 kg per day for a body weight of 64 kg during early lactation. Le Houérou & Hoste (1977) used for their rangeland production and carrying capacity review in relation to rainfall in the Mediterranean Basin the following conversion factors: One mature sheep (40 kg) = 300 Feed units per year, and 1 kg of dry matter = 0.33 feed unit. This means a requirement of 900 kg dry matter per head of sheep per year or about 2.5 kg per day. Other authors give figures of the same magnitude (Brown 1954, USDA 1976) and a figure of 2 $kg.day^{-1}$ per head of sheep had been used as an approximation for grazing capacity estimates (Kernick 1966, Baig 1977). Adopting this figure, the requirement for all livestock in the desert rangelands amounts to $3.31 .10^9$ $kg.year^{-1}$ for the 1971 situation. On the same basis a demand figure for 1965 can be worked out: $6.66 . 10^9$ $kg.year^{-1}$. The livestock figures, however, from which these demands were calculated are based on census data and certain assumptions for the percentage of animals feeding on the desert rangelands (see chapter XII). For the sake of comparison a second set of demand figures was worked out, based on very low estimates of the percentage of animals feeding on the desert ranges (table VII–20). The figures thus arrived at show (table VII–20)

Table VII–20. Calculated estimates of supply and demand of forage for the desert rangelands of Iraq, comprising the Central and Lower Jezira and Western and Southern Deserts.

Calculated estimates of supply and demand	Dry matter ($kg.10^9.year^{-1}$)
Supply of forage by the perennial (mainly shrub) vegetation in the early seventies	2.91
Demand of forage in 1971 based on livestock estimates as given in chapter XII*	3.31
Demand of forage in 1971 based on very low livestock estimates**	2.52
Demand of forage in 1965 based on livestock estimates as given in chapter XII*	6.66
Demand of forage in 1965 based on very low livestock estimates**	5.09

* Total livestock numbers from table IV–14 and based on the assumption that in Iraq camels for 90%, sheep for 67%, goats for 33% and donkeys for 10% take their forage from the desert rangelands.

** Total livestock numbers from table IV–14 and based on the assumption that in Iraq camels for 75%, sheep for 50%, goats for 25% and donkeys for 10% take their forage from the desert rangelands.

that supply and demand in the early seventies are of the same magnitude. Under average or good rainfall the ephemeral vegetation may supply all forage required during two or three months or so allowing rest for that period to the shrub vegetation. In such years the shrub vegetation may recover from earlier damage. In a poor rainfall year, however, the emphemerals' contribution is negligible and the total current year's production, and most likely more than that, is needed to fill the food-requirements of the livestock. On the basis of the available data with all uncertain factors, this is the most optimistic picture that can be given for the supply-demand balance in the early seventies.

The drop in total sheep numbers for Iraq from 11 040 000 in 1965 to 5 582 000 in 1971 is an unlikely one (table IV–14). The 1971 figure most likely is at the lower side. The 5.6 million sheep were actually counted by the enumerators or mentioned to them by the owners. No owner will give too high a figure for fear of taxation. An overestimate is thereby ruled out. An underestimate is likely, due to (i) too low figures given by the owners for fear of taxation, (ii) the overlooking of grazing areas by the enumerators, and (iii) livestock coming in from neighbouring countries. An actual higher demand for 1971 combined with lower production of forage than suggested in table VII–20, for reasons mentioned earlier, would show a far more unfavourable balance, which in fact may be the true one.

VII.F.4. *Selected evidence for overgrazing*

Table VII-20 shows for 1965 a required amount of forage twice as high as for 1971. Even if this requirement was estimated too high, the desert vegetation of Iraq at that time must have been able to provide far more forage than the amount calculated for 1971. Degradation of the production capacity of the food resources of the desert as a result of overexploitation is the obvious explanation. The desert rangelands once must have had a far higher food production capacity. The following observations support this view:

— During the reconnaissance survey, deteriorated stands were the most common ones, encountered all over the desert. Large aread had to be ranked 'poor' for their present production capacity, which once must have been far better. Kaul & Thalen in UNESCO (1975) listed over seventy sites fit for direct improvement, where such situations were observed.

— In 1938 the calculated forage demand from the desert rangelands was of the same magnitude as calculated for 1971. The vegetation was able to produce the required amounts in view of the increase in number of livestock in the years following. However, in 1971 the whole desert could be utilized being accessible because of a large number of newly drilled wells and introduction of tankers and other means of transport. In 1938 only part of the desert could be used at all seasons because of lack of water supply in may areas, at least for part of the year. In other words, it is likely that only part of the desert rangelands around 1938 produced the same amount of forage as the whole desert by 1971.

—The intensity of decreased forage production under increased grazing was clearly observed and quantified in a few cases. An example of an *Artemisia herba-alba* covertype is given in table VII–21 and Fig. VII–13. In mid-June 1971 the dry matter production ratio for very heavy, heavy and slightly grazed stands was found to be approximately 1:2:10. Kaul & Al–Mufti (1974) investigating the effect of grazing on very good stands of *Artemisia* suggest an aerial biomass ratio of 8:2:1 for the species under light, heavy and very heavy grazing. Le Houérou (1972b) reports for Algeria that any specific plant community produced on average 4.5 times more after two to five years of protection than the same community under heavy permanent grazing. Extreme figures reach eight to ten times for continually grazed and ungrazed stands of the same community. Table VII–20 and the preceding discussion, although because of lack of accurate data in simplified and sometimes only qualitative terms, provide the evidence for overutilization by livestock of the desert shrub rangelands over the last 25 years of so. The data and discussions presented in the chapters VIII and IX will show that overgrazing is not the only cause of vegetation degradation.

Table VII–21. Comparison of height, density, crown cover, aerial biomass and current season's growth in three *Artemisia herba-alba* stands under different grazing intensities. Mid-June 1971, near Rutba, Western Desert (from Thalen 1972, see Fig. VII–13).

	Very heavily grazed		Heavily grazed		Slightly grazed (protected)	
Number of analysed 1 m^2 quadrats	64		64		32	
Height (cm) of *A. herba-alba*	10		26		35	
Density (no.m^{-2})						
A. herba-alba	2.2		3.0		3.2	
other perennials	0.7		0.0		0.1	
Crown cover (%)	2*		7*		33	
Aerial biomass (kg.ha^{-1})	Fresh	Dry	Fresh	Dry	Fresh	Dry
A. herba-alba	191	94	586	308	3150	1896
other perennials	56	23	–	–	negl.**	negl.**
Current season's growth in mid-June (kg.ha^{-1})	–	80	–	150	1610	795

* estimated by visual observation.

** negl. – neglectable.

Figure VII–13. View of two physiographically comparative areas, slight depressions with a silty soil, near Rutba in the Western Desert. Both areas are under a cover of mainly *Artemisia herba-alba*, (above) slightly grazed, protected, and (below) very heavily grazed. Quantitative characterization of the stands in table VII–21.

VII.F.5. *Comparison of Iraq with other areas*

Improved grazing can be achieved by a management programme aimed at a grazing pressure not exceeding the grazing capacity.

Grazing capacity refers to the capacity of the land to produce forage over a longer period of time. It may be defined as:

'The number of animals that produces the greatest return without damage to the physical resources and in concert with other values received from the land.'

Optimum grazing capacity suggests the most profitable stocking rate (Heady 1975).

Grazing capacity is often indicated as carrying capacity. However, as rightly pointed out in Heady (1975): carrying capacity includes more than a maximum number of animals. It relates to all possible resource values of the rangeland and should be used in relation to particular management objectives *e.g.* recreational carrying capacity.

The establishment of a data base to work out grazing capacities for the various parts of the desert shrub rangelands of Iraq will be discussed in chapter XIII. At present such a data base is lacking, but a comparison can be made between the current situation in Iraq and calculated grazing capacities elsewhere.

Le Houérou (1972b) produced a curve outlining the actual and desirable grazing pressure against average rainfall for North Africa. For the 100–250 mm rainfall areas a desired grazing pressure of more than ten hectares to approximately four hectares per head of sheep are given. Van der Veen (1967) reported on a grazing trail with sheep in the 'Syrian Steppe'. Comparing the effect of three stocking rates (9.0, 6.0 and 4.5 ha per sheep) in a wadi with a vegetation comparable to the types found in the Iraqi Jezira, he came to the preliminary conclusion that 4.5 ha per adult ewe on the basis of yearround grazing with some supplementary feed for extreme drought or cold periods, does not lead to serious depletion of grazeable herbage. It should be kept in mind that this experiment was carried out under forage conditions better than the average conditions in the desert areas of Iraq. Kernick (1966) estimates for *Rhanterium epapposum* range in Kuwait under a seasonal rainfall of 50–60 mm a stocking rate of 5 ha per sheep from November to the end of February and 1.75 ha per sheep from the end of February to the end of June.

Grazing capacities for Iraq will be of the same magnitude. Kaul & Al-Mufti (1974) arrived at 0.9 ha per sheep for an excellent *Artemisia herba-alba* range to 5–7.5 ha per sheep for a poor range. This, however, relates to an improved situation as a result of several years of protection against overgrazing. In working out a grazing capacity, the difference between 'actual' (= under the current conditions) and 'potential' (= under improved management) should be made. An overall figure for the desert rangelands of Iraq may well be around 6–8 ha per sheep for the 'actual', against twice that many for the 'potential' situation.

An other approach to work out the overall potential productivity in relation to the actual situation is by comparing productivity in Iraq with productivity in isoclimatic regions. It has been shown that a good proportion of variation in productivity in the ecosystems under discussion can be directly related to precipitation (Noy-Meir 1973; le Houérou & Hoste 1977). The data on rangeland production and annual rainfall relations in the Mediterranean Basin was recently reviewed (le Houérou & Hoste 1977). The authors found that each millimetre of rainfall produced four kilograms of above-ground phytomass or two kilograms of consumable dry matter per hectare per year. Novikoff *et al.* (1975) found an average of 2 $kg.ha^{-1}.year^{-1}$ per millimetre rain in southern Tunisia on sandy soils under a rainfall of 15–182 mm per year.

For the desert shrub rangelands of Iraq a figure of approximately 120 $kg.ha^{-1}.year^{-1}$ of dry matter production was calculated for the perennial vegetation only. This means about one kilogram, or somewhat less, of dry matter per millimetre rainfall, not including the contribution of annuals and geophytes. The contribution of this ephemeral vegetation can not be estimated without further data, but it is unlikely that this component could raise the observed overall one kilogram per millimetre rain to the four kilograms found by le Houérou & Hoste (1977) for the Mediterranean Basin to which in this respect the desert rangelands of Iraq also belong. It is another indication of the poor actual condition of the desert shrub rangelands of Iraq.

CHAPTER VIII

FUEL, THE EFFECT OF CUTTING

VIII. FUEL, THE EFFECT OF CUTTING

VIII.A. Introduction

Cutting and uprooting shrubs for fuel is a common practice, reported from arid and semi-arid lands all over the world. It has been mentioned by several authors as an important cause of vegetation deterioration and land degradation, in particular from a number of countries in South West Asia and North Africa (cf. Warren & Maizels 1977).

Freitag (1971), in the first overall descriptions based on personal observations of the vegetation of Afghanistan, mentioned progressive destruction of tree-and shrub

Figure VIII-1. Site, where recently charcoal was burned (near Shitathe, Western Desert). Note car tracks right foreground and remaining low hummocks where shrubs were cut in the middle part of the photo.

vegetation types for firewood and charcoal production. Near towns and villages all accessible larger woody species have often been destroyed. Species of *Artemisia, Astragalus, Acantholimon, Cousinia* and *Ephedra* and in semi-desert areas also the herb *Alhagi sparsifolium*, are now being collected and, especially in autumn, in large quantities transported to the settlements, while in the desert areas the nomads' need for fuel has its effects.

Van der Veen (1964) in a discussion on the process in Syria, remarks that: 'The virtual disappearance of perennial sub-shrubs in the neighbourhood of watering points (wells) and along migration tracks is partly due to overgrazing but for a large part due to uprooting'. He emphasizes the danger of cutting in combination with heavy grazing, whereby regeneration of a once destroyed shrub vegetation becomes difficult.

Le Houérou has discussed the process for North African countries in several papers. In le Houérou (1969) the author refers to his earlier work in Morocco, Algeria, Tunesia and Lybia and concludes: 'The removal of wood for fuel has been calculated at one kilogram per person per day; that is, an average of 0.5 to 1 hectare of woody vegetation is destroyed per person each year'.

More or less similar examples as the ones quoted above can be given for other countries and regions, e.g. Indo-Pakistan (Kaul 1970), Central Asia (Petrov 1970), Saudi Arabia (Heady 1972) and Kuwait (Kernick 1966).

The destructive process of uprooting and cutting shrubs for fuel in Iraq has been mentioned by almost all authors dealing with the vegetation of the desert areas in one way or another (Gillett 1948 a, b; Long 1956; Schwan 1957; Anderson 1958; Guest 1966; Winkworth 1969; Kaul & Thalen 1971a; Thalen & Al-Mufti 1973). More than general remarks or a rough qualitative picture are not given. In the following parts an attempt is made to present a somewhat more detailed account (VIII.B), incorporating results of cutting experiments (VIII.C).

VIII.B. Fuel cutting in the desert shrub rangelands

VIII.B.1. *Use of the fuel*

Wherever one travels in the Iraqi desert areas, women, animals (usually donkeys) and trucks transporting loads of shrubs for firewood may be encountered. Fire is not only necessary for cooking a meal or preparing a kettle of tea, but also for heating during the cold nights and 24 hours a day in the short but extremely cold winter period. Firewood from the desert thus serves a variety of people in and around the desert areas. Three categories may be distinguished:

- Woody material is collected and locally processed into charcoal (Fig. VIII–1), an important fuel for the preparation of some popular Iraqi meat dishes, which require roasting of the meat. A 'kabab' or 'tikèh' consumed in Baghdad or another large city, may have been prepared on the heat of desert shrubs.

Figure VIII-2. The daily picture of women with donkeys transporting fuel to their homes in the desert villages and towns. In his case mainly *Haloxylon salicornicum* – shrubs had been collected and were brought to Nukhaib in the Western Desert.

- Firewood is collected by people of all the settlements bordering the desert areas. Near the major villages in the desert, As Salman, Busaiya, Zubair, Rutba, Shabicha, etc., not a single shrub has survived this process. When the distance to the woody vegetation becomes too far for the women to walk and carry the load, donkeys are used and even occasionally trucks, tractors, etc. (Fig. VIII–3).
- In addition to forage and water Bedouins consider amongst others the firewood supply as a factor in choosing their camping site. However, water ranks first and the result is that the Bedouin women may have to walk over a kilometre, or ride on their donkeys even much farther to transport the fuel to the campsite (Fig. VIII–4). The only alternative fuel is dry cameldung. This is used as far as the stock lasts, but for relatively few Bedouin families it is the major source of fuel. Cameldung in most cases is only a welcome addition to the shrubs as a source of fuel.

It is hard to arrive at a weight estimate of the amount of firewood used by a Bedouin family over a certain period of time. There is a strong seasonal influence. In the period from November to February, firewood is not only used for cooking, but also for heating. Thicker branches will be more effective than the thinner ones, etc. The amounts mentioned by Bedouins in interviews varied widely.

VIII.B.2. *Extent of the cutting*

It is difficult and at the present state of knowledge impossible to make an *accurate* estimate of the area yearly denuded by uprooting shrubs for fuel. For the three groups mentioned above an estimate of the overall daily firewood supply with its seasonal variations would be necessary. Such estimates are not available. In the following a first *rough* estimate is given, based on field observations and literature data of comparable situations.

During a several days' stay at Nukhaib in February 1972, it was observed how six to eight donkeyloads of shrubs to be used as firewood were brought into the village daily (Fig. VIII–2). In addition firewood was occasionally brought in by cars (Fig. VIII–3), and by women on foot. An estimate of 500 kg total daily is at the lower side. During cutting experiments, roughly about 500 kg of *Haloxylon salicornicum* shrubs on a hectare were found. For the fuel supply of a small village as Nukhaib alone (appr. 100 inhabitants) at least one hectare of desert shrub vegetation is uprooted daily. In a year this adds up to an area with a radius of over one kilometre around the village where all larger shrubs have disappeared.

On one occasion, when discussing the firewood issue with a Bedouin, the bundle of firewood just collected and transported to the tent by one of his wives was shown to us. It weighed about 60 kg (low estimate) and was mentioned, in

Figure VIII-3. Trucks are occasionally used to transport shrubs collected in the desert areas for fuel over large distances.

Figure VIII-4. The most extensive cutting of shrubs for fuel is done by the Bedouins, who usually have no alternatives. Near every tent piles of shrubs can be seen, which sometimes had to be carried over long distances.

February, to be a sufficient supply for about two days for the small family. An estimate of 25 kg per family per day for the cold season seems, therefore, certainly justified.

Van der Veen (1964) made a calculation for Syria, based on some realistic estimates. On the assumption that about five *Haloxylon* shrubs are needed to boil an average kettle of tea, and that approximately 40 000 Bedouin families prepare one kettle of tea a day each in winter and spring, the author figures that about 40 000 000 *Haloxylon* shrubs are yearly burnt for tea only, equalling a thousand hectares of dense shrub vegetation. For heating, bread-baking etc. the figure, he suggests, may be multiplied by ten.

Le Houérou (1969) in his study on the Tunesian steppes, gives a figure of 5 kg per day per family as an average for the whole year and refers to work carried out by FAO in the Mediterranean, giving a figure of 1 kg per person per day. Later studies in Tunesia have shown that 1.5–2 kg per day per person as a minimum is a more realistic figure (Novikoff 1975). Le Houérou, on the basis of 1 kg per person per day, calculates that some 300 000 ha are yearly denuded for the total steppe area of Tunesia, assuming that no regeneration takes place.

In relation to the above figures, an estimate of 2 kg fuel taken per person per day seems a realistic figure for the desert shrub rangelands of Iraq. Assuming a Bedouin population of 150 000 (see table IV–15) and using the figure of 500 $kg.ha^{-1}$ firewood from shrubs (see table VIII–1), in one year about 200 000 ha are being cut in the desert shrub rangelands by the Bedouin population alone. A stand with an aerial biomass of 500 $kg.ha^{-1}$ would be qualified moderate to good in the terms given in the previous chapter. The total area assessed as good in the study area was approximately 1 750 000 ha, and assessed as moderate approximately 5 250 000 ha. Considering these figures the conclusion can be drawn that in a period of less than ten years an area equal to the area presently occupied by all good stands is being cut and that about 35 years would be needed to clear off all good and moderate stands together. The shrubs cut for fuel need some five to ten years to develop from germination to cuttable height (chapter VI; van der Veen 1964). If regeneration had not been hampered by overgrazing and no increased erosion had resulted from the denudation of the soil, it would be well possible for the desert shrub vegetation to withstand the impact of cutting. Cutting and uprooting itself, therefore, is not a serious problem, except near permanent settlements and for some selected species preferred as fuel (see VIII.B.3). However, in combination with the other utilization processes taking place, overgrazing (see chapter VII) and cultivation (see chapter IX), fuel cutting often leads to a total denudation as can be seen all over the Southern and Western Deserts and Jezira. This aspect of combined effects of utilization is separately discussed in chapter XI. In this context cutting and uprooting is a serious problem.

VIII.B.3. *The species involved*

Species of the Iraqi deserts suitable as firewood may be grouped in decreasing order of usefulness: (i) trees, (ii) large shrubs, (iii) thick branched low shrubs, and (iv) thin branched shrubs.

(i) Trees are almost absent in the Southern Desert, Western Desert and Jezira. The places where trees are found are usually known to people inhabiting these areas or frequently travelling them. A small clump of *Acacia iraquensis* is found near Shabicha in the Southern Desert. The species is for Iraq only known from this location. For Kuwait a lone *Acacia* tree of the same species was found (Firmin & Kernick 1964). There is no evidence that this tree has been more widely distributed and common in earlier times. The only (small) 'tree' species rather frequently encountered nowadays are *Zizyphus nummularia* and *Tamarix* sp. Guest (1966) discussed the '*Zizyphetum nummulariae*'. His remarks still hold more than ten years later. It is likely that the absence of *Z. nummularia* in the Jezira is the result of the pressure of a much greater Bedouin population over a longer period compared to the Western and Southern Deserts. Two factors are mentioned why the species has managed to survive locally:

- the thorny and bushy nature of *Z. nummularia* and its common associates, *e.g. Lycium depressum*,
- the widespread popular belief that *Z. nummularia* is sacred or haunted. This places the *Zizyphus* trees of the desert in the same light as the few sacred oaks in the foothill zone.

The species is mostly found in rather remote places at sites where runoff water comes down or accumulates, along the border of depressions and along watercourses. Well known thickets are found in the Busaiya area. About 37 km north of

Figure VIII-5. Rather dense stand of *Zizyphus nummularia* at Bir "Ruahan" between Samah and As Salman.

Busaiya on the track to Nasiriya, wadi 'Sidr', where *Zizyphus* is still found abundantly, is named after the species. Another well known stand of *Zizyphus*, mixed with *Lycium depressum* is found at 'bir Ruahan', about halfway the Samah – As Salman track (Fig. VIII–5). The species was regularly encountered during the survey trips of 1970–1973. Most observations, however, pertain to scattered single trees or groups of a few trees. The species was found deep in the south along the Ansab - Samah track about 25 km north of Ansab in a depression. The locations mentioned and the one shown in Fig. VIII–5 are presently the most extensive and dense stands of 'trees' found in the Iraqi desert areas. *Zizyphus nummularia* may have been a substantial firewood producer once, its resource value in this respect is presently reduced to almost zero. *Tamarix* spp. are mainly found as components of riverine open forest in the Mesopotamian Plain, but also in low open stands in saline depressions in the desert (e.g. Abu Dibbis depression, Shitatha oasis). *Tamarix aphylla* is found extensively around Zubair. The trees were planted as wind breaks and irrigated together with the small plots where vegetables were grown. By the time these plots were abandoned because of increasing salinity of the ground (irrigation-) water, the trees were well established with deep roots. Guest (1966) mentioned a plant less than two metres tall with roots penetrating over five metres down. Eucalypts (*Eucalyptus* spp.) have been introduced and grow locally near wells in desert villages under ample water supply from these wells. At such places other tree species were also encountered, but they are of negligible value as firewood producers.

(ii) The above observations show that firewood in the Iraqi deserts has to be provided by shrubs. Two *large shrub species* should be mentioned in this respect. *Haloxylon ammodendron* (syn. *H. persicum*) and *Calligonum comosum*. Of these, *H. ammodendron* has been collected so intensively that the species has almost disappeared in Iraq. Guest (1966) lists the *Haloxyletum ammodendri* first in his account on the associations of the desert and found it quite common.

> 'This community has not previously been described from Iraq – no doubt because it is only found in a few remote and extremely sandy desert places difficult of access: also because it has largely been destroyed in the more sandy places where it was formerly to be found in abundance.
> In Iraq we have found this community at various points (Ichrishi, Shagra, Umaighir, W. of Darraji) along the Euphrates sand-belt, which varies in width from about 5–25 km. and runs for some 320 km. in a north-westerly direction from the ridge of Makhazuma, some 110 km. S.W. of Basra, to about 60 km S.S.E. of Najaf, in the vicinity of Shinafiya. There seems little doubt that the greater part of this belt of dunes was once an open Haloxylon thicket, though the treelets have now been eradicated from the more accessible places by the Baduin and the fuel gatherers of the delta. Another locality where we have found vestiges of this community is in the Lower Jazira among the scattered sand dunes along the lower reaches of the Wadi Thirthar – here, in 1954 we actually found charcoal burners digging up the trees, root and all: also in the sand hills immediately to the south of Millaih Thirthar. It has been reported from the small sand dune area at the foot of the Jabal Hamrin, between the Tigris and Adhaim rivers and north-east of Samarra, but we have not yet seen any specimens from that locality.'

About twenty years later hardly any specimens of the species were found. Bedouins ranked the species highest on the list of desirable firewood species, but they always added that the species was not any more found. At the places where in 1971–1973 the Euphrates sand dune belt was studied, at the crossings from Tel Lachm and Ur to Busaiya, only *Calligonum comosum* was seen. In wadi Ubayidh, a few kilometres east of Athaman, a few sprouting stumps were found and Kaul (pers. comm.) encountered the species in the Lower Jezira at a few places. The role of *Haloxylon ammodendron* as a firewood producer for the population of the desert seems passed. The common associate of the species in sandy areas, *Calligonum comosum*, is still growing estensively in the Euphrates sand dune belt, but this species is also extensively harvested locally (Fig. VIII–6) and may in due time follow the fate of *Haloxylon ammodendron* if no counteraction is taken.

(iii) Low shrubby species are many in the desert rangelands, as shown before. Their selection as firewood mainly depends on (a) whether they are easy to obtain, i.e. widely distributed and easy to handle (no spines or thorns); (b) whether they form thick, woody, branches which last long in the fire. *Haloxylon salicornicum* and *Rhanterium epapposum* both are very suitable, especially the first. Zohary (1940) in 1932 and Gillett (1948a) fifteen years later observed 'droves of donkeys bringing *Rhanterium epapposum* into Basra for sale as fuel'. Given the choice,

Figure VIII-6. Stand of *Calligonum comosum,* in the sand dune belt between Ur and Busaiya in the Southern Desert, where shrubs have been cut for fuel.

Bedouins prefer *Haloxylon salicornicum* for fuel. Other species are occasionally collected, such as *Haloxylon articulatum, Ephedra alata, Halogeton alopecuroïdes* and *Nitraria retusa*, but are certainly not preferred and also less common.
(iv) As least usefull, but nevertheless sometimes collected, group of species for fuel, may be listed: *Artemisia herba-alba, Achillea* spp., *Salsola* spp. and *Artemisia scoparia*. All of these have been collected, some of them even by donkeyloads, to serve as fuel. In these cases, however, thick branched woody species were completely absent in the wide surroundings and the Bedouins was left no choice. Grazing

in such areas is usually restricted to spring grazing on an open, ephemeral vegetation.

VIII.C. Response to experimental cutting

VIII.C.1. *Cutting of 'Haloxylon salicornicum'*

Areas where shrubs have been uprooted for fuel are usually easily recognised and very common all over the Iraqi desert. Where little or no hummock formation around the plants had taken place before cutting, a stump of a single stem or the upper few centimetres of the root can often be observed (Fig. VIII–7). At places with a loose topsoil, accumulated around the plants, often the remains of hummocks can be seen with a few branches still sticking out.

Experimental work was carried out to investigate the influence of different cutting treatments on the amount of regrowth, especially of *Haloxylon salicornicum*, the species presently most widely collected for fuel. These experiments had, by and large, to be conducted without control over the situation in the field

Figure VIII-7. Remnants of a once dense stand of well developed *Haloxylon salicornicum* plants, cut for fuel. The hummocks, unprotected by the plant cover, have disappeared through wind erosion (Southern Desert 1972).

Figure VIII-8. Denuded area, resulting from three repeated cuttings of the shrubs, *Haloxylon salicornicum* and a few *Ephedra alata,* in just over a year (area IV from table VIII-1). Stick at the foreground and person at the background are two cornerpoints of the cut area to the right.

and considerable difficulties were faced in repeating treatments and observations at the desired intervals.

Table VIII-1 shows the results on weight basis of a cutting experiment over 20 months in a good stand of *Haloxylon salicornicum* near Wadi Ubayidh between Nukhaib and Faidhat al Jeez. By the end of October 1971 all plants in an area of 32 x 64 m^2 were cut at surface level and weighed individually, after the

Table VIII–1. Effect of repeated cutting with various intervals on a stand of *Haloxylon salicornicum* near Nukhaib (Western Desert). The areas I, II, III and IV were adjacent areas of 512 m^2 each.

Date of cutting and recording	Aerial biomass (fresh) in per cent of aerial biomass at first cutting. In brackets the recorded fresh weight in kg.			
	Area I	Area II	Area III	Area IV
Oct. 1971	100 (24.4)	100 (20.3)	100 (32.7)	100 (26.9)
June 1972	–	44 (8.9)	–	46 (12.5)
Oct. 1972	–	–	75*(24.4)	10 (2.7)
May 1973	32 (7.9)	17 (3.5)	5 (1.7)	1 (0.4)

* The relatively high aerial biomass and weight percentage of regrowth in area III is biased by the influence of a few very large plants. Disregarding these five plants, would bring down the percentage to 46.8. For the effect of plant size on regrowth see table VIII–2.

Figure VIII-9. Regrowth of *Haloxylon salicornicum* in May 1973 after cutting in October 1971 (area I from table VIII-1).

height and crown diameters had been measured and the exact position mapped. The area was then subdivided in four areas (I, II, III and IV) of 512 m^2, each of which was subjected to a different cutting treatment carried out at the end of June and the end of October 1972. The final response to the different cutting regimes was recorded at 26 May 1973. Three times cutting in just over a year (area IV) reduced the crowncover to almost nil (Fig. VIII–8).

Area 1, only cut in October 1971, had extremely well recovered in May 1973 (Fig.VIII–9). In Fig. VIII–10 based on actual mapping in the field, the reduction in crown cover for the two extremes, area I and area IV, is illustrated. The twice cut areas, II and III, hold an intermediate position.

It was experienced that large plants had a much higher resistance against (repeated) cutting than plants of smaller size. This is elaborated in table VIII–2 and Fig. VIII–11. These data are based on observations in 15 quadrats of

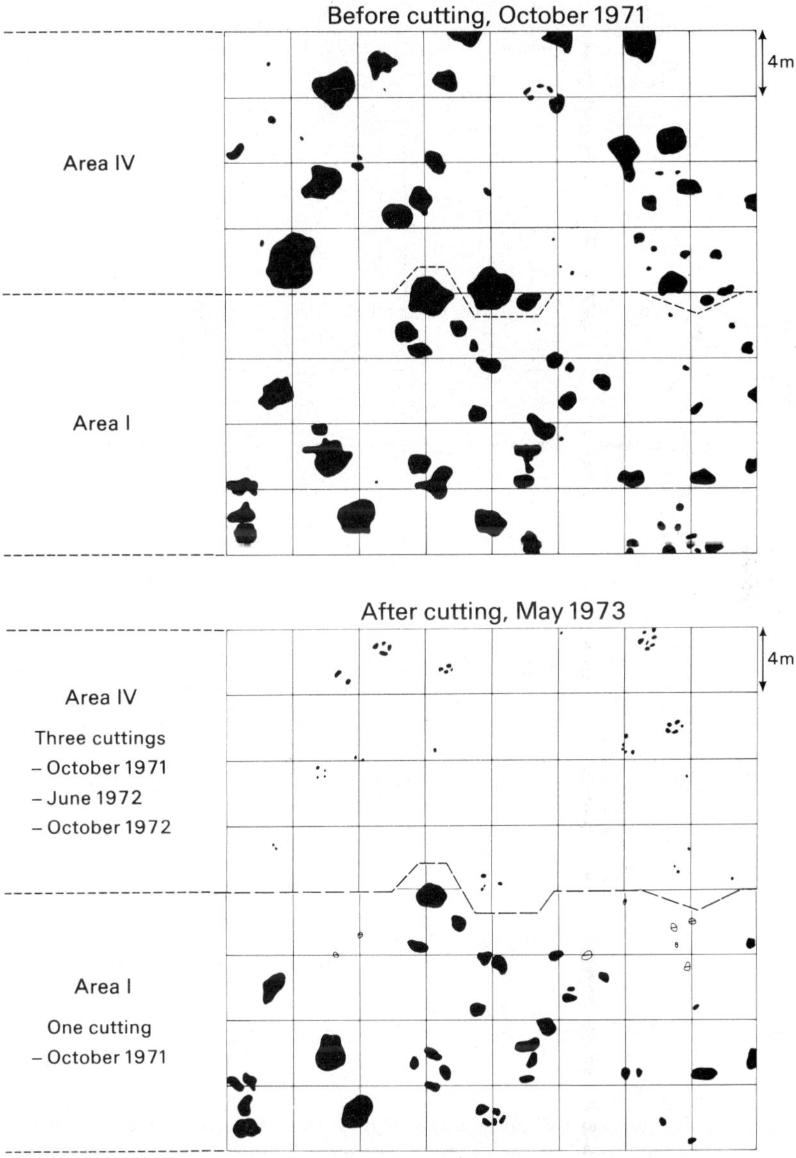

Figure VIII-10. Crown cover charts of part of the cutting experiment in a stand of *Haloxylon salicornicum*, see table VIII-1.

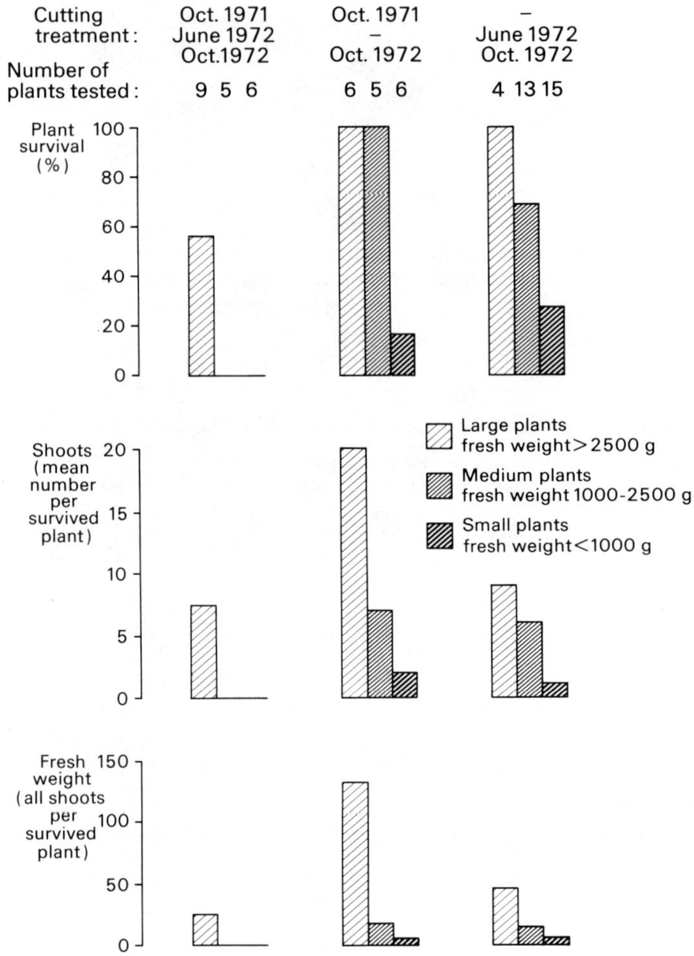

Figure VIII-11. Response of *Haloxylon saticornicum* plants of three weight classes to different treatments of repeated cutting; recorded May 1973, about 20 months after first cutting.

8 x 8 m^2, randomly laid out in an area nearby the place of the above mentioned experiment. In this area plants of different sizes were found. Fig. VIII–11 shows amongst others that three times cutting at surface level, in October 1971, June 1972 and October 1972, killed all plants with a fresh weight of the aerial parts below 2.5 kg and four out of nine of the larger plants. Only one small plant out of six survived cuttings in October 1971 and October 1972, but all medium size and larger plants did. Table VIII–2 shows the regrowth characteristics (number of new shoots developed in the branches cut and total fresh weight of these shoots) on an

individual plant basis. The general trend of larger plants having a higher resistance is here clearly illustrated. The four smallest plants died after the first cutting. Another four plants of medium size are killed by the second cutting, while five of the six largest plants are the only survivors of the third cutting. The reducing vigour in these five plants is shown in Fig. VIII–12. Not only the number of shoots decreases drastically, but also their average length and the variation in this length. Only a few small shoots remain.

The repeated experimental cuttings, and their results as discussed, are very similar to the cutting of shrubs for fuel and subsequent grazing of the new shoots. The

Table VIII–2. Response of *Haloxylon salicornicum* plants of different size to repeated cutting at the given dates. The plants are arranged in order of decreasing crown cover. A dash indicates no sprouting at all and death of the plant.

No.	Plant characteristics before first cutting at 26 Oct. 1971				Plant characteristics following repeated cutting at the given dates					
	Crown cover	Height	Aerial biomass fresh	Current years growth fresh	Number of new shoots			Total fresh weight new shoots (g)		
					26 June 1972	30 Oct. 1972	27 May 1973	26 June 1972	30 Oct. 1972	27 May 1973
	(m^2)	(cm)	(g)	(g)						
1	11.00	70	13 990	6 517	49	23	3	2 490	435	3
2	7.50	45	6 680	5 528	151	23	–	3 517	185	–
3	6.25	60	5 820	3 630	97	30	4	1 710	290	8
4	6.00	55	4 420	2 380	78	6	21	2 393	121	73
5	5.63	64	6 970	4 042	71	12	7	1 517	282	5
6	5.25	70	7 060	3 827	54	9	4	2 055	650	20
7	4.38	75	5 930	3 353	34	2	–	780	7	–
8	4.75	76	5 600	3 498	35	5	–	1 019	95	–
9	2.25	35	1 960	1 317	39	–	–	586	–	–
10	2.00	40	2 010	1 203	9	3	–	570	78	–
11	1.75	62	4 180	2 431	7	4	–	267	11	–
12	1.75	50	1 400	1 091	26	8	–	742	42	–
13	1.50	35	1 260	875	10	–	–	164	–	–
14	1.00	50	2 080	1 161	13	2	–	335	45	–
15	0.63	42	370	234	3	–	–	73	–	–
16	0.50	25	180	126	5	–	–	13	–	–
17	0.12	36	115	83	–	–	–	–	–	–
18	0.12	34	100	71	–	–	–	–	–	–
19	0.10	22	50	36	–	–	–	–	–	–
20	0.06	20	25	15	–	–	–	–	–	–

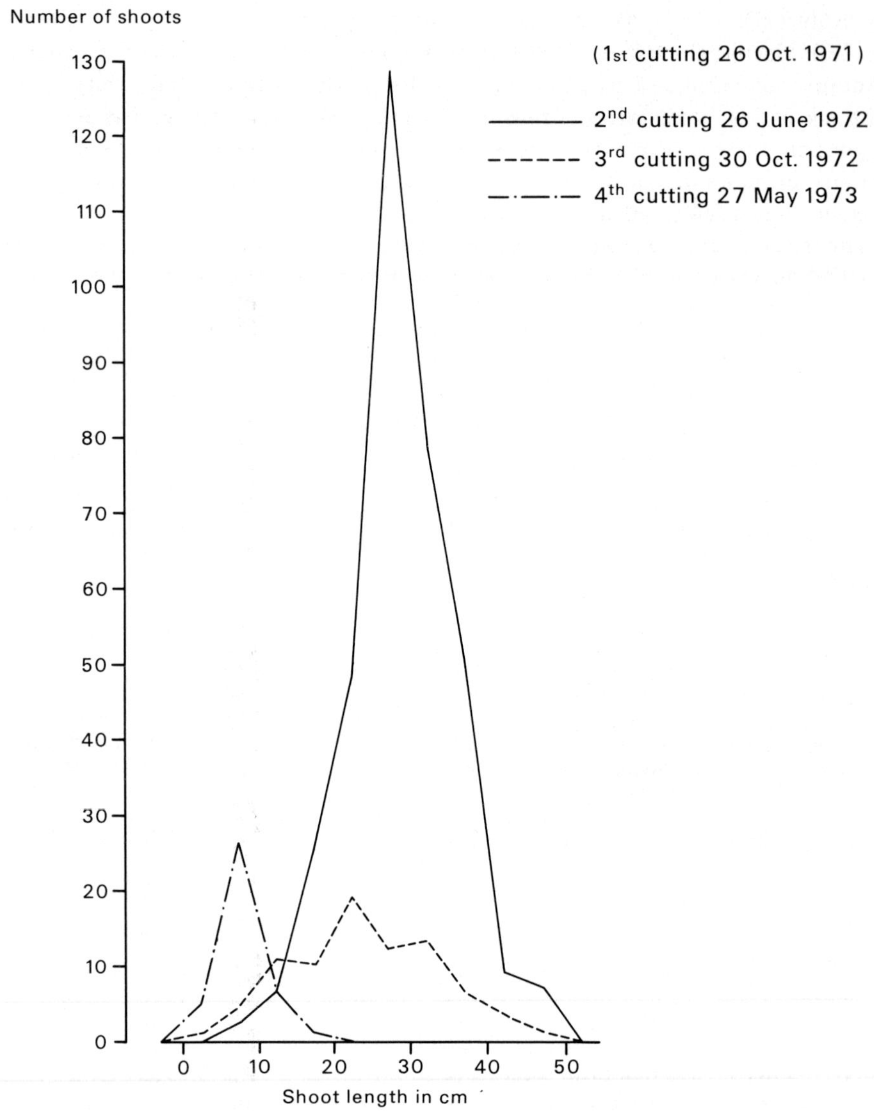

Figure VIII-12. Shoot length measured in the regrowth of five large *Haloxylon salicornicum* plants at three successive cutting dates.

preference of livestock for these fresh green shoots was several times observed in the field. The fact that most of the cutting is done as near as possible to the wells and settlements, places frequently visited by the herds, cause an even more frequent and greater impact than simulated by the above experimental cutting.

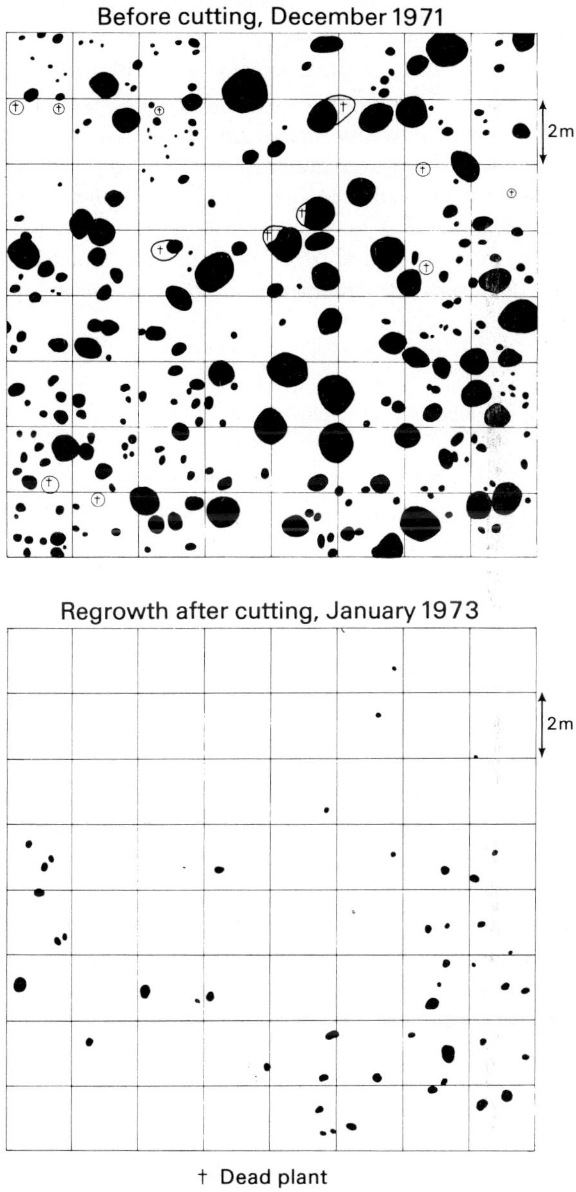

Figure VIII-13. Crown cover charts of part of an area where a cutting experiment was carried out in a stand of *Rhanterium epapposum* near Samah, Southern Desert.

The regrowth data on *H. salicornicum* provide evidence on the following aspects of the impact of fuel-cutting on the species:

(i) A *one-time cutting at surface level,* especially when hummock formation has taken place, *may induce the development of new growth* in case of larger plants and be advantageous to the development of the shrub. The regrowth in area III, mentioned before, shows this and to a minor extent the regrowth in area I.

(ii) *Small plants*, with a fresh weight in October below 1000 g , *will often die* as a result of cutting at surface level.

(iii) *Repeated cutting* with an intensity of twice a year and more with several months in between the separate cuttings, *will kill all Haloxylon salicornicum shrubs in a few years.*

VIII.C.2. *Cutting of 'Rhanterium epapposum'*

Systematic experimental cutting as carried out on *Haloxylon salicornicum*, proved too complicated to carry through on *Rhanterium epapposum*, mainly for logistic reasons.

In the field the impression had been gained that *Rhanterium* had far less resistance to cutting than *Haloxylon*. An experiment proved this. At 18 and 19 December 1971 an area of 16 x 32 m^2 in a good *Rhanterium* stand was cut, after the height, crowndiameters, and exact position of each individual plant had been determined. The regrowth was determined 13 months later. Half of the area is shown in Fig. VIII–13 and Fig. VIII–14. A few large plants had produced some new sprouts and some small plants were seen, too large to be one year old seedlings. These small plants probably were an aberration, resulting from inadequate cutting the year before when some small plants must have been overlooked. A stimulating effect of cutting was, therefore, not seen in *Rhanterium epapposum*. On the contrary, a cutting only once at surface level caused the death of virtually all plants. Fuel cutting in a *Rhanterium* vegetation seems in all cases highly destructive. In one case was seen how sheep and camels grazed the young seedlings, developing in a stand formerly cut. As mentioned before, the combination of utilization practices causes the most serious destruction of the vegetation cover.

VIII.C.3. *Cutting and uprooting*

In the experiments discussed above, cutting of shrubs was in all cases at surface level. Fuel collection by Bedouins, villagers, etc. is different, as was observed in the field and mentioned to us in discussions on this point. The thickest part of the stem is usually in the ground, or rather in the upper 20 cm of the hummock and the shrubs are cut some 10–20 cm below the surface (Fig. VIII–15). To investigate the consequences of surface cutting and 'uprooting', an experiment was carried

Figure VIII-14. Result of experimental cutting in a good stand of *Rhanterium epapposum* near Samah, Southern Desert, photographed in January 1973, 13 months after cutting of the whole area at surface level in December 1971. Persons stand at the two corners of the treated area (see Fig. VIII-13).

Table VIII–3. Effect of extra deep cutting (uprooting) of shrubs against cutting at surface level, demonstrated for two groups of plants of *Haloxylon salicornicum* in a depression near Wadi Ubayidh at Faidhat al Jeez (in brackets the standard deviation).

29 October 1972, before cutting (n = 10 for both groups)	Shrubs cut at surface level	Shrubs cut extra deep (uprooting)	
Height (cm)	57 (± 10)	48 (± 14)	
Crown cover (m^2)	2.8 (± 1.3)	3.2 (± 1.0)	
Total fresh weight (kg)	2.3 (± 1.3)	2.6 (± 1.0)	
Gain by extra deep cutting (kg)		0.4 (± 0.2)	
26 May 1973, seven months after cutting (n = 8 for both groups)			**Difference**
Number of new shoots per shrub	25 (± 13)	13 (± 8)	12*
Fresh weight of regrowth per shrub (g)	147 (±120)	62 (± 63)	85*
Shoot length (cm)	13.2 (± 2.9)	9.4 (± 2.7)	3.8*

* Significant at $p < 0.05$

Figure VIII-15. Shrubs cut by fuel gatherers, ready for transport to the village at the background; near Abu Dibbis in a stand of *Haloxylon salicornicum.*

Figure VIII-16. Results of a cutting experiment of *Haloxylon salicornicum,* recorded seven months after treatment. Plant right: cutting at surface level; plant left foreground: extra deep cutting (uprooting); small plant left background; no cutting.

out. In a well developed stand of *Haloxylon salicornicum* two groups of ten plants each of about equal size were selected, measured and cut. Of one group all thick branches in the top layer of the hummock were pulled out, broken and weighed. The treatment of this group compared favourably with a more or less similar treatment given to *Haloxylon* shrubs by Bedouin women, collecting fuel nearby. The response was measured seven months later, in May 1973. (Fig. VIII–16 and table VIII–3). Some difficulties were encountered in recovering the treated plants, which had been marked with small heaps of stones, and two plants from each group were not found again. The upper half of table VIII–3 shows that the treated plants had an average height of about 50 cm, a crown cover of about 3 m^2 and a total fresh weight of about 2.5 kg. They can thereby be classified as good, medium size shrubs (see table VIII–2), a category commonly collected for fuel. The lower half of the table demonstrates that the extra deep cutting (uprooting), yielding comparatively little extra wood, resulted in a reduction of regrowth by one half, an equal reduction in total shoot fresh weight and a reduced shoot length (Fig. VIII–16). This provides strong evidence that the impact of cutting shrubs for fuel is in actual practice more destructive than indicated by the experiments discussed before, even without subsequent intensive grazing, an aspect of combined effects, discussed separately in chapter XI.

CHAPTER IX

CULTIVATION, THE EFFECT OF PLOUGHING

IX. CULTIVATION, THE EFFECT OF PLOUGHING

IX.A. Introduction

The River Euphrates is the only perennial river crossing the study area. The river is a source of irrigation water, but only for a narrow strip of land along its streambed. The use of groundwater for irrigated agriculture in the area is, with minor exceptions for local supplementary irrigation, considered uneconomical. Dry-land farming or rainfed agriculture is in general in the long run only economically profitable in areas with an average annual rainfall exceeding 250 mm. Specialists do not agree on this point, but a value of 250 mm must be considered at the lower side (table IX–1). The average annual rainfall in the Southern and Western Deserts and Lower and Central Jezira ranges from about 50 mm to 250 mm (see IV.B.5). This should virtually preclude cultivation from the study area.

The reality is different. Ploughed, sown, harvested and abandoned fields can be found all over the Iraqi desert rangelands at locations where the local environmental conditions have been judged conducive for growing a grain crop. The northernmost part of the area, the Upper Jezira and the transition to the Central Jezira has almost completely been under cultivation at one or another time or is still under this land use. Examples can also be found elsewhere in the area, be it confined to

Table IX–1. Boundary of dry-land farming as given by various authors.

Author	Required minimum rainfall for cereal cultivation under dry-land farming	Remarks
Wirth (1962)	300 mm	The author mentions 100–150 mm for extremely favourable soil and relief conditions and the 400 mm isohyet as probably the proven long-term best boundary.
Brichambaut & Wallén (1963)	200–300 mm	Figures are a rough indication; a detailed water balance is worked out, incorporating available water and potential evapotranspiration during the period of active growth.
Al-Kasab (1966)	300 mm	Boundary in the Jezira, based on an assessment of average yields, incorporating a certain number of crop failures.
Gangopadhyahya in UNESCO (1975)	appr. 300 mm	An approximate boundary for the Jezira, worked out on the basis of climatological water budget conditions more of less coïncided with the 300 mm isohyet.

local depressions and wadi bottoms, as far as the border with Saudi Arabia in the south where rainfall is less than 100 mm. Before a crop can be grown the natural vegetation has to be removed. Large parts of the desert rangelands, and by definition the best grazing areas, have been affected by this practice. Because of its destructive, often irreversible effect on the local natural vegetation, a separate treatment here is in every respect justified. The situation in the Jezira is different from the situation in the Western and Southern Deserts and the two are separately discussed below in IX.C and IX.D, respectively, after a brief description of the general process has been given.

IX.B. The general process

In arid areas plant growth is most closely linked to available soil moisture. The best developed natural vegetation is found at places receiving the highest amount of water that can be stored in the soil, either through rainfall or accumulation of runoff water or both. These are also the places with the best opportunities for growing a crop. For this purpose the natural vegetation cover of such areas is removed first (Fig. IX–1). Shrubs are uprooted or cut by using simple tools or heavy equipment (Fig. IX–2) and the area is ploughed and sown under a grain

Figure IX–1. Good stand of *Haloxylon articulatum* and *Artemisia herba-alba* in a slight depression, being uprooted to clear the soil for cereal cultivation.

Figure IX–2. Since about 1950 heavy equipment owned by contractors or groups of families is being used for cultivation in the desert. The area in the foreground has just been cleared from the shrubs. The woody plants have been collected and may be used as fuel.

crop. The crops best adapted to the harsh conditions and used for this purpose are hard wheat (*Triticum durum*) and barley (*Hordeum sativum*). Their development coincides with the period during which soil moisture is often available and selected varieties can even survive a certain drought (Brichambaut & Wallén 1963). Barley prevails in the relatively lower rainfall areas. Some 80 kg wheat or 100 kg barley are sown per hectare; sometimes mixtures are used. In exceptionally good rainfall years the yield can be some 1500–2000 kg per hectare; 600–800 kg per hectare is considered a normal harvest in the Western Desert (Sheikh Naif pers. comm.), about the same as in the Jezira (Al-Kasab 1966)*. Barley usually gives a considerable higher yield than wheat. Both crops, but especially barley, are used as green fodder when the grain harvest is expected to be a failure. The harvested grain is usually made into barley or wheat flour and used to bake 'gubs', the local flat, round bread. By-products are used as supplementary fodder. Once the crop has been harvested, a bare area remains, often with a loosened topsoil from which the stones have been removed. The furrows made by the plough are still there (Fig. IX–9 and Fig. IX–10) and often run in a downslope direction (see shape of the plots in Fig. IX–8 and Fig. IX–11). Water erosion and wind erosion start. The

* For the whole of Iraq over the period 1961–1970 the average yield for wheat was 640 kg. ha^{-1} and for barley 880 $kg.ha.^{-1}$ (Statistical Pocket Book 1960 1970, Central Statistical Organization 1972).

topsoil with a relatively high organic matter content and comparatively rich in nutrients is easily blown away by wind action or washed away after flooding. Surface sealing and sheet erosion occur and may continue until hard layers are exposed. Where stony layers exist not too deep in the profile, a desert pavement may be the result (Buringh 1960). In places with deep, fine textured, soils often gully erosion can be seen. Ploughing means that the relatively best grazing areas are stripped from their natural, productive and protective vegetation and, thereby, exposed to a high erosion hazard. Occasional ploughing of small areas in an attempt to obtain a grain crop is an old practice in the area. It has for centuries been a sort of 'nomadic cultivation', rather harmless because of the small areas involved and the primitive equipment used. Ploughing was not deep and a good deal of the deeper rooted shrubs survived and later sprouted again, as demonstrated in the previous chapter. This picture has changed considerably, especially since mechanization was introduced.

IX.C. Cultivation in the Central and Upper Jezira

The Jezira has a gradient in annual rainfall from about 120 mm in the south near the River Euphrates (Hit 120 mm, Ana 122 mm) to about 400 mm in the north (Mosul 390 mm, Balad Sinjar 403 mm). This gradient in amount of rainfall goes together with an increased rainfall reliability and also of annual mean number of rainy days, from less than 50 in the south to an estimated 100 in the north. Between the two extremes, along the gradient, two land use types meet, rainfed agriculture and one or another form of nomadic pastoralism. Cropping, when successful, yields per unit area more than livestock raising, but is also far more uncertain. From the distribution pattern of ancient settlements, clearly distinguished in the field as 'tels', an idea can be formed of how man in the past judged the possibilities for rainfed cropping. These tels, in fact mostly the ruined remnants of small farmers villages, now visible in the landscape as low mounds, indicate an occupation of the land by an early farmers population, probably dating back to the second millennium B.C. The villages were destroyed during the early Islamic times and tels are found in a band of some 150–200 km wide, stretching from the Tigris River over Tel Afar and Baaj with a few as far south as Hatra. In terms of rainfall these early settlements are confined to an area north of the 300 mm isohyet. Shifts in tribal areas in former days moved these people from the land, which was then again mainly used for grazing. Periods of drought were of particular significance in this respect. For centuries the Upper Jezira has been an area of struggle between sedentary cultivators, under regular government control and protection, and nomadic pastoralists, mainly of the Shammar, Aneze and Dulaim tribes. This part of the Jezira was excellent grazing land and fights between the tribes over grazing and water rights were common. Settlement spreaded southward extensively in the 19th and 20th centuries, causing subsequent increase in ploughed area. Three distinct periods can be separated. From 1850 to World War I the area around Jebel Sinjar and Tel Afar, roughly in the 350–450 mm rainfall zone, was occupied again, mainly by non-Arab tribes. In a second period, after World War I and according to Al-Kasab lasting, to 1942, Arab groups pushed north by the Shammar from the Central Jezira, occupied the areas north and south adjacent to the earlier settled tribes. For the present account the third period, started after 1951, is most significant.

Supported by the Government who distributed the land to the families and offered some medical and education facilities, Bedouins, mainly of the Shammar tribe, settled in the northern Jezira, in the area between Hatra and Baaj. Al-Kasab (1966) mentions the establishment of 400 new villages between 1951 and 1965 where in total some 47200 people were settled. These developments fall outside the study area or at its northern fringe and can be justified from the point of view of the most profitable land use. However, they induced another development further south, well into the Jezira rangelands. Following a few good rainfall years in the early fifties, cultivation pushed far south and were in 1952 in total some 250 000 ha cultivated, in 1963 this had been extended to well over 900 000 ha. With the traditional equipment this could have never been achieved. The use of heavy machinery, owned by contractors living in the cities, made this possible as can be seen from the details given in Al-Kasab (1966):

From 1948 to 1964 the number of tractors in the Jezira increased from 39 to 2000, the number of combine harvesters from 19 to 1650.

The old fashioned plough worked about 0.25 ha a day, the new tractor plough over 4 ha and considerably deeper.

Mechanized seeding increased the area treated in one day twenty-fold compared with hand seeding.

Some 4 to 5 ha a day could be mown by machine against 0.2 ha by hand. A substantial part of the best grazing lands of the Jezira was ploughed, but the cultivation of larger areas did not result in equally increased yield. On the contrary, average yields per unit area decreased rapidly and it soon became clear that cultivation had been pushed too far south. A sequence of bad years revealed the misjudgment and in 1963 the Iraqi authorities fixed a southern boundary for rainfed agriculture. The line was not established on climatological grounds alone, but rather to indicate where the possibility of a paying mixed farming system of animal husbandry and rainfed cropping ended. In 1965 the line had to be put further south to incorporate amongst others 14 already established villages. A new boundary was fixed by law in 1971, again slightly modifying its position.

Fig. IX–3 gives a summary of the developments discussed. The data on the 1971 – boundary were insufficient to incorporate this line. Until recently, only a time consuming ground survey or an extensive aerial survey could have shown the present southern border of rainfed cropping. Thanks to the very large field sizes (often adjacent fields are ploughed together by the same contractor), the first LANDSAT-satellite imagery of this part of the world revealed it in one glance

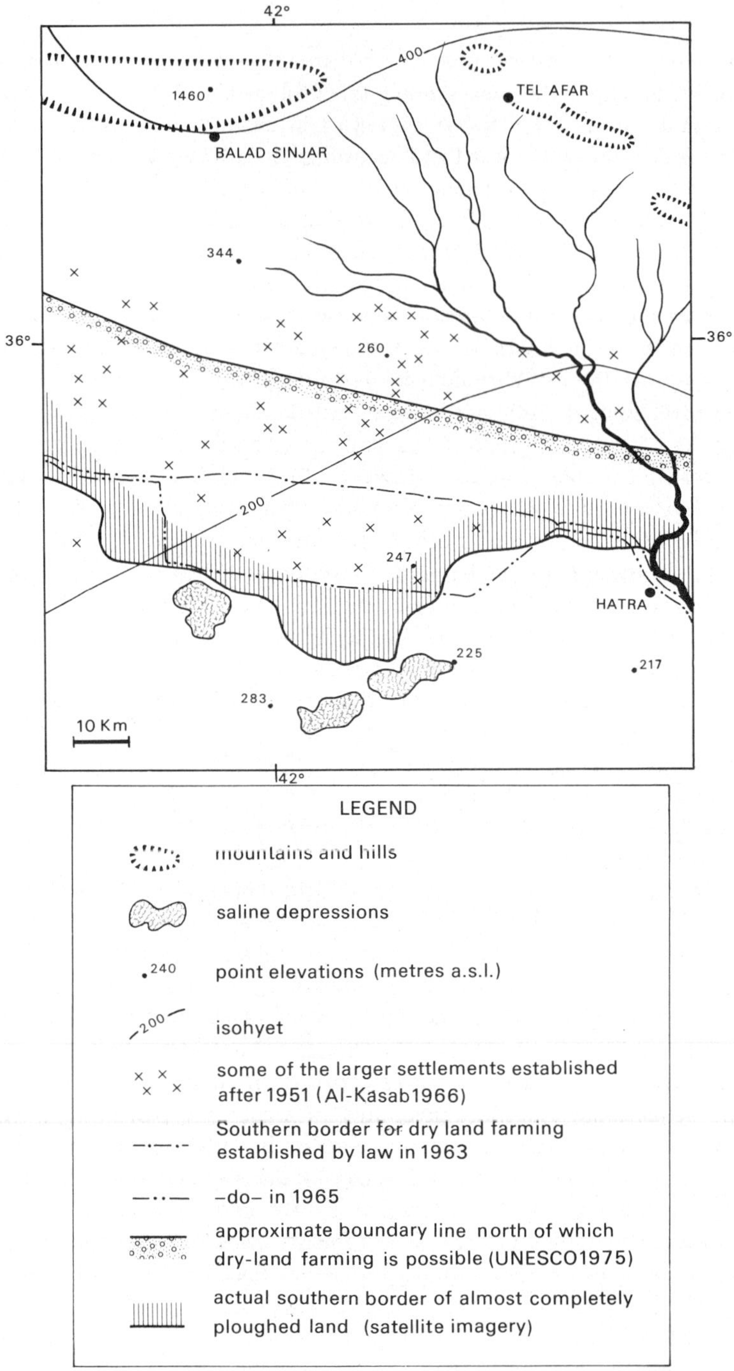

Figure IX–3. Map of the northern part of the Jezira, showing the southward extension of dry-land farming.

(Fig. IX–4). That in a relatively short period of time the ploughed area in the northern half of the Jezira was extended further south than ever, can be clearly explained from an agricultural or economic point of view. The discussions on the boundary mentioned before, either the theoretical or the practical boundary, are mostly concerned with the uncertainty and the low grain yields, rather than with the effect of the ploughing on the natural vegetation and the potential of grazing as an alternative use of these areas (e.g. Wirth 1962; Al-Kasab 1966; Brichambaut & Wallén 1963; with Kaul 1973 as an exception).

For an evaluation of this aspect information is required on (i) the (semi-) natural

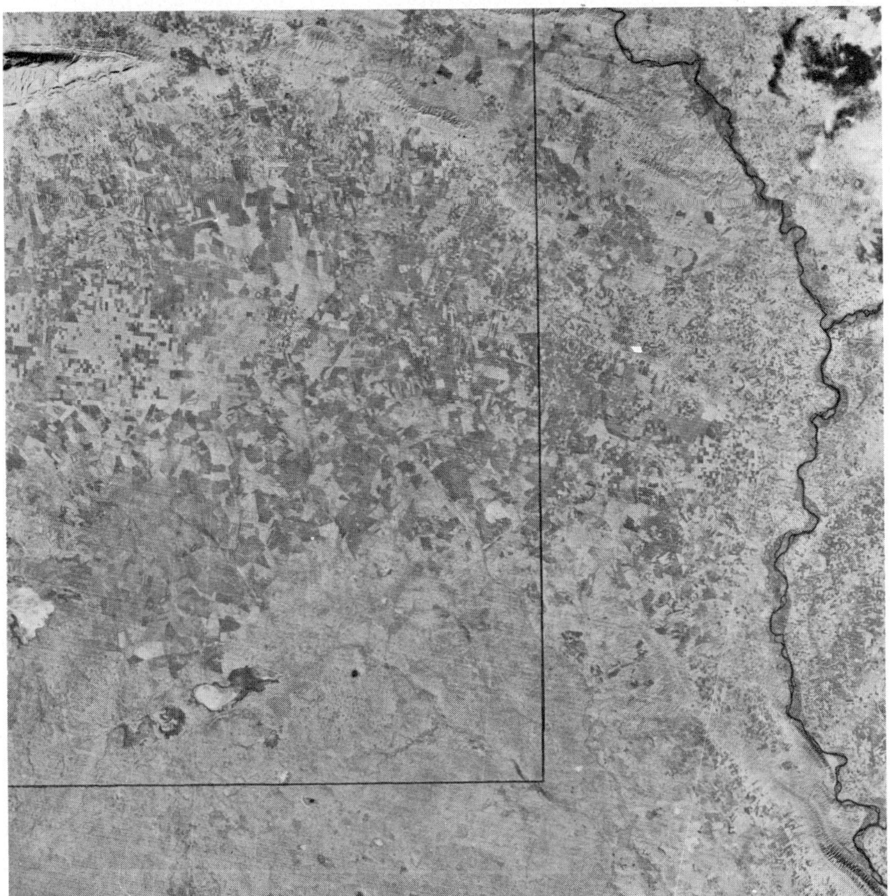

Figure IX–4. LANDSAT-1 satellite image of 22 November covering part of the Central and Upper Jezira (Band 7, infrared). At the right the Tigris River is seen with the tributaries Lesser Zab and Greater Zab. Jebel Sinjar is at the upper left corner, Jebel Makhul at the lower right. The area partly mapped in Fig. IX–3 is indicated.

vegetation before ploughing and (ii) vegetation development after ploughing. Data on the first point are extremely scarce. The area is often described as an excellent grazing area, but no details on the vegetation are given. Naval Intelligence Division (1944) reported for the area around Hatra:

> 'Much gras grows in spring and affords good grazing for the camels and sheep of nomads, mostly Shammar Jarba and Dulaim' (pp. 77/78)

and

> 'In its neighbourhood there have always been water and plentiful grazing for camels, and in ancient times the caravan trade needed protection from desert raids' (p. 80).

Older information is available from Ainsworth (1841) who travelled from Mosul to Hatra. He quoted an earlier traveller (De Lamartine) of the area who camped with Arabs near Hatra ... 'for eight days, the pasturage being very abundant', and writes on his return to Mosul:

'. . . we struck right across the grassy plains towards Mosul'. In a last page Ainsworth gives the 'geographical botany of the tracts which we have travelled', where he mentions amongst others: 'There are scarcely any spots that are actually deprived of all vegetation'. 'Oat grass is by far the most abundant of the graminеous plants. This single species covers whole uplands of miles in extent, to the exclusion of everything except a few flowering plants, which at this season of the year were the *Ranunculus Asiaticus*, and certain species of *Hieracium* and *Crepis*.' 'Other grasses were also met with, among which *Hordeum pratense* and a delicately-panicled *Poa* advanced upon the most sandy spots.'

'In the drier parts of the plains, grasses became more rare and lichens more common, but these tracts were clothed with a more prominent vegetation of under-shrubs of wormwood; among which the most common species were *Artemisia fragrans* and *A. absinthium*'.

In the Upper Jezira little is left of the vegetation indicated by Ainsworth. Not only the ploughing has been destructive, but especially the cultivation in combination with continued intensive grazing. The settled Bedouins still keep some livestock and stubble grazing is still an important practice. However, with the grain crop on the land the animals must be grazed elsewhere and near the settlements now only relatively unpalatable species or no vegetation at all is found. Only an extensive floristic search can show whether the components of the vegetation before the cultivation are still present. Observations in the Hatra exclosure (see chapter X) give no reason for optimism.

Development of a new vegetation cover after cultivation, the second point of interest, is a slow process and the resulting vegetation is very poor from a utilization point of view. The only common large plant species found abundantly are *Artemisia scoparia* and *Peganum harmala*. Especially the first species covers extensive areas in the northern Jezira. Both species are virtually useless as forage producers, *Artemisia scoparia* because of its low production of palatable material, *Peganum harmala* because of its being poisonous when green. The species were

Figure IX–5. Vegetation dominated by *Artemisia scoparia* and *Peganum harmala* on once cultivated, but long since abandoned fields in the Upper Jezira.

found to be typical invaders when a once cultivated area is abandoned (Fig. IX–5 and Fig. IX–6). Both species are also common near human settlements. Guest (1966) calls *Peganum harmala* the ruderal plant par excellence of the desert, being invariably found along caravan routes, near camping sites, in the vicinity of wells, etc. The distribution of this species, considering the rather heavy seeds, may well be by man, e.g. with the grain used for seeding. *Artemisia scoparia* seeds are easily distributed by wind. The two species together, in dominance, determine the community type shown in table IX–2. Total crown cover in comparison to other stands need not be very low (Fig. IX–6) but the forage available from this type is virtually nil. The species shown in table IX–2 have extremely little forage value, either because of their morphological characteristicss (*Lagonychium farctum*, *Cornulaca* spp., *Alhagi mannifera*) or their highly aromatic smell (*Achillea conferta*).

A common phenomenon observed in the formerly cultivated areas in the Southern and Western Deserts is the vigorous growth and good development of the few surviving woody species such as *Haloxylon salicornicum* and *H. articulatum*, especially the last species. Batanouny & Zaki (1969) have worked on this in Egypt and found that 'ploughing leads to changes in physical properties of the soil in such a way that it accelerates water permeation. Ploughing increases penetrability and thus modifies the depth and the lateral extension of the root system. It also accelerates the growth of the shoot system.' Absence of competition for water could be another factor. Several years of successful cropping can have a serious

Table IX–2. Characteristics of four stands representative for the vegetation in formerly cultivated areas in the Jezira (Community type XIV, *Artemisia scoparia* – *Peganum harmala* distinguished in chapter V).

Stand number	51		47		46		48	
Sampling date	20 March 1973		23 March 1973		24 March 1973		23 March 1973	
Location	28 km from Baaj along the track to Traifawi		1 km from the Hatra ruins, east of the Mosul-road		17 km south of Hatra police-post along the Bareed-track		near northern boundary exclosure, 1 km from gate, outside	
Environment	Slight depression, last cultivated at least two years before sampling		Flat, long abandoned cultivated field, under grazing		Gently sloping plain, last cultivated over a year before sampling		Gently sloping plain, abandoned cultivated field	
Total aerial biomass ($g.m^{-2}$ dry)	20		15		23		8	
Estimated total available forage ($kg.ha^{-1}$ dry)	10		3		23		2	
Total crown cover (%)	12.6		8.5		14.3		2.8	
Total density ($no.m^{-2}$)	1.5		2.1		3.7		0.6	
	Crown cover (%)	Density ($no.m^{-2}$)	Crown cover (%)	Density ($no.m^{-2}$)	Crown cover (%)	Density ($no.m^{-2}$)	Crown cover (%)	Density ($no.m^{-2}$)
Species:*								
Peganum harmala	8.6	1.2	6.0	1.1	4.7	1.2	1.0	0.3
Artemisia scoparia	3.8	0.3	0.8	0.3	2.4	0.4	1.1	0.2
Lagonychium farctum	–	–	1.5	0.7	1.4	0.5	0.7	0.1
Salsola spp.	–	–	0.1	negl.	5.8	1.6	negl.	negl.
Cornulaca spp.	–	–	negl.	negl.	negl.	negl.	negl.	negl.
Achillea conferta	0.1	negl.	negl.	negl.	–	–	–	–
Teucrium polium	–	–	0.1	negl.	–	–	negl.	negl.
Alhagi mannifera	0.1	negl.	–	–	–	–	–	–
Scrophularia deserti	–	–	–	–	–	–	negl.	negl.

* negl. = value less than 0.05

Figure IX–6. *Peganum harmala,* the densely growing species to the right, is the most common invader on once cultivated areas. At this site (near Traifawi in the Jezira), not only the old field to the right, but also the surrounding shrub vegetation is invaded by the species, which is considered poisonous to the livestock and has no grazing value.

effect on the soil fertility, as was demonstrated by decreased yields in the Syrian Jezira (Brichambaut & Wallén 1963). Some larger woody species would be expected in the Iraqi Jezira, even after cultivation, but they are conspicuous by their absence. The processes of fuel cutting or overgrazing discussed in the previous chapters are a likely explanation. If seedlings could develop undisturbed by grazing and older plants were not cut for firewood, some regeneration would be expected. Based on evidence of the vegetation of remote or sacred places in the northern Jezira, Gillett (1948 b) suggested that the natural condition of this zone once was a kind of savannah or open forest with trees (*Pistacia, Prunus*), shrubs and grasses. Guest (1966) supported this view with evidence from similar areas in Jordan, where *Pistacia* scrub was found 'uptil as recently as the 1914–18 war'. Observations by Poore & Robertson (1963) on range types in Jordan provide similar information. For the distinguished '*Artemisia* – brush' type where *Artemisia herba-alba* is the dominant, found on rolling hills and valleys in the steppe region, the following is mentioned:

> 'The vegetation of this type is produced and maintained by overgrazing. Evidence of the original vegetation is scanty, but it was probably, in the west, Sub-Mediterranean woodland (*Quercus calliprinos*), and on the Steppe-Forest margin an open woodland of *Pistacia*. In the steppe region proper it almost certainly contained a higher proportion of palatable species and may once have possessed a humus-rich steppe soil.'

In Iraq the abrupt transition from forest (or remnants of it) to a treeless area, now largely ploughed, seems improbable. The above remains speculative. Long-term, detailed observations on vegetation development in exclosures may at one time give clues for an undisputable picture of what once has been. It could, at the same time, be a decisive argument for counteracting the present deterioration processes.

The changes in forage production in the Upper Jezira have had their influence on the situation in the central part. Guest (1966) discusses the degradation of the vegetation of the steppe, distinguishing after Gillett (1948 b) between the Dry Zone (200–350 mm annual rainfall) and the relatively Moist Zone (350–500 mm). For the Dry Zone his observations agree with the data of the early seventies. Only remnants of perennial shrub vegetation are found in localities far from permanent water and 'large tracts have been reduced to an almost bare expanse dominated by *Poa bulbosa* and other short grasses, with a colourful showing of small annual herbs in spring'. The change in vegetation has mainly taken place in the twentieth century. Guest compares the statements of several travellers on *Haloxylon articulatum*, common and conspicuous in 1910 (Handel-Mazzetti 1914b) and only occasionally found in the fourties. In the course of the present study the species was encountered several times between 1970 and 1973, however certainly not 'common and conspicuous'. A quotation from Guest (1966) summarizes the impressions for the Central Jezira, also as obtained during the 1970–1973 observations and communications with Bedouins:

> 'This confirms the testimony of the older tribesmen in the Jezira today when they recall that the vegetation used to be much thicker in their boyhood, the ground being covered as a thicket of SHIH (*A. herba-alba*), QAISUM (*Achillea*) and RIMTH (*Haloxylon*) while HAMDH (salt bushes in general) was abundant, and that most of the more valuable shrublets have disappeared during the course of their own lifetime.'

Figure IX–7. Wadibottom, partly ploughed several years ago, flooded by runoff water from adjacent areas during a heavy shower.

IX.D. Cultivation in the Lower Jezira and Western and Southern Deserts

Cultivation in the Lower Jezira and Western and Southern Deserts is concentrated along the wadis and in depressions, areas which receive an additional amount of rainwater through runoff. The less than 125 mm rainfall annually is normally insufficient to build up the soil moisture required for the growth of a grain crop. In the area precipitation occurs mostly as short, intense showers (see Fig. IV–5) resulting in a flooding of the wadi bottom by accumulated runoff water. An area with an average annual rainfall of not more than 100 mm can thereby be flooded by several centimetres water a few times a year (Fig. IX–7). Closer to the wadi channel the water can be many decimetres deep and in and near the channel a torrential stream of more than one metre deep can be flowing for a duration of several hours to a few days. Spreading of the water is, especially in larger depressions, being done by furrows, ploughed to guide the water from the surrounding

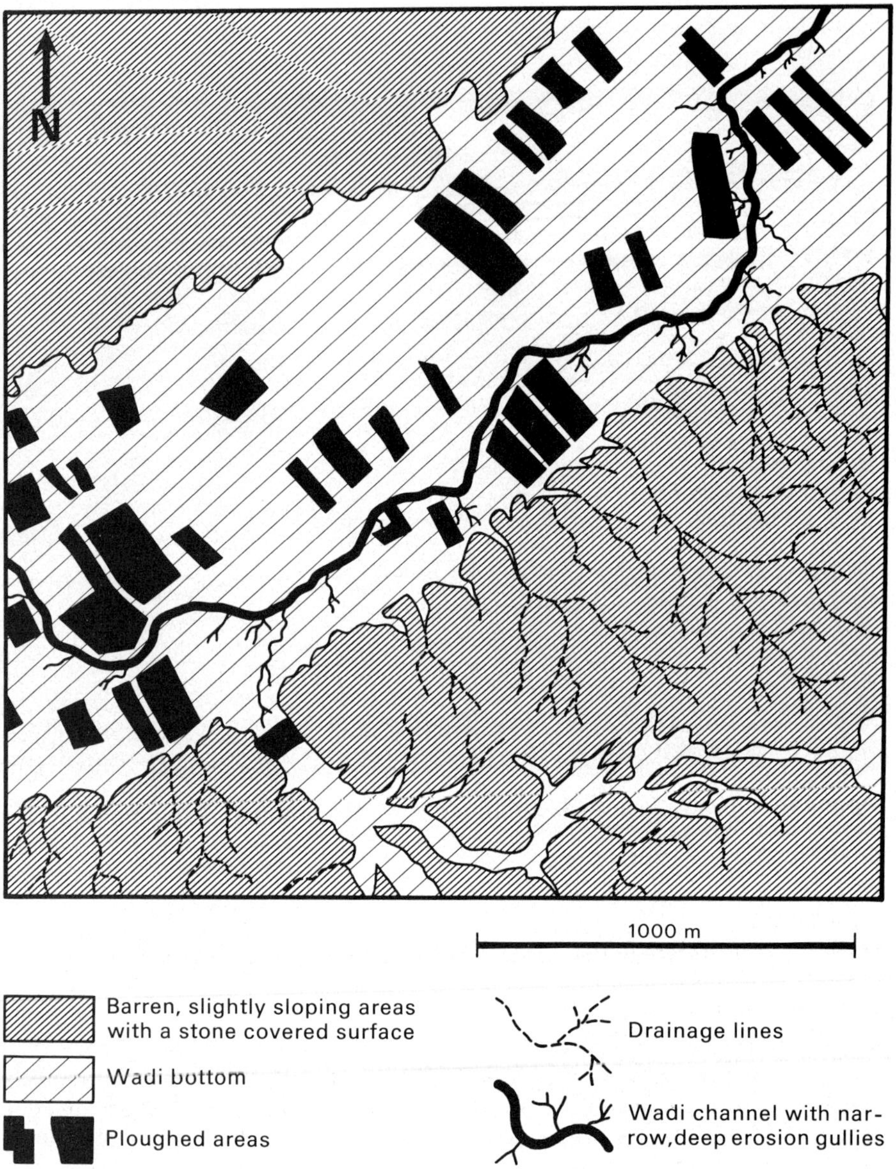

Figure IX–8. Pattern of cultivation in a wadi as seen on aerial photography (Wadi Ubayidh, 1962).

area to the fields (Wirth 1962). Short canals are sometimes constructed diverting the water from the streamchannel of the wadi to lower situated areas nearby.

Depressions and wadibottoms, if not cultivated bear the best developed vegetation of the desert rangelands, consisting of tall shrubs, sometimes even small trees, sub-shrubs and a herbaceous layer with perennial grasses and forbs. This can be deduced from remnants found at remote and inaccessible places and observations in exclosures (e.g. the Rutba exclosure, see X.6). For centuries these wadis and depressions together with the water sources have determined the migration routes of the Bedouins with their flocks. The processes of overgrazing, fuel cutting and cultivation have been most intensive here, in the vegetation of the areas with the highest production potential. The direct effect of the ploughing, destruction of the existing vegetation cover, is serious, but far more serious on the long run is the effect on the soil. Cultivation in these areas results in an almost bare surface with shallow furrows, often leading downslope. Surface sealing takes place. Heavy rainfall forms a crust of a few millimetres thick. By raindrop impact the aggregates of fine material, especially silt, are broken down and the particles in suspension are washed into the pore space of the sand grains skeleton. The volume of the surface pores is reduced and after drying a hard cemented crust is formed. Rainwater penetration becomes difficult and accelerated sheet erosion takes place. In addition the shallow furrows can lead to gully formation, especially near the wadi streamchannel, where the action of the streaming water is most violent and lasts longest (Fig. IX–8). Surface sealing also

Figure IX–9. Vegetation in a slight depression in the Western Desert, with a rather heavy soil, after several times ploughing. Only two species were found, *Achillea fragrantissima* and some *Zilla spinosa*, both of low grazing value, but nevertheless grazed.

Figure IX–10. Vegetation near Wadi Tibal (Western Desert) in an area once cultivated, but long abandoned since. *Artemisia scoparia* and *Capparis spinosa* are the dominants, both of low grazing value.

makes the establishment of a new vegetation very difficult. Seedbed conditions are poor and litte moisture penetrates to the subsurface soil. When plants succeed in establishing themselves, there is an immediate danger of being grazed by livestock or, in case of shrubs, being cut for fuel. The vegetation of abandoned cultivated fields is, therefore, extremely poor and consists mainly of unpalatable species. Fig. IX–9 and Fig. IX–10 show typical examples. The crown cover of the grazed *Achillea*-vegetation shown in Fig. IX–9 is only 1.5% and its aerial biomass dry 17 $g.m^{-2}$. Non-cultivated areas nearby had a perennial shrub cover of well over 20% and an aerial biomass of over 100 $g.m.^{-2}$. Fields are usually abandoned after one or more years of crop failure. Over a period of approximately twenty years (1950–1970), in Faidhat al Jeez near Nukhaib, a lower situated area near wadi Ubayidh, there had been five good years and six years with a complete crop failure (Sheikh Naif pers. comm.). In a sequence of years with yields above the average there is a tendency to increase the ploughed area, the same effect as discussed above for the Jezira. In the early sixties there were three consecutive good years in this area and a comparison of aerial photographs of 1962 and 1963 (by chance available) of the nearby Athaman-area clearly showed this increase (Fig. IX–11). Ten years later no recent ploughing was seen in this area. From the exposed roots of some *Haloxylon articulatum* plants who had for some time survived the ploughing, but died afterwards from fuel cutting, the erosion was obvious. A hard surface crust was formed and along the wadi channel many small gullies were to be seen. The species most commonly found in such areas include *Artemisia scoparia, Achillea fragrantissima, Haloxylon articulatum* (usually in a

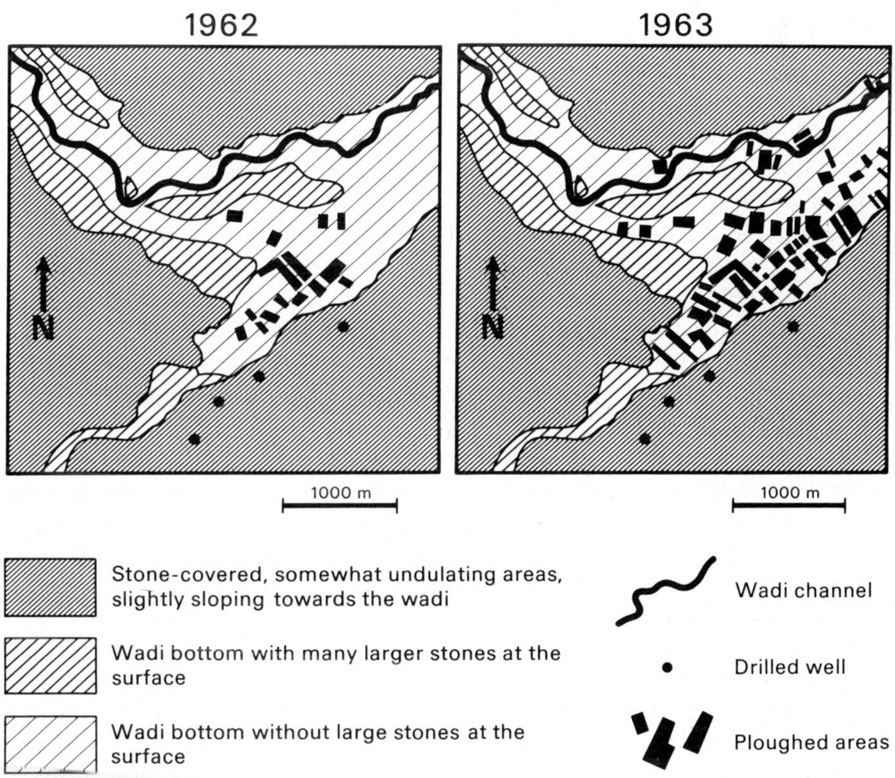

Figure IX–11. Increase in area under cultivation after above average rainfall years (see text), as seen on sequential aerial photography (Athaman, Western Desert).

few, well developed specimens, see Batanouny & Zaki 1969, referred to before) *Capparis spinosa* and the almost unpalatable *Zilla spinosa*.

Van der Veen (1964) followed the vegetation development on an experimental small field ploughed by tractor for barley cultivation in Syria in 1960 on the border of an old and recent floodplain, under an average annual rainfall of about 150 mm. The situation may be considered representative for a good part of the Lower Jezira and somewhat better (higher rainfall) than in the Southern and Western Deserts. He found that four years after ploughing, during three of which rainfall had been favourable, the grazing value of the land was still extremely poor and he estimated that it would take some 5–15 years for the vegetation to regenerate. In such a lapse of time soil conditions might change to such degree that little or no vegetation at all will develop. This is proved by the extensive, poorly vegetated and often completely bare areas in depressions and along wadis in the Southern and Western Deserts and Lower Jezira.

IX.E. The overall effect

The overall effect of ploughing for rainfed cereal cultivation as practised in the area under discussion can be summarized in a few points:

The immediate effect of ploughing is the destruction of the perennial vegetation of the (potentially) best grazing areas in the desert rangelands. In the northern part of the Jezira these are the areas with the highest average rainfall and the most reliable rainfall, in the rest of the area these are the places receiving an additional amount of water through runoff from the adjacent areas.

Crop failures are common and fields are abandoned after one or a few years.

Soil conditions in the abandoned fields, especially of the surface layer, are not favourable for germination and development of a new vegetation and re-vegetation is poor or absent. Surface sealing is common and leads to accelerated erosion by water and wind.

The few plants succeeding to re-establish in the formerly ploughed area are subject to heavy grazing and fuel cutting and consequently the resulting vegetation consists mainly of species useless for both these types of utilization.

The deteriorating effect of cultivation for the rangelands has increased tremendously since about 1950 with the introduction of modern farm machinery such as tractors and combines, allowing to plough larger areas and much deeper than before and harvest more efficiently. A commercial aspect previously non existent, with outside interests, has been introduced.

The hazardous cultivation has in a few decaded destroyed a good deal of the best rangelands in the area and degraded them into areas with little or no grazing value at all.

CHAPTER X

EXCLOSURES, THE EFFECT OF PROTECTION

X. EXCLOSURES, THE EFFECT OF PROTECTION

X.A. Location of exclosures and earlier observations

Local elimination of exploitation processes through the establishment of exclosures is an obvious method to create a situation for investigating the impact of utilization on the vegetation. A study of the progress reports of the Range Management Division and personal communication with the staff of that Division revealed the establishment of several exclosures, ranging in size from 0.25 ha to 100 ha, during the period 1953–1958 (Thalen & Al-Mufti 1973). Anderson (1958) listed nine and indicates the initiation of studies on recovery and succession. However,

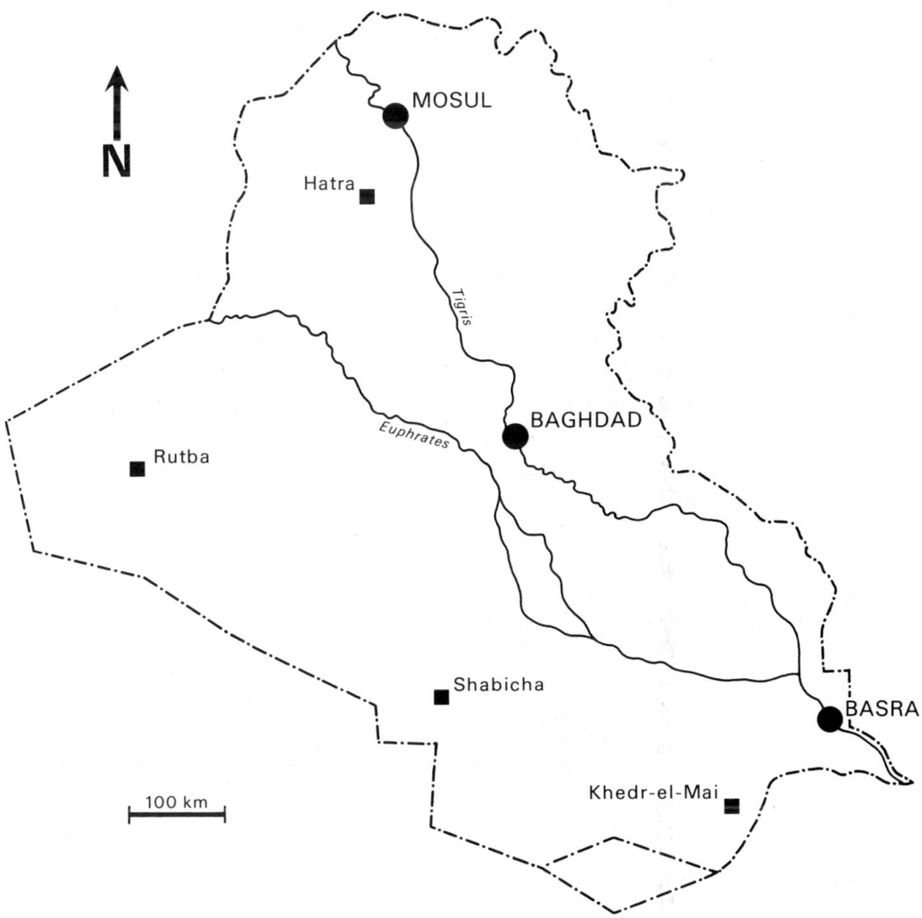

Figure X-1. Location of the four exclosures studied.

no work has been reported. Long (1956) visited 'the exclosures of Fakka, Bsaia, As Salman, Ash Shabicha, Nukhaib and Rutba' and found after less than a year of almost complete protection the results very encouraging. Agnew c.s. spent two days each at the protected areas of Khedr-el-Mai and Shabicha at the end of March 1960 (Agnew 1960). His data are the only quantitative record of the vegetation in exclosures, collected prior to 1970. By that year four exclosures were in existence: the ones near Khedr-el-Mai and Shabicha on which Agnew had reported, and large exclosures near Rutba in the Western Desert and near Hatra in the Jezira (Fig. X–1). Some qualitative impressions of exclosures obtained during a consultant mission to Iraq are given in Kaul (1969a). The Rutba exclosures has been used to study light grazing of *Artemisia herba-alba*, as shown in chapter VII (see also Kaul & Al-Mufti 1974). The situation in the Shabicha exclosure, a protected area abandoned in 1971, has been recorded in Thalen & Al-Mufti (1973). The observations on the four exclosures mentioned above, made during the period 1970–1973 are discussed below.

X.B. The Khedr-el-Mai exclosure

X.B.1. *Establishment and early data*

In 1957 the then Range Management Section of the Division of Animal Husbandry fenced an area of 400 donums (100 ha) near the Khedr-el-Mai police post and well known wells (29°46'N 46°50'E) in the Southern Desert. A guard, appointed in 1959, made full protection effective. The exclosure is situated in a slight depression with a sandy-gravelly topsoil. Data from soil samples taken at various locations inside and outside the fence are given in table X–1.

The only early record of the vegetation, soon after the time of establishment, is found in Agnew (1960). He writes:

Table X-1. Selected properties of the soil at vegetation sampling sites inside and outside the Khedr-el-Mai exclosure. Mean values and range for five samples.

		Inside		Outside	
		0–30 cm	30–60 cm	0–30 cm	30–60 cm
Texture	Sand (%)	88 (84–94)	85 (81–90)	85 (82–90)	88 (84–90)
	Silt (%)	7 (5–10)	7 (6–10)	9 (7–10)	7 (6–11)
	Clay (%)	5 (1– 6)	7 (3–10)	7 (3– 8)	5 (1–10)
EC_e (mmho.cm^{-1})		0.2 (0.1–0.2)	0.2 (0.2–0.3)	0.3 (0.2–0.4)	0.6 (0.2–2.3)
Lime (%)		17 (15–18)	17 (15–20)	18 (15–20)	15 (11–18)
Gypsum (meq.100g^{-1})		nil	nil	nil	48 (nil-176)

'The vegetation of the whole area consists of scattered perennial bushes (mean density 0.52/sq. metre). Principally of *Haloxylon salicornicum* with some *Agathophora alopecuroïdes**, *Heliotropium persicum, Cornulaca monacantha, Anabasis setifera* and *Anvillea garcini*. Perennials of low stature occupied the spaces between the shrubs. These included the most abundant plant in the area *Aristida plumosa***, with a mean density of about 8 plants per square metre, and the following species: *Haplophyllum sp., Helianthemum lippii, Fagonia bruguieri* and *Paronychia arabica*'.

Subsequently the annuals are listed and the quantitative data on perennials that could be obtained in the short time available, are presented. The limited number of copies of Agnew's report warrant the summarized presentation of these data in table X–2. Agnew gives as principle conclusion:

'that as yet the protection has not changed the composition of the vegetation to any marked extent. This would be expected in view of the short period of complete protection, but a superficial note on the spectacular increase in cover might lead to the opposite, erronious, conclusion'.

Several visits were paid to the Khedr-al-Mai exclosure between 1970 and 1973 and quantitative data on the various aspects of vegetation development inside the exclosure as compared to outside were recorded.

The area is rather inhomogeneous in its vegetation cover, which may be due to differences in the (top) soil and differences in the amount of runoff water received from adjacent areas. This means that comparing an area randomly choosen inside the 100 ha fenced with an area of similar size outside, anywhere in the Khedr-el-Mai depression, could give a very biased picture. An approach of compa-

Table X–2. Cover and density of the main perennial species recorded inside and outside the Khedr-el-Mai exclosure in March 1960, after a few years protection (compiled from Agnew 1960).

	Cover (%)*		Density (no.m^{-2})**	
	Outside	Inside	Outside	Inside
Stipagrostis plumosa	0.53	18.57	8.18	7.74
Haloxylon salicornicum	–	–	0.14	0.26
Helianthemum lippii	–	–	0.41	1.46
Halogeton alopecuroïdes	–	–	–	present
Heliotropium bacciferum	–	–	–	present
Haplophyllum sp.	–	–	–	present
Anabasis setifera	–	–	–	present
Fagonia bruguieri	–	–	present	–

* Measured with two line intercepts of 32 m each.
** Measured with 50 randomly placed quadrats of 1 m^2.

* Syn. *Halogeton alopecuroïdes*
** Syn. *Stipagrostis plumosa*

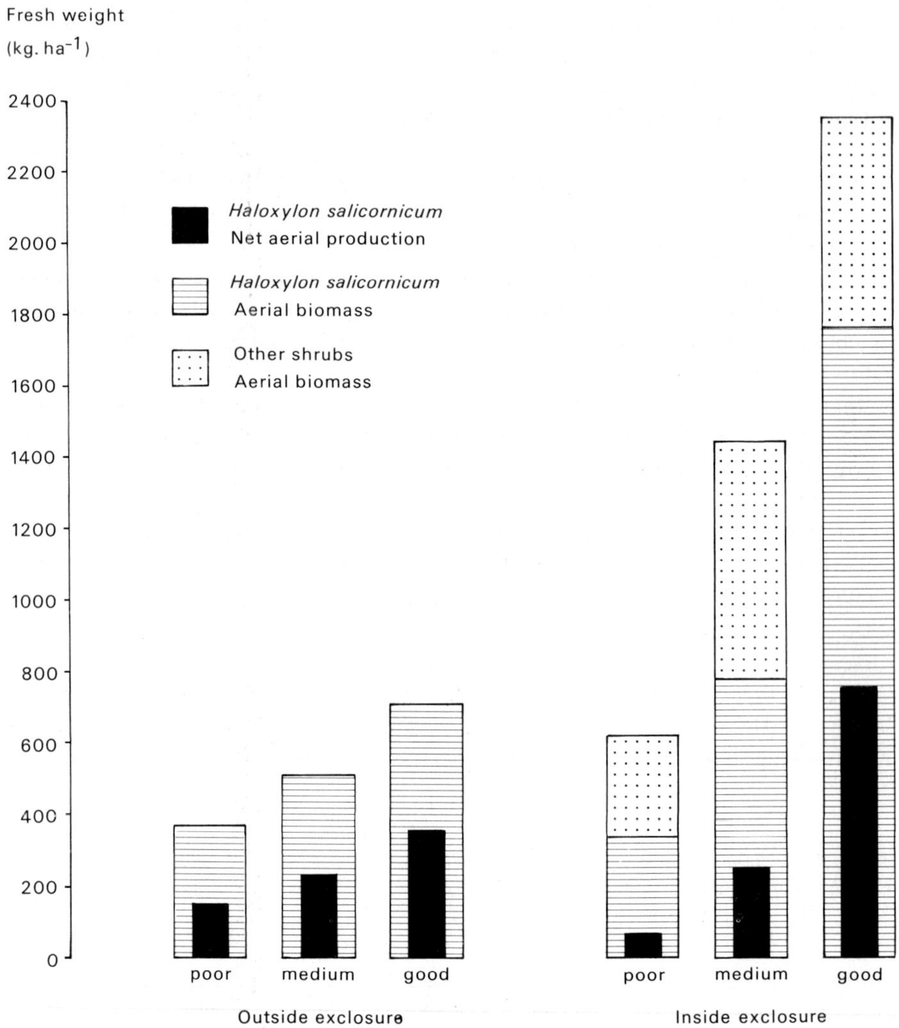

Figure X-2. Aerial biomass and net aerial production (of *Haloxylon salicornicum* only) inside and outside the Khedr-el-Mai exclosure, May 1971, after about 13 years of full protection. Twelve areas of 64 m^2 each, considered representative, inside and outside, were measured and ranked in three classes according to the crown cover.

ring *adjacent* areas inside and outside the fence, with similar soil- and water conditions, as far as it could be judged without detailed analysis (see table X–1), was therefore adopted.

X.B.2. *Aerial biomass and production*

In May 1971, November 1972 and April 1973, aerial biomass in adjacent areas inside and outside the fence was determined. Total biomass, on fresh weight basis, inside ranged roughly from two to five times higher than outside the fence (table X–3, Fig. X–2). Inside an average value of appr. 1.5 ton.ha^{-1} was found, against outside less than 0.5 ton.ha^{-1}. From Fig. X–2 it can be seen that the difference is amongst other things due to the development of species other than *Haloxylon salicornicum* inside the fenced area. *Halogeton alopecuroïdes* is the most dominating one in this respect. In the poor and medium quality sites examined table X–3 this species contributes almost a similar amount to the total aerial biomass as *Haloxylon salicornicum*.

It was observed that inside the exclosure *Haloxylon salicornicum* plants generally had developed to a large size. Diameters over two metres were no exception. However, the plants seemed to put on little new growth. This was confirmed by aerial production data collected on a per plant basis at three locations inside and outside the fence in May 1972 (Fig. X–3). On a fresh weight per plant basis, outside the fence production was almost twice as high as inside. The outside new material contained more moisture; on a dry weight basis a seven to four ratio for outside-inside aerial production was found for individual *Haloxylon salicornicum* plants.

Regular grazing and perhaps also moderate cutting (see VIII.C.1.) is apparently beneficial and even necessary for continued growth of this species. A number of

Table X–3. Aerial biomass inside and outside the Khedr-el-Mai exclosure in 1972 and 1973, after about fifteen years of full protection of the exclosure ($\overline{x}$–mean, s–standard deviation).

Nature of the sample	Date of collection	Aerial biomass fresh (kg.ha^{-1})			
		Inside		Outside	
		$\overline{x}$	s	$\overline{x}$	s
At five locations inside and outside the fence a cluster of 10 randomly placed quadrats of 4 m^2	29/30 Nov. 1972	1600	534	538	251
In an area near the western fence 5 clusters of 10 randomly placed quadrats of 4 m^2	24 April 1973	1256	424	255	100
In an area near the southern fence 5 clusters of 10 randomly placed quadrats of 4 m^2	25 April 1973	1215	262	329	117

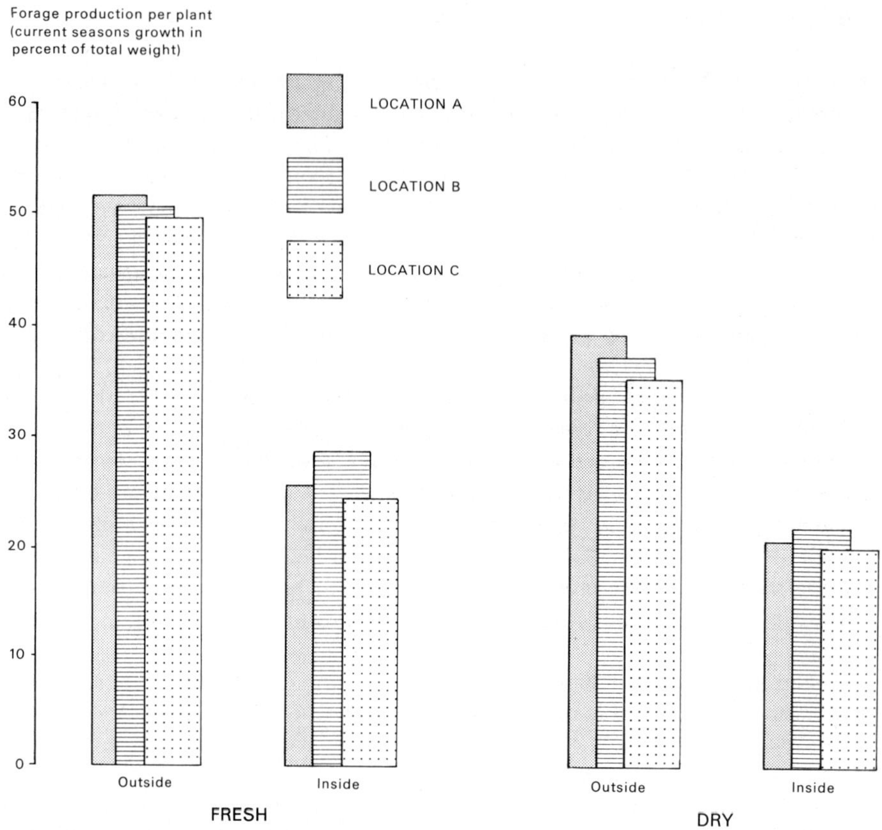

Figure X-3. Current season's growth as a measure of forage production, determined for six plants inside and outside the Khedr-el-Mai exclosure at three locations in May 1972.

dead and almost dead large *Haloxylon* plants in the exclosure supported this hypothesis and an experiment was carried out to test this further. In the exclosure a number of *Haloxylon* plants of about equal size were selected, tagged and subjected to either no cutting, half* cutting or total cutting. Table X–4 shows the response for the three groups, of 10, 10 and 9 plants, half a year after the treatment. Sprout length in the total cut and half cut plants was three and two times that of the non-cut plants, respectively. For another set of three groups, of resp. 5, 5 and 6 plants, the effect of no cutting, half cutting and total cutting was recorded 11 months after the treatment (table X–4). The two cutting treatments resulted in an average increase in new growth of about 40% for the total cutting and 115% for the half cutting. This was produced in sprouts of approximately

* The whole plant cut at half its height.

Table X–4. Response of plants of *Haloxylon salicornicum*, after twelve years of complete protection in the Khedr-el-Mai exclosure, to experimental cutting at different levels, recorded at six months and eleven months after the treatment (x-mean, s-standard deviation).

Group	Treatment 16 December 1971	Crown volume before treatment (m^3)		Response to treatment, 28 May 1972: Dry weight new growth (g)		Length of longest sprout (cm)	
		$\bar{x}$	s	$\bar{x}$	s	$\bar{x}$	s
I (n=10)	no cutting	0.45	0.15	22.8	10.0	4.9	1.9
II (n=10)	half cutting	0.51	0.18	11.6	9.3	10.8	3.6
III (n=9)	total cutting	0.40	0.17	12.4	9.9	15.3	4.7
				Response to treatment, 28 November 1972			
Ia (n=5)	no cutting	0.50	0.15	41.4	34.2	11.6	3.9
IIa (n=5)	half cutting	0.48	0.27	89.0	61.3	22.6	7.1
IIIa (n=6)	total cutting	0.44	0.19	58.3	48.1	23.7	6.5

twice the length of those of the uncut plants. In spite of the observed individual variation in response to the various treatments (see standard deviations in table X–4), these observations provide strong evidence that regular grazing or cutting is favourable for the development of *Haloxylon salicornicum*.

X.B.3. *Cover and density*

In May 1971, a crown cover of 6.1% outside against 14.2% inside the exclosure for the larger shrub species was found, result of subjective sampling of a few large plots. More detailed, objective, observations were made in November 1972 and April 1973, following the procedure of sampling stands by 50 random quadrats of 4 m^2 (see V.B.3.).

Table X–5 shows that for the sampled areas in all three cases crown cover inside the fence was 2 to 3 times higher than outside, as a result of differences in *Haloxylon salicornicum* cover, and, not less, in the cover of *Halogeton alopecuroïdes*. The latter species hardly contributes to the cover outside, while inside the fence cover contributions from five to over ten per cent were recorded. *Anabasis setifera* was not highly represented in the sampled areas, but seemed locally to

Figure X-4. About the same part of the Khedr-el-Mai exclosure photographed at two different dates, showing the difference in the vegetation outside (foreground) and inside the fence. Above: March 1960, after only one year 'more or less' and one year full protection; below: twelve years later, March 1972.

Table X–5. Comparison of cover and density of the most abundant perennial species, at adjacent locations inside and outside the Khedr-el-Mai exclosure, observed in 1972 and 1973 after about fifteen years of full protection of the exclosure. Crown cover was calculated from individual crown diameters.

	29/30 November 1972				24 April 1973				25 April 1973			
	Mixed sample of five locations				Appr. middle of western fence				Appr. middle of southern fence			
	Crown cover (%)		Density ($no.m^{-2}$)		Crown cover (%)		Density ($no.m^{-2}$)		Crown cover (%)		Density ($no.m^{-2}$)	
	Inside	Outside	Inside	Outside	Inside	Outside	Inside	Outside	Inside	Outside	Inside	Outside
Haloxylon salicornicum	8.13	4.65	0.22	0.31	5.93	1.84	0.18	0.17	13.84	5.19	0.24	0.35
Halogeton alopecuroïdes	7.30	0.90	0.28	0.44	10.89	0.89	0.48	0.25	5.00	0.26	0.22	0.06
Anabasis setifera	0.59	0.01	0.01	0.01	0.01	0.06	0.01	0.02	0.31	0.03	0.02	0.01
Cousinia sp.	–	–	–	–	0.00	–	0.01	–	–	–	–	–
Helianthemum lippii	1.19	0.64	0.43	0.61	0.71	0.63	0.65	0.48	1.55	1.98	0.79	1.51
*Heliotropium bacciferum**	–	–	–	–	0.26	0.02	0.04	0.02	0.27	0.08	0.03	0.01
Cornulaca spp.	0.44	0.03	0.03	0.01	0.07	0.06	0.03	0.01	0.20	0.03	0.03	0.02
*Stipagrostis plumosa***	2.04	0.49	0.65	0.61	2.29	3.09	0.75	1.37	3.24	2.52	1.10	1.20
Anvillea garcini	–	0.01	–	0.01	–	–	–	–	–	–	–	–
Total	19.69	6.73	1.62	2.00	20.16	6.59	2.15	2.32	24.41	10.09	2.43	3.16

* Syn. *H. persicum, H. ramosissimum*
** Syn. *Aristida plumosa*

develop better inside the fence. Large numbers of the shrublet *Helianthemum lippii* were found outside as well as inside the exclosure. The species has little grazing value, except in spring, and its distribution and development is perhaps more governed by seedbed conditions than by the influence of grazing. Moreover, its occurence is very local, which may give a biased picture with the sampling procedure adopted.

The perennial grass *Stipagrostis plumosa* (Syn. *Aristida plumosa*) is very widespread in the Khedr-el-Mai depression. It has a high grazing value (Bor 1968). Agnew, in 1960, measured for this species a cover of 18.57% inside the exclosure and a density of 7.74 plants.m^{-2} Outside he found approximately the same density (8.18), but a cover of only 0.53% (table X–2, Fig. X–4). Agnew noted already that the species was more abundant on the low, flood levelled, area than on the hummock areas and wondered what would happen to this species in the future. The answer is given in table X–5. With an exception for the November 1972 outside observation, the cover of *Stipagrostis plumosa* twelve years later inside as well as outside the exclosure was between two and just over three per cent. Density inside ranged from 0.65 to 1.10 plants.m^{-2}, against 0.61 to 1.37 plants.m^{-2} outside the exclosure. This means that between 1960 and 1972 about six *S. plumosa* plants out of seven disappeared, inside the exclosure as well as outside. This process of dying of *S. plumosa* was still going on in 1972 and 1973. Out of some hundred carefully examined plants inside the exclosure, only just over fifty showed clear signs of life. Because of the time required to examine this, in table X–5 all plants, dead (?) and alive, are included.

The cover differences observed by Agnew in 1960 for the species and again in November 1972 are easily explained as the result of recent heavy grazing outside the fence (Fig. X–4, Fig. X 5). As regards the decrease in density and cover, inside as well as outside the exclosure, between 1960 and 1972/73 different explanations are possible, *viz.*

(i) A severe drought, e.g. a year without any rainfall, may have caused the death of part of the *S. plumosa* plants. Inside and outside the fence the species would be equally affected. There are no data to support or reject this possibility.
(ii) Outside the exclosure the abundance of *S. plumosa* as observed by Agnew must have attracted the attention of the graziers and their herds of sheep and camels. This must have resulted in a heavy grazing pressure, especially in view of the availability of nearby watering points.
(iii) Inside the exclosure, and to a less extent also outside, hummock formation around the shrubs was observed, combined with a lower density of *S. plumosa* at such sites. The species in 1972 and 1973 was not found in the pattern clearly shown in Agnews photograph of 1960 (Fig. X–5 above) but rather in local concentrations, mostly at sites away from the sandy hummocks. Of 300 quadrats of 4 m^2, sampled in 1972 and 1973, 85 contained no *S. plumosa* at all, while densities of over 20 per 4 m^2, even up to 34 per 4 m^2 were found in others. The

Figure X-5. Comparative, general view of the vegetation in about the same area inside the Khedr-el-Mai exclosure. Above: March 1960; below: March 1972.

contagious distribution (clumped or aggregated) of the species in thereby well illustrated.

Although no further proof can be given, it is likely that (iii) played the important role for *Aristida* decrease inside the fence and (ii) outside, perhaps with (i) as an overall extra factor for both in a particular period.

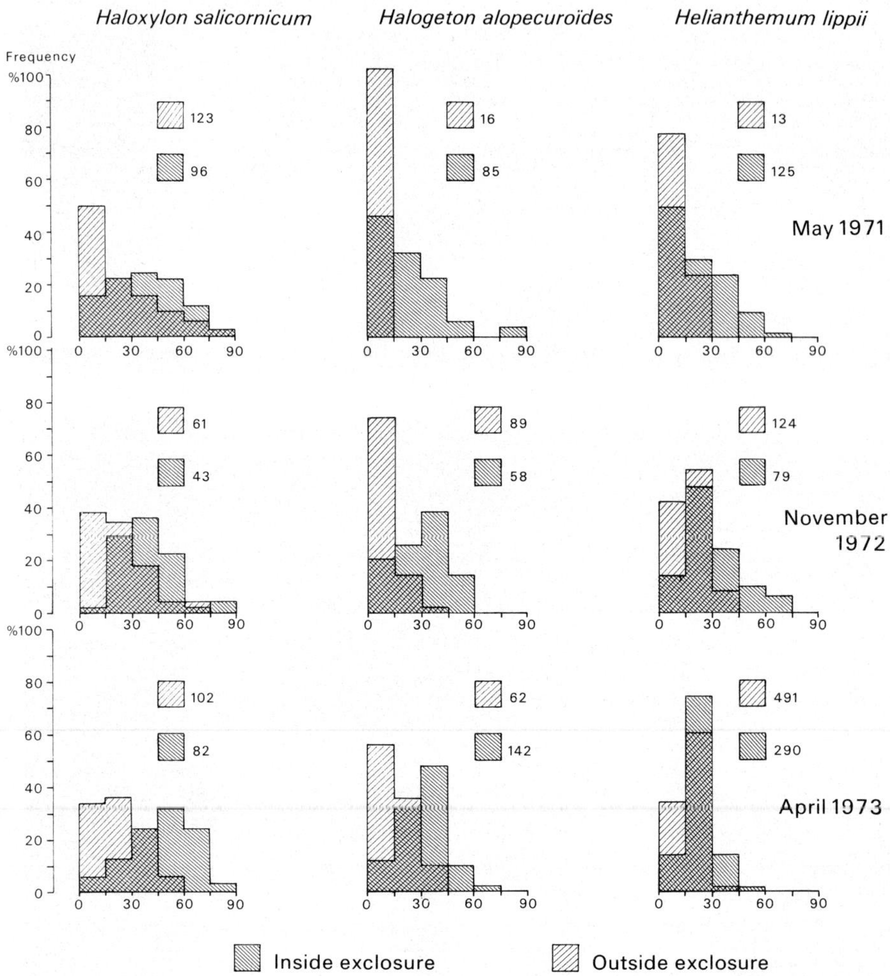

Figure X-6. Relative height distribution in six classes of the species *Haloxylon salicornicum*, *Halogeton alopecuroïdes* and *Helianthemum lippii*, measured inside and outside the Khedr-el-Mai exclosure at three different dates. Figures indicate the number of plants measured.

X.B.4. *Successional trends*

For the shrub species no clear differences in species composition in- and outside the exclosure were observed, but large differences in cover and density (table X–5). All individual plants had been measured, which allowed a grouping in height classes. To a certain extent these classes, certainly for the lower heights, represent age classes. The relative height distribution in percentages for the two shrubs *Haloxylon salicornicum* and *Halogeton alopecuroïdes* and the shrublet *Helianthemum lippii* is shown in Fig. X–6. For *Haloxylon salicornicum* a distinct different pattern was found inside and outside, more or less similar in the three observations:

Inside the exclosure the classes between 30 and 60 cm are dominant. Higher plants up to the 75–90 cm class were observed at all three sampling dates. Few small plants, however, were found and hardly any seedlings in 1972 and 1973. Some of these small plants will die before reaching the greater height classes. The general picture, therefore, is one of an aging stand with a very limited replacement of old dead shrubs.

Outside the fence the opposite picture was found. Large numbers of seedlings and small plants. The greater the height, the lower the number of plants encountered. The grazing apparently does not allow the plants to reach the height observed under protection; at the same time however conditions conducive for the development of young plants are maintained. A well regenerating stand is the result.

Halogeton alopecuroïdes inside the exclosure at the sites sampled in May 1971 showed a picture similar to that found for *Haloxylon* outside the fence: a well developing stand with many young plants. At the sites sampled in 1972 and 1973 a lower number of young plants was found, and the 30–45 cm class dominated. Outside the fence the species apparently has no chance to develop. Seedlings are produced, a few can develop to about half a metre height, but then disappear. Although the species seems not very palatable, the only explanation for the observed inside-outside difference is grazing, in this case most likely grazing pressure by camels on the young plants.

The observed pattern for *Helianthemum lippii* is rather similar to that of *Halogeton* and the cause may well be the same: irregular intensive grazing pressure on the small plants, especially in spring when they grow amidst the ephemerals.

From the above observations some conclusions can be derived on vegetation development under protection in this area:

In a not too far degenerated stand of *Haloxylon salicornicum* as found presently in the Khedr-el-Mai depression outside the exclosure, the *Haloxylon* plants in the first years of protection will develop to full grown specimens with diameters to two metres and a height of nearly one metre. After some years however, production of new growth will decrease, to an extent that very little is being produced, including hardly any viable seeds. Other large shrub species in the meantime will establish and develop, such as *Halogeton alopecuroïdes* and *Anabasis setifera*. In due time they may locally take over as dominant from *Haloxylon salicornicum* (see table X–5 the area sampled at 24 April 1973).

In between the shrubs a good perennial grass cover can develop, which however needs a certain grazing pressure to maintain its viability. In spring, after the rains, the existing seed sources guarantee a lush ephemeral vegetation.

X.C. The Rutba exclosure

In a wide wadi in the Faidhat-al-Mesad area in the Western Desert, about 15 km south of Rutba, an area of approximately 6500 donums (1625 ha) was fenced in

Figure X-7. The Rutba exclosure in spring 1970, after about eight years of protection.

Figure X-8. Vegetation differences near the fence of the Rutba exclosure, partly due to differences in grazing intensity, partly to not yet disclosed differences in the physical environment.

1962. In 1966 another 1200 donums (300 ha) adjacent to the first exclosure, were added. Local personnel of the Range Management Section, posted at Rutba, ensured full protection. In the years of observation, between 1970 and 1973, only the small flock of sheep owned by the organisation was regularly grazing in the area. In 1970 the longest protected area was bearing a very good low shrub cover of *Artemisia herba-alba*, with *Haloxylon articulatum* and *Achillea fragrantissima* as co-dominants. After the rains a wealth of annuals did show e.g. *Hordeum glaucum, Avena barbata, Plantago spp., Malva parviflora, Helianthemum spp., Tribulus terrestris, Erodium cicutarium,* and *E. deserti.* Perennial grasses and herbs were also quite abundant locally, with *Carex stenophylla* and *Poa sinaica* in the highest densities (Fig. X–7).

Unfortunately the lay-out of the exclosure is such that it occupies vitually the whole wadi. This makes a comparison of vegetation development outside and inside the fence rather useless because of the non-comparability of the physical conditions (Fig. X–8). No data on the vegetation cover at the time of fencing were available and the only remaining possibility was a comparison with the vegetation in similar habitats nearby. Results of such a comparison have been given in table VII–21.

The protected area itself is not homogeneous. Vegetation cover and density vary considerably. In June 1972 a sampling was carried out in a poor, fair and excellent stand to evaluate especially aerial biomass and forage as a result of ten years protection. The results (based on Kaul & Al-Mufti 1974) are given in table X–6. A comparison with biomass and forage production data mentioned before (see chapter VII) shows the tremendous scope for increased productivity under proper management. The 1600 kg dry forage per ha estimated for the excellent stand, could feed two sheep per hectare on a yearlong basis, not taking into account even the annuals' contribution. Incorporating the annuals component would certainly allow this grazing pressure for the wadi under the for Rutba average rainfall of only 120 mm per year.

Table X–6. Properties of the vegetation at three locations of different potential inside the Rutba exclosure, after about ten years of almost complete protection (June 1972, based on data in Kaul & Al-Mufti 1974).

	Poor stand	Fair stand	Excellent stand
No. of 1 m^2 quadrats measured	128	96	64
Average height (cm)	32	35	39
Density ($no.m^{-2}$)	0.5	1.1	2.7
Crown cover (%)	8.5	16	44
Aerial biomass dry ($kg.ha^{-1}$)	800	1800	5200
Available forage dry ($kg.ha^{-1}$)*	250	500	1600

* Only *Artemisia herba-alba*, estimated from the relationship established between aerial biomass and available forage, measured on 30 plants.

X.D. The Shabicha exclosure

The village Shabicha is situated in a large depression in the Southern Desert at 30°47'N, 43°40'E, about 300 m a.s.l. Such large depressions are common in the limestone areas of this part of Iraq. They may have been formed by solution of underlying limestone, followed by soil accumulation (alluvial and loess) in the newly formed internal drainage regions. Soil permeability could decrease through deposition of fine particles after flooding, resulting in water accumulation after every shower (cf. Agnew 1960). Vegetation in the Shabicha region is by and large confined to these depressions and to suitable sites in runoff areas where some soil has accumulated. The most common perennial plant species observed include *Artemisia herba-alba, Achillea fragrantissima, Lagonychium farctum, Peganum harmala* and *Artemisia scoparia*. In the flat plains *Haloxylon salicornicum* is found locally where sandy soil has accumulated. The spiny *Astragalus spinosus* and *Zilla spinosa* were seen scattered all over the area.

At the northern edge of the Shabicha depression, near a stony slope, an area of four donums (1 ha) was fenced, most likely in 1955. The exclosure was maintained irregularly over the years. Agnew found in 1960 the fence destroyed and attempted no comparative inside-outside studies because 'the vegetation was in general similar inside and outside the fence'. At the time the entire depression was green as a result of a severe flood earlier in the year and groundwater was at a depth of only one metre in the wells. Agnew measured a groundcover of 56% in the central part of the exclosure and 80% on low hummocks. Kaul passed the exclosure during a rangeland reconnaissance trip at 20 November 1968. He reports:

> 'Near Shubaija* a small fenced plot of the Range Management Division was visited. In spite of rather poor protection as evidenced by droppings of goats and sheep, the vegetation stand here compared to the unprotected area was good and the main species forming the vegetation were *Prosopis farcta***, *Achillea fragrantissima, Cynodon* spp., *Carthamus oxycanthus* and *Citrullus colocynthis'* (Kaul 1969a).

In December 1971 the exclosure was found abandoned since a few months. Although apparant grazing might have levelled off part of the difference in vegetation, an inside-outside comparison was nevertheless made (table X–7).

The date of observation was not appropriate for detailed quantitative data on grasses and forbs. It was seen that these were almost completely absent outside the exclosure. Inside the former fence locally scattered plants of *Plantago* spp., other forbs and the grass *Cynodon dactylon* were found. Density of *Achillea fragrantissima* and *Lagonychium farctum* together inside was 6.58 against 2.31 outside. The density of young *Achillea* plants was inside four times higher than outside (table X–7). The species is known as relatively unpalatable, probably due to its strong aromatic smell and taste. Under protection *Achillea* may reach a height of 50 cm and occasionally even more, as was observed in the Rutba fenced area. 47% of the plants inside and 74% of the plants outside were lower than 10 cm, clearly showing the heavy grazing pressure. Plants inside were still slightly higher than outside where not a single stump was found with a height exceeding 20 cm. The density of 0.97 and frequency percentage 26 illustrate the grouped occurrence of the young plants. The differences observed between the 1971 inside and outside situations can only be explained as the result of differences in grazing pressure. The estimated 2000 sheep in the village in 1971 found sufficient forage in the depressions etc. only in spring and early summer. Supplementary feed and transport over long distances was otherwise necessary. This indicates a very heavy grazing pressure for the Shabicha depression which has caused:
(i) a virtual disappearance of the already scarce, mostly annual grasses and forbs;
(ii) a reduction in the density of the perennial woody species, for *Achillea* with a factor five to six, for *Lagonychium* with a factor two (table X-7).

For both the species, *A. fragrantissima* and *L. farctum* Agnew for 1960 gives a more limited distribution than observed in 1971 inside the fence. Propagation of

* Shabicha
** Syn. *Lagonychium farctum*

Table X–7. Comparison of percentage frequency and density for three perennial species inside and outside the Shabicha exclosure. Data from 38 quadrats of 1 m^2 inside and 38 quadrats of 1 m^2 outside, recorded about 15 years after irregular protection, December 1971, (modified from Thalen & Al-Mufti 1973).

		Frequency (%)			Density (no.m^{-2})		
		Inside	Outside	Difference	Inside	Outside	Difference
Achillea fragrantissima	young plants	26	16	10	0.97	0.24	0.73
	older stumps	68	29	39	2.56	0.39	2.17
Lagonychium farctum	emerging shoots	26	16	10	3.05	1.68	1.37
Peganum harmala	whole plant	5	5	0	0.05	0.05	0.00

A. fragrantissima is mainly through seeds and differences in seed production and dispersal, germination or seedling establishment and development could explain the observed inside–outside difference. In view of the existing seed sourse, the easy seed dispersal, certainly over such short distances as here under discussion, and the similar soil conditions inside and outside the fence, the grazing pressure on the newly established seedlings seems the most likely factor inhibiting better development of the species in the depression outside the exclosure. *L. farctum* is generally found on deep alluvial soils with the groundwater not too deep, although the species has a considerable flexability through its deep and extensive root system. Its grazing value for sheep is very poor to zero, but new shoots and leaves are selectively grazed occasionally. Near Shabicha the species was found in good numbers with a height of a few decimeters at the most, while under undisturbed conditions the species may form thorny thickets two to three metres high (Guest 1966). Its spiny nature is a protection against complete eradication under heavy grazing pressure.

X.E. The Hatra exclosure

Near Hatra in the Jezira an area of several hundreds of hectares was fenced in 1966. Some preliminary observations on vegetation development were made in April 1971 and March 1973. The terrain and vegetation are rather heterogeneous and it proved difficult to find representative sites for a quantitative comparative vegetation description. Some results are given in table X-8. The overall impression was that vegetation changes were taking place very slowly. An increase in density and thereby cover of the two small species *Poa bulbosa* and *Carex stenophylla* was generally observed. Inside the fenced area well developed specimens of *Andrachne telephioides* and *Scrophularia xanthoglossa* were seen, species virtually lacking outside the fence. The area around Hatra is mostly ploughed land, fallow land, or land that has been cultivated some time ago and now bears a cover of weeds of low palatability *e.g. Lagonychium farctum, Alhagi maurorum* and *Peganum harmala*. Lack of sufficient seed source may, therefore, well be one of the important reasons for the initial slow recovery of the vegetation in this part of the Jezira, which receives an average annual rainfall of about 250 mm. Also in view of the discussion in IX.C. an intensive study of the vegetation development in the Hatra exclosure would be of great value.

X.F. Overall impression

The observations on the vegetation in the exclosures, in comparison to the vegetation outside under continuous use, lead to the following general conclusions.

Under protection from overgrazing the aerial biomass and production of the shrub vegetation increase and vegetation types with a very high shrub cover and a productive herbaceous layer can develop. Light to moderate grazing may be necessary to maintain the productivity of some of these shrub species as clearly shown for *Haloxylon salicornicum.*

Succession is in the shrub vegetation a very slow process; it may take fifteen to twenty years or more before distinct differences in shrub species composition (in terms of presence or absence of species) inside and outside exclosures can be seen. There is sufficient evidence, however, to show that under protection succession is taking place. Considering the representativeness of the Khedr-el-Mai area for extensive areas in the Southern and Western Deserts, this suggests that the widespread occurrence of *Haloxylon salicornicum* (in often almost monospecific stands) in the study area may well be the result of long time heavy utilization. To what extent this also holds for other species and which species under protection

Table X–8 Density of the most common species observed inside and outside the fence at two locations of the Hatra fenced area, after five and seven years of protection, respectively.

	Density (no.m^{-2})			
	Area I (18 April 1971)		Area II (23 March 1973)	
	Inside	Outside	Inside	Outside
Poa bulbosa	126	110	n.r.	n.r.
Carex stenophylla	212	111	n.r.	n.r.
Artemisia scoparia	0.53	0.18	0.55	0.24
Andrachne telephioides	0.10	––	0.14	––
Scrophularia xanthoglossa	0.25	0.01	0.10	0.03
Lagonychium farctum	0.24	0.02	0.09	0.12
Salsola sp.	––	––	0.68	0.01
Achillea conferta	––	––	0.09	––
Erodium spp.	0.13	0.19	––	––
Peganum harmala	––	––	––	0.27

will take over in due time can only be worked out with further longterm studies (over decades) in exclosures. Only with such data an indication can be found of the vegetation types that once may have existed in the area, before the overall human impact.

CHAPTER XI

COMBINED EFFECTS, THE OVERALL PICTURE

XI COMBINED EFFECTS, THE OVERALL PICTURE

XI.A. Fuel cutting and grazing combined

The utilization processes in the desert areas of Iraq directly affecting the natural vegetation, grazing, fuel gathering and local cultivation, have been treated separately, but in reality take place in the same areas with accumulative destructive effect. Grazing alone, as it is generally practiced in the area under discussion, will not destroy the vegetation. The herds will move on before all vegetation has disappeared. At least the thick woody parts of the shrubs survive and will sprout again. Only continued, severe overgrazing may cause the death of all plants. What generally happens is a shift in species: the more palatable and nutritious ones will decrease and may totally disappear, less palatable ones will, at least relatively, increase in importance and new, often relatively unpalatable, species may appear.

Fuel gathering is different, more destructive, but usually limited to relatively small areas. It affects all woody plants and may result in the local disappearance of the shrub vegetation. The positive effect of the perennials on the annuals, as analyzed and demonstrated by Agnew & Haines (1960), was already mentioned before. Denudation of an area under perennials, especially shrubs, may in due time well result in a decreasing ephemeral cover, with as final stage a barren desert.

The added effect of fuel cutting and grazing is often disastrous for the vegetation. As a result of cutting at surface level the woody perennial will die or the stump and roots will survive and sprout again. Repeated grazing of these sprouts will in most cases still cause the death of the shrubs (Fig. XI–1). Grazing of young sprouts from the remaining stumps of plants once cut for fuel is very likely to occur. The places for Bedouin camps are in many cases fixed (e.g. near wells), not to speak of the few villages and towns. Near such sites fuel cutting takes place most extensively while there also concentrations of livestock can be found. In the Khedr-el-Mai exclosure the added effect of fuel cutting and subsequent grazing was simulated for *Haloxylon salicornicum* (table XI–1). All (six) plants, subjected to one cutting only, survived, against only two out of eight subjected to an additional cutting simulating intensive grazing afterwards. The results suggest an even heavier impact of repeated cutting than the results of the Nukhaib repeated cutting experiment discussed before in VIII.C.1. (table VIII–1).

Once the perennial shrubs have been cut, the forbs and grasses, usually found in the highest concentrations near these plants, lack their protective cover and are easily and intensively grazed. Organic matter content of the hummock and soil around the original shrubs decreases and thereby the moisture holding capacity, which again makes the soil conditions less suitable for germination. Moreover, the lack of a vegetation cover may cause serious wind and water erosion.

Table XI–1. Response of *Haloxylon salicornicum* to simulated fuel cutting with and without subsequent grazing, after about 12 years of complete protection, in the Khedr-el-Mai exclosure ($\overline{x}$ – mean, s-standard deviation).

	Treatment		Crown volume before treatment (m^3)		Response to treatment, 28 November 1972.				
	Actual	Simulating			Dry weight new growth (g)		Length of longest sprout (cm)		Number of plants alive
			$\overline{x}$	s	$\overline{x}$	s	$\overline{x}$	s	
I (n = 5)	no cutting	no fuel cutting or grazing	0.50	0.15	41.4	34.2	11.6	3.9	all five
II (n = 6)	one cutting at surface level at 16 December 1971	fuel cutting without subsequent grazing	0.44	0.19	58.3	48.1	23.7	6.5	all six
III (n = 8)	cutting at surface level at 16 December 1971 repeated at 18 May 1972	fuel cutting followed by grazing of the new sprouts	0.39	0.18	8.0	16.4	5.1	9.6	two out of eight

XI.B. Cultivation combined with grazing and/or fuel cutting

Uprooting all existing vegetation, when ploughing particular areas, to grow a grain crop is the process with the most destructive impact for the natural vegetation and its environment. Even more than for the combination fuel cutting and grazing, in the case of cultivation the effect of subsequent grazing and, in later years, also fuel

Figure XI–1. Deterioration of a once good *Haloxylon salicornicum* cover by fuel cutting and repeated grazing. Fuel cutting has only left the root system and a few centimetres of the woody stems intact. Subsequent regular intensive grazing does not allow the plants to recover to their original size and may cause the plants death in course of time (see right foreground). The now unprotected topsoil is easily removed by wind (and probably water) action. In the photographed area the top 10-20 cm have already disappeared exposing the upper parts of the root system and giving the plants this particular appearance.

Table XI–2. Comparison of two stands growing under similar environmental conditions except that one is only lightly grazed while the other is found in an area at least once ploughed and under continuous heavy grazing and fuel cutting (see Fig. XI–2).

Area and type of utilization	Bank of Wadi Ubayidh, 1.5 km east of Faidhat-al-Jeez. No recent fuel cutting, light grazing.		Bank of Wadi Ubayidh in front of Faidhat-al-Jeez well. Cultivated till two or three years ago, continuous heavy grazing and fuel cutting.	
Date of sampling	4 June 1973		2 June 1973	
	Density ($no.m^{-2}$)	Crown cover (%)	Density ($no.m^{-2}$)	Crown cover (%)
Haloxylon salicornicum	0.33	22.3	0.03	0.7
Haloxylon articulatum	–	–	0.01	0.5
Artemisia herba–alba	–	–	0.04	negl.
Artemisia scoparia	0.04	negl.	0.27	0.1
Peganum harmala	–	–	0.06	0.3
Achillea fragrantissima	–	–	0.01	negl.
Astragalus spinosus	0.01	negl.	0.01	negl.
Zilla spinosa	0.01	negl.	–	–
All species	0.39	22.4	0.43	1.6
Total aerial biomass, dry ($kg.ha^{-1}$)	1312		102	
Estimated total available forage, dry ($kg.ha^{-1}$)	525		20	

cutting in the same areas has to be added. For the northern part of the Jezira this aspect of combined effects was touched upon in IX.C. An example for the Southern and Western Deserts is shown in Fig. XI–2 for an area near a well. A similar area nearby had an exceptional good cover of *Haloxylon salicornicum.* The cultivation was followed by fuel cutting and continuous grazing and this species, and also others establishing in the newly created open area, had little chance. Table XI–2 gives a comparison of the good stand of *Haloxylon salicornicum* with the vegetation resulting in the area under continuous heavy use, including cultivation. The remark must be made that it is likely that the almost monospecific stand of *Haloxylon salicornicum* in itself already represents a degradation stage and a response to occasional fuel cutting and continuous, but not too heavy, grazing. The table shows that the cover has been brought down from over 20% to 1.6% and the estimated available forage from over 500 $kg.ha^{-1}$ to only 20 $kg.ha^{-1}$. It may be regarded the last step in vegetation degradation by overexploitation in the absence of any proper management. Erosion in such areas is often serious.

Figure XI–2. Remnants of the vegetation in a shallow depression near a well after cultivation and subsequent grazing and fuel cutting. A few shrubs apparently survived the ploughing and around these ones new hummocks were formed. Such, yet intact, specimens of *Haloxylon articulatum* can be seen in the background. The ones in the foreground were already cut for fuel. All more palatable species have disappeared. A few plants of *Peganum harmala* and *Astragalus spinosus* only survive.

XI.C. Summary of the impact of utilization

XI.C.1. *The processes*

To what extent the processes discussed above cause land degradation depends on a variety of abiotic and biotic environmental factors, e.g. the vegetation type that is affected, the underlying soil (moisture holding capacity, fertility, depth), local relief (runoff) and the degree and time of utilization in relation to the prevailing climatic conditions (drought years). Heavy grazing not only affects the vegetation by the taking of forage, but perhaps even more by trampling and altering the condition of the top soil. As stated in Warren & Maizels (1977): 'Hooves are far more damaging than mouths, . . .' The structure of the top soil is damaged, aggregates disintegrate, litter is reduced and blown away and after a heavy shower the remaining fine particles are washed in the pore space. A crust is formed which seals the surface and leads to an increased runoff and erosion at the next shower or strong wind. In the process so many variables are involved that a statement on the impact of heavy grazing can only be of a general nature. For the impact of the various processes discussed an overall summary has, nevertheless, been attempted and is given in table XI–3. The utilization types are arranged in an increasing order of impact, from the beneficial effect of light grazing to the devastating combination of cultivation followed by fuel cutting and grazing. In Fig. XI–3 a,b,c,d the process of degradation of the desert shrub rangelands is shown in its extreme appearance. It should be emphasized that the figure does not show successive stages photographed in the same area, but a sequence of stages that might be found in course of time in one area.

XI.C.2. *The pattern of the processes*

After the opening up of the desert mainly with the help of equipment provided by modern technology (see IV.G and IV.H) the entire area is now under a continuous use and the utilization processes take place in a range of intensities all over the desert rangelands. There are hardly any rest periods for the vegetation. The system of migration of man and his livestock within a well defined large area as dictated by tribal customs and watering rights has been replaced by a movement into the area from a large number of fixed points, at the fringe as well as inside the deserts. At all these places a permanent water supply is guaranteed and they form the nuclei of the desertification process. Around each perennial source of water an area, the so called 'piosphere' exists where the degrading effect of overexploitation diminishes with the distance from the centre (Lange 1969). Warren & Maizels (1977) used the term 'phryganosphere' for areas around settlements where firewood is collected. The extent of such areas under direct influence of utilization varies with terrain and vegetation conditions. Warren & Maizels (1977) quote examples of a radius of five to eight kilometres in the Sahel. In Iraq larger areas were observed. Bedouins mentioned in interviews a movement with the flocks to distances of at least 10 km and as far as 20 or 30 km from the wells. At one occasion a flock was found at a distance of six hours walking with the sheep from the nearest well. Assuming a direct influence-sphere with a radius of 15 km would require some 340 watering points equally spaced to cover the entire area under discussion. For a radius of 20 km it amounts to less than 200 watering points.

Table XI–3. Qualitative summary of the effects of exploiting the desert shrub rangelands of Iraq by grazing, fuel cutting, cultivation and some combinations thereof.

Utilization	Effect (generalized)		
	Shrub vegetation	Grasses and forbs, environment	Examples and evidence
1 LIGHT TO MODERATE GRAZING	Hardly any effect. Some less palatable species may increase their cover at the cost of high palatable ones. Growth of new shoots can be induced.	Hardly any effect. Possibly a slight shift towards less palatable species and a positive effect on germination by breaking the surface sealing, distribution of seeds and creating of some open space.	Rutba fenced area, Fig. X–7. The fact that grazing could until recently continue for centuries without much damage.
2 MODERATE TO HEAVY GRAZING	Change in species composition. Relatively high palatable species disappear, only less palatable ones will survive.	Reduction in cover, mainly through too early or too long selective grazing of the high palatable species. Limited, usually local, erosion hazard.	Khedr-el-Mai exclosure, table X–5. Disappearance of *Salsola rigida* (Figs. VII–3 and VII–4). Disappearance of grasses; table VII–17.
3 CONTINUED VERY HEAVY GRAZING	After a lapse of time only a few scattered specimens of species of low to non-palatability will survive.	Reduction in cover upto total disappearance in case of too early, severe grazing (no seed source left). Serious erosion hazard.	The areas not in the immediate vicinity (see 8) but at some distance of wells and villages, e.g. around Busaiya and As Salman. Not separable from 5.
4 FUEL CUTTING	Complete eradication of species with high fuel value; local eradication and thinning out of lower valued species.	Reduction in cover by local disappearance of the protecting and good seedbed providing shrubs. Increased erosion hazard.	Eradication of *Haloxylon ammodendron* (VIII.B.3.) local disappearance of *Rhanterium epapposum* (Figs. VIII–13 and VIII–14); Agnew & Haines (1960).
5 FUEL CUTTING FOLLOWED BY GRAZING	Complete eradication of all species with high fuel plus forage value. Thinning out of less palatable species with high fuel value. Survival of non-palatable species without fuel value.	Reduction to total eradication, depending on the intensity of the cutting and grazing. Increased to extremely serious continuous erosion hazard.	All denuded areas around villages and often visited wells. *Haloxylon salicornicum* data in tables VIII–1 and XI–1 and Fig. XI–1.
6 CULTIVATION	Temporary denudation. A few plants of high resistant species may survive and develop luxurious growth, likely to be grazed or cut soon.	Reduction to almost zero; only after a long time re-established. At least a temporary, extremely serious erosion hazard.	Table IX–2, Figs. IX–5, IX–6 and XI–3b.
7 CULTIVATION FOLLOWED BY GRAZING	Eradication of virtually all plants. An occasional unpalatable shrub may sprout and escape the grazing.	Reduction to almost zero; under sheep and goats grazing virtually no re-establishment. Permanent extremely serious erosion hazard.	For the shrubs the situation in the Upper Jezira. The stands shown in Fig. IX–9 and IX–10.
8 CULTIVATION FOLLOWED BY FUEL CUTTING AND GRAZING	Complete eradication of all species.	Complete eradication of virtually all species. Only a few relatively unpalatable plants may remain. Continued extremely serious erosion hazard.	Table XI–2, Figs. XI–2, XI–3c and XI–3d. The most common picture in the near surroundings of often visited wells, and of villages.

Figure XI–3. a,b,c,d. Four stages in the overexploitation – denudation – erosion process.

a. The vegetation cover of a wide depression after a good rainfall season. *(Artemisia herba-alba, Malva parviflora* and *Plantago* spp., are the dominants). The only exploitation is an occasional light grazing. This picture may once have been the general appearance of large areas in the Iraqi deserts, but can now hardly anymore be found.

b. A similar area as in a) after repeated cultivation, followed by grazing. Almost all plants have disappeared. The furrows resulting from the ploughing will remain visible for many years. After heavy showers run off will collect in these funnels, soon forming small streams. This stage can be observed all over the Iraqi deserts.

c. Soil erosion in full progress at the fringe of a formerly cultivated area. Note the few *Artemisia scoparia* plants, established after the cultivation, and the remnants of a once rather dense cover *(Artemisia herba-alba, A. scoparia, Haloxylon salicornicum)* still found at the transition of the stony plain and the depression.

d. Extreme case of accellerated erosion. Deep gullies have been formed. Reclaiming such areas for grazing is difficult and costly. This (final) stage is not often seen in Iraq. Sheet erosion causes greater losses than gully erosion. Usually the soils in the desert are shallow and erosion leads to the exposure of the parent rock or extensive otherwise stony areas.

There are many more, although during the fieldwork many were seen to be out of order. The wells are not equally spaced but occur in clusters. On the other hand in such a calculation the use of tankers to distribute the water (see Fig. IV–25) is not incorporated. A more accurate picture can, therefore, not be given. However, it may be clear that under the existing conditions few areas can escape the continuous impact of use by man and his livestock. During the fieldwork this general picture was confirmed. Very few areas were found to be relatively unused or little used. The degradation processes are not confined to a well defined area, e.g. to a zone bordering an already earlier overused area. The only exception in this respect are parts of the Upper Jezira under cultivation. Desertification in the rangelands of Iraq is not a 'desert creep' but rather a 'rash of degraded areas' as pointed out by Warren & Maizels (1977) for other areas threatened by desertification.

XI.C.3. *Conclusions*

The information presented in the previous chapters and the summary picture in the foregoing sections lead to the following conclusions:

– The overall impact of an increased exploitation of the vegetation of the desert shrub rangelands in Iraq has reached a highly destructive phase.
– Areas once under a productive vegetation cover have been denuded and the soils of these areas are blown away and washed away in erosion processes.
– Over extensive areas natural regeneration of a productive vegetation has become difficult or impossible, e.g. through an alteration of the physical environment (increased runoff, removal of the top soil, formation of desert pavement or of a hard crust, surface sealing), limiting the possibilities for germination and seedling development, through the virtual absence of a seed source of the valuable species, and through a continuous grazing of the newly established plants.
– Overexploitation of the desert shrub rangelands is the cause of a decreasing benefit and an ongoing degradation of the land and its production potential.

Intensive management is required to stop further degradation, increase chances for recovery and eventually restore the productivity at its highest level. In the present situation, in the absence of well organized management programmes, degradation will continue and in a few more decades the once productive desert shrub rangelands may have been converted into a barren, dead desert. As stated in the first lines of chapter I, such management programmes require, apart from a knowledge of the resources and the impact of the utilization processes, organization and coordination. Inventory and evaluation should be a first step in the management programmes. Selected procedures and techniques dealing with the inventory and evaluation of desert shrub rangelands are discussed in the last part of this study, after (in chapter XII) the economic value of the desert shrub rangelands for Iraq has been shown.

PART FOUR

Management

CHAPTER XII

ECONOMIC JUSTIFICATION

XII. ECONOMIC JUSTIFICATION

XII.A. Introduction

The conclusion that large areas of the desert shrub rangelands of Iraq are in an advanced stage of degradation automatically leads to the question how improvement could be achieved. In this respect the Iraqi situation is not unique. 'What can be done to shift from destruction to improvement' is a heading in Williams *et al.* (1968) discussing this pertinent question for the world's rangelands. For the countries neighbouring Iraq reference can be made to Kernick (1966), Heady (1972), Poore & Robertson (1963) and van der Veen (1964), for the Middle East as a whole to Ibrahim (1967), Pearse (1970), and Clawson *et al.* (1971).

Most of the users seem already well aware of the degradation. In communications during the fieldwork, Bedouins and other graziers mentioned the scarcity of forage in comparison to earlier times. On occasion a direct request for assistance to improve the situation was made. The increasing meat prices have been noticed by another part of the population, in their cities often far away from the reality of the desert rangelands (Ahmed & Al-Roumi 1967; see also Kaul & Thalen 1971a).

Arguments related to the functioning of the rangeland ecosystem could also be used, e.g. the danger of land deterioration beyond recovery. Since desertification has become a worldwide concern after the United Nations conference in Nairobi in 1977, such arguments are gaining ground (Warren & Maizels 1977). For Iraq they can be found in the previous chapter.

A prerequisite for action is that both, the user of the land and the owner of the land believe in the necessity for improvement. This can be based on different arguments as given above, but economic ones are usually most convincing. In this chapter, therefore, an attempt to evaluate the resource values of the rangelands of Iraq in general, and those of the desert shrub ranges in particular, is presented.

XII.B. The concept 'range resources'

Natural resources in general have been briefly characterized as 'wealth supplied by nature' (Dudley Stamp 1966) and in a more lengthy definition as 'those uncaptured natural stores which are useful to mankind in any way' (Allen 1959). The list of natural resources is not a fixed one because human assessment of what is useful and thereby constitutes a resource is constantly changing.

In recent years the ideas about natural resources related to rangelands, and also about rangeland management in relation to the resource values, have broadened as a result of the ever increasing attention paid to the potential multiple use of such areas. While earlier definitions of range management stressed only the livestock production side, more recently aspects of wildlife, recreation and especially conservation of the vulnerable water-soil-plant-animal compex are also mentioned

(Hedrick 1966; Pechanec 1967; Lewis 1969; Harris 1977.) In this context the natural resource values of the rangelands i.e. the range resources, in the modern concept, are many. A number of possible resource values was already indicated in I.C.3. when pointing out the value of shrubs in arid lands. For rangelands in general one could think amongst others of the following resource values:

- production of forage for livestock
- providing a habitat and producing food for wildlife
- production of firewood and small timber
- production of medicinal plants
- preservation of plants and animals for breeding and domestication
- preservation of natural and semi-natural ecosystems for study
- conservation of water and soil
- offering possibilities for recreation
- offering scenic beauty.

Other values may be thought of, but from the above list it can be seen that an assessment of the total resource value of the rangelands in a certain area, in the modern concept, will always be very complicated and often impossible. Most of the resource values mentioned above apply to the desert shrub rangelands of Iraq.

XII.C. Problems in evaluating the range resources

An assessment of the value of the range resources of Iraq should preferably be given in a form that allows comparison in one way or the other with other resources. Contribution to the national income is such a form when an economic evaluation is attempted. However, most of the values, indicated above as part of the total resource value, cannot be expressed in such economic terms. There is a lack of data, but certain resource values can never be evaluated in economic terms because of their abstract and subjective nature.

Although in Iraq the resource values, pertaining generally speaking to environmental quality, are increasingly being realized, nevertheless there are no data available on which an assessment in economic terms could be based. The same can be said from a number of products from rangelands, which can sometimes be seen on local markets, e.g. charcoal, medicinal plants and meat from gamebirds. The only renewable natural resource value of the rangelands of Iraq for which an estimate in quantitative terms can be attempted is the contribution of the natural vegetation to the national income through livestock production and livestock products. This aspect is discussed below.

The grazing areas of Iraq in the broadest sense comprise various types of land. With the exception of relatively minor areas occupied by fresh water, villages and towns, virtually the whole area of the country contributes to livestock forage supply, at least during part of the year In this respect the country can be classified as follows (see Fig. XII–1):

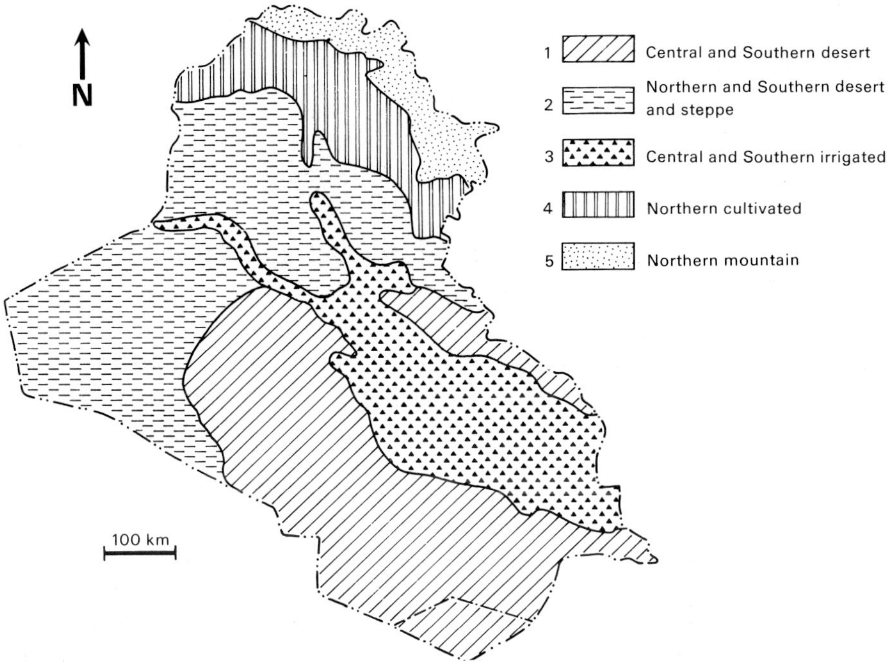

Figure XII-1. Agro-climatic regions in Iraq, mainly with respect to forage production (after Springfield 1957).

- Deserts and dry steppes virtually without cultivation (1) and (2)
- Irrigated cultivated land in the plains (3)
- Rainfed agricultural lands in the less arid steppes and lower mountains (4)
- 'Pasturage' in the oak forest and mountain pastures above the upper limit of oaks (5)

The first group, desert and dry steppes, was object of the present study. The Southern Desert, Western Desert and Jezira comprise about 55% of the country's total area and constitute in area and total forage amount by far the most important rangelands.

Crucial point in a discussion on the contribution of rangelands to the national income through livestock production is the estimate of the part of the forage contributed by native plants in relation to the contribution of cultivated fodder-crops and other cultivated plants. Such an estimate is difficult to make for the arid areas where plant production directly depends on – and vaires with – the

Table XII–1. Forage contributed in Iraq by native and cultivated plants in the five regions delineated in Fig. XII–1 according to Springfield (1957).

Region (see Fig. XII–1)	Per cent forage contribution	
	Cultivated plants	Native plants
1. Central and Southern Desert	0	100
2. Jezira and Western Desert	2	98
3. Central and Southern irrigated	30	70
4. Northern cultivated	20	80
5. Northern mountainous	10	90

yearly highly variable amount of rainfall. Springfield (1957) has made an estimate, dividing Iraq in five agroclimatic regions (Fig. XII–1). Without mentioning the method or calculations applied he presents the figures reproduced in table XII–1. It is the only estimate of its kind available. The figures given for the study area may be high for the actual contribution by native plants, but an estimate of 80% for the total amount of forage contributed by this group is certainly well below the actual situation. Ahmed (1968) referring to Springfield (1957) estimated it at 88%. The estimated part of this amount taken from the desert rangelands is discussed in XII.E. For a quantitative assessment, the contribution of livestock products to the national income can be summarized in the following points, subsequently discussed with their trends of development:

- changes in the livestock population
- goods and services consumed or exported, *viz.*
  - meat (sheep, cattle, buffaloes, goats, camels)
  - hides, skins, casings, etc. (species as above)
  - wool (sheep) and hair (goats and camels)
  - milk and milk products (buffaloes, cattle, sheep, goats, camels)
  - dung for manure and fuel.

XII.D. Contribution of the livestock industry to the national income

XII.D.1. *Changes in livestock population*

Under the same heading in IV.F.4. the available data on the total livestock population of Iraq were discussed and the figures presented in table IV–14. For sheep, goats, camels and donkeys, the livestock primarily depending on the forage supply of the desert rangelands and other natural ranges, the figures are repeated in table XII–2. The changes in stock can be translated in economic terms by using average ex-farm prices for each livestock type (Haseeb 1964). This aspect has been worked out and taken into account in calculating the figures presented in table XII–5.

XII.D.2. *Export of live animals*

Data on the number of live animals exported by Iraq are available from the yearly issued surveys of the Central Bureau of Statistics (table XII–3). Until 1959 the country was an important and regular exporter of live animals, especially sheep, goats and camels, mainly to other Arab countries. Syria, Jordan, Lebanon and Saudi Arabia were the major importers from Iraq. From 1959 onwards livestock production was directed to satisfy the domestic demands. Haseeb (1964) compared the published export statistics of Iraq with the published import figures of Syria and Jordan and found a significant understatement of livestock exported, especially sheep, in the Iraqi statistics, showing the illicit exports. The official figures of table XII–3 clearly show the rapid decrease to almost zero in exported live animals in a period of 20 years. In 1953, net exports counted for 6.8% of the total livestock addition to the national income. In 1961, this figure was 0.8% (calculated from Haseeb 1964). In 1969 total export of live animals was for the value of I.D. 34 783.–, while for the total imports about four times that amount (I.D. 137 935.–) was spend (Central Statistical Organization 1970). In 1974 food and live animals imports reached about I.D. 200 million, nearly 10% of the official oil receipts (Europa Publication Ltd. 1975).

Table XII–2. Estimated total number of sheep, goats, camels and donkeys in Iraq for selected years, from available census data. For the sources see table IV–14.

	(thousands)			
Year	Sheep	Goats	Camels	Donkeys
1971	5 582	1 794	38	497
1965	11 040	1 846	202	542
1956	9 221	2 640	193	493
1950	7 490	1 754	279	326
1938	5 525	2 224	82	–

Table XII–3. Average *yearly* figures for the periods indicated of the number of animals examined by veterinary authorities and granted certificate for export. (Source: 'Annual Abstract of Statistics', Central Bureau of Statistics, Ministry of Planning, several volumes).

Livestock type	1949–1950	1954–1958	1959–1963	1964–1968
Sheep (males)	954 805	556 276	33 203	4 774
Bulls	27 391	9 873	1 257	117
Horses	5 186	3 170	2 218	3 084
Mules	3 228	1 304	261	37
Goats (males)	216 724	158 559	1	–
Camels (males)	621	953	221	–
Buffaloes	1 120	–	–	–
Total animal units (thousands)	262.8	151.3	11.1	4.8

XII.D.3. *Goods and services consumed or exported*

The decrease in exported live animals, shown above, goes together with an increased slaughter and export of livestock by-products, due to the increased domestic demands. Meat consumption and meat prices were studied by Ahmed & Al-Roumi (1967), Ahmed (1968) and Sykes *et al.* (1969); slaughter receipts by Haseeb (1964), Jones *et al.* (1967) and Kassir *et al.* (1967).

It has been shown that there is very low consumption of protein rich foods particularly in the low income families. Per capita yearly consumption of meat in the low income group, compared to the high income group in 1966 was 15.2 kg against 50.9 kg. For milk these figures were 16.7 kg and 47.5 kg, respectively. It was indicated that it is highly significant that the low income group will increase its consumption of meat, milk and other dairy products provided they have the means to do so (Sykes *et al.* 1969). Iraq experienced a period of comparatively rapid development and population growth. Increase in per capita income during 1953–64 was calculated at 6.5% per year, while the population increased at a rate of about 3% (Ahmed & Al-Roumi 1967). If the trends continue the livestock industry will have to meet the demands of a higher wage earning increasing population (van der Veen 1962; Central Statistical Organization 1970; Al-Hashimi 1972).

Ahmed (1968) has shown that per capita real income is the most important factor in explaining mutton and beef prices. The number of slaughtered animals has been increasing rapidly (table XII–4). In spite of this increased production, however, and stopping of export, meat supply has lagged behind its consumption by about two per cent yearly. Ahmed & Al-Roumi (1967) calculated an average yearly rate of increase for mutton and beef prices of about seven per cent during 1964–1975.

Price index figures for meat and milk products increased from 100.0 to 1048.4 during 1939–1965, while the general index number increased to 506.2 only over the same period. The price index for meat in Baghdad markets increased to 156.45 in 1968 during six years from the base year 1962, the foodstuff wholesale price index to only 110.96 over the same period (Kaul & Thalen 1971a).

As a result of increased slaughter, the added value by meat, hides, skins, casings etc. to the national income almost doubled between 1953 and 1961 (Haseeb 1964). The milk products contribution increased by about 50% over the same period (for breakdown see Haseeb 1964) The total value of these two groups of products accounts for about 80% of the total contribution made by livestock. Wool and hair production increased proportionally to the increase in number of sheep and goats and camels respectively. Its values range from about 5% to 10% of the total livestock contribution. Values calculated for dung (manure and fuel) are the lowest of all factors to be considered and of little importance for the present discussion.

Table XII–4. Animals slaughtered and meat produced in Iraq over a twenty year period, 1951–1970 (Calculated from: 'Annual Abstract of Statistics', Central Bureau of Statistics, Ministry of Planning, several volumes).

	Number of animals slaughtered* (thousands)			Meat produced** (1000 kg)		
Year	Sheep	Goats	Camels	Total from (1), (2) and (3)	All meat, (4) + buffaloes and cattle	(4) as per cent of (5)
	(1)	(2)	(3)	(4)	(5)	(6)
1970	2 892	749	17	71 625	114 480	62.6
1969	2 794	563	15	66 575	106 700	62.4
1968	2 901	563	13	68 010	106 405	63.9
1967	2 816	516	13	66 010	102 568	64.4
1966	2 526	650	15	62 520	97 209	64.3
1965	2 213	620	15	55 810	92 470	60.4
1964	2 276	602	12	56 350	89 309	63.1
1963	2 312	603	12	57 085	87 276	65.4
1962	2 184	819	12	57 765	85 234	67.8
1961	2 263	742	17	58 940	87 827	67.1
1960	2 286	654	14	57 630	87 250	66.1
1959	2 110	732	6	54 080	82 668	65.4
1958	2 054	638	6	51 550	77 743	66.3
1957	1 895	744	5	49 090	75 994	64.6
1956	1 780	613	2	45 095	69 703	64.7
1955	1 657	515	3	41 315	63 528	65.0
1954	1 439	508	3	36 850	56 726	65.0
1953	1 149	433	3	29 925	44 947	66.6
1952	864	251	2	21 495	33 307	64.5
1951	863	245	2	21 235	32 768	64.8

* Adjusted to include illicit slaughtering and slaughter of animals in homes and rural areas (sheep, goats and cattle by 20% upwards, buffaloes and camels by 10% upwards, see Haseeb 1964).

** The following values were used for average carcase weight of dressed meat: 20 kg for sheep, 15 kg for goats, 100 kg for cattle and 150 kg each for buffalo and camels (see Haseeb 1964, note p. 63).

XII.D.4. *Total contribution through livestock*

Agriculture, including livestock, contributed about 30% to the national income around 1950. This value had decreased to just above 20% in 1960 and 19% in 1969. Its contribution to the national income, excluding the revenue from crude

Figure XII-2. The desert shrub rangelands of Iraq contribute to the national economy mainly through livestock production. Slaughtering outside abattoirs as shown in this photograph of a desert village, has to be taken into account in an assessment of this contribution.

Table XII–5. Contribution of the livestock and livestock products to the national income for the years 1953, 1957 and 1961; calculated from Haseeb (1964).

	1953	1957	1961
Total livestock contribution to the national income (1000 I.D., constant 1956 prices)	27 959	41 527	42 260
Total livestock contribution to the national income (per cent of total, including oil revenues)	10.6	11.7	9.0
Livestock contribution to the agricultural sector, i.e. agriculture, forestry and fishing (per cent)	32.7	35.6	43.5
Percentage of the total contribution added per livestock type			
Sheep	43.5	41.1	43.1
Goats	8.1	8.7	7.8
Cattle	35.4	37.7	35.9
Buffaloes	11.9	11.7	12.2
Other	1.1	0.8	1.0

oil, was about twice as high: 63.0% in 1950, 42.5% in 1961 and 40.1% in 1969, as can be calculated from Haseeb (1964) and Central Statistical Organization (1970). The tremendous increase in oil revenues after the nationalization in 1972 (1964: US $ 353 million, 1969: US $ 483 million, 1974: US $ 7200 million) makes calculations on the present contribution to the national income little illustrative for the actual rôle played by the desert shrub rangelands in the country's total agricultural product.

A summarized picture *for a period for which detailed data were available* of the total contribution to agriculture and to the national income, of livestock and livestock products by species, is given in table XII–5. The totals show that the livestock industry contribution to the total agricultural contribution increased from 32.7% in 1953 to 43.5% in 1961. Although the figures in table XII–5 are not from recent years, they nevertheless serve the purpose to show the relative importance of livestock production in the total agricultural product. More recent statistics would first require a thorough analysis including adjustments, as done by Haseeb (1964) for the data used above, a task beyond the scope of this study.

XII.E. Contribution of the desert rangelands to the national income

Livestock completely or partly dependent on the vegetation of the desert rangelands includes camels, sheep, goats, donkeys and very few horses and mules, in order of dependence. Figures on the rate of dependence have to be rough estimates, as no census data on this aspect have ever been collected. An indirect indication on the distribution of livestock over holdings and cultivated areas against the natural ranges can be obtained when the census figures from 1952/53 and 1958/59 on livestock are compared with the estimated totals for those years. In this respect the two censuses showed some similarity, especially for sheep, goats and cattle that are easiest to cover. The figures indicated that about half of

Table XII–6. Estimated number of livestock in Iraq, dependent on the forage supply of the desert rangelands, for five selected years. Figures were calculated from table XII–2 (except the estimated (extrapolated) figure marked with*), based on assumptions discussed in the text (XII.E.).

Year	Livestock numbers (thousands)				Animal units (see table IV–14)	
	Sheep	Goats	Camels	Donkeys	(thousands)	Difference with 1938 (%)
1971	3 721	598	34	50	906	– 5
1965	7 360	615	182	54	1824	+ 90
1956	6 147	880	174	49	1608	+ 68
1950	4 993	585	251	33	1420	+ 48
1938	3 683	741	74	30*	958	0

the sheep and one third of the goats were kept outside holdings and cultivated areas. This was taken into account and also the fact that livestock from the cultivated areas also partly and irregularly feed on the desert rangelands. Considering further the percentages given in table XII–1, it was estimated that on an average in Iraq camels for 90%, sheep for 67%, goats for 33% and donkeys for 10% depend on the vegetation of the desert rangelands. Applying these estimates to table XII–2 a new table was compiled, indicating the estimated number of livestock in Iraq fully dependent on the forage supply of the desert rangelands (table XII–6). From this table we see that from 1938 to 1965 the total animal units increased by 90%, dropping dramatically afterwards. A comparison of the animal units in table IV–14 and table XII–6 shows that according to the estimates in 1950, 1956, 1965 and 1971, for 37, 32, 36 and 27%, respectively, of the animal units the feed came from the desert rangelands. This is roughly one third, dropped to just over one quarter in the last few years. Although necessarily a rough approximation the figure seems certainly not on the high side compared to estimates of others, e.g. Springfield (1957, table XII-1) and Ahmed (1968). Having arrived at these figures and applying the percentages given in table XII–5 of the contribution of livestock to the national income, it can be stated, that the desert rangelands of Iraq, through livestock production alone contributed 10–15% of the agricultural contribution to the national income during the period concerned. To see this figure in right perspective in relation to other agricultural products. It may be mentioned that the income from the vegetation of the desert rangelands during that period:

- was about one and a half times the total income from fruit and dates together,
- was in the same range as the income from all vegetables grown in the country,
- was about one third to one half of the total income from field crops.

XII.F. Conclusions

From the above figures and discussion, the following can be concluded:

- The desert rangelands of Iraq constitute an important natural resource in a complex of values. A number of these values, especially those related to environmental quality cannot yet be properly evaluated.
- Based on the contribution through the livestock industry alone, the desert rangelands produce about 10–15% of the total agricultural contribution to the national income.
- Livestock numbers have largely increased over the period 1938–1965 as far as available census data indicate, followed by a sharp drop to about the 1938 level in the early seventies.
- The demand for livestock products is sharply increasing, mainly as a result of population growth and higher per capita income, the latter to a great extent due to increasing oil revenues. The number of sheep slaughtered annually rose

by 235% between 1950 and 1970, the total amount of meat produced by 250%. Export of live animals has virtually stopped.

Attention for the desert shrub rangelands, be it only as an important source of forage for the livestock industry is thereby well justified from a purely economic point of view.

CHAPTER XIII

EVALUATION, SURVEY AND MEASUREMENT

XIII. EVALUATION, SURVEY AND MEASUREMENT

XIII.A. Introduction: Land evaluation and land use planning

From the conclusions on the effects of the ongoing utilization as given in chapter XI and the justification for improved management as given in chapter XII, it will be clear that for a wise use of the land resources of the desert shrub rangelands of Iraq an inventory and assessment of the value of the land for a specific use are required. The need for such an evaluation, brought about by changing pressures on the land, and not only rangeland, is a common phenomenon in many parts of the world, including developing as well as developed countries, and has been given due attention during several decades. Procedures and techniques for the evaluation of soil and/or land suitability were developed, but the systems in the different countries were often limited in their applicability elsewhere (FAO 1974; Ashraf 1976). There was a need for a systematic approach, a generally applicable methodology for *land evaluation*, including a definition of terms and concepts. The basis for such an approach was given in Beek & Bennema (1972). An expert consultation on 'Land Evaluation for Rural Purposes' convened by FAO in cooperation with institutes in the Netherlands in 1972 'devised a framework into which national land evaluation systems could fit' (Brinkman & Smyth 1973). These guidelines, discussed and tested, resulted a few years later in the FAO framework for land evaluation (FAO 1976). This generally applicable approach is now being employed worldwide for land use planning and continually refined (Beek 1978; Zonneveld i.p.).

Land evaluation in itself does not bring about any change in land use; to be effective it should be part of an overall land use planning process. The function of this process is 'to guide decisions on land use in such a way that the resources of the environment are put to the most beneficial use for man, whilst at the same time conserving those resources for the future' (cf. FAO 1976). The process involves a sequence of activities and decisions, which can roughly be formulated in the following ten steps (FAO 1976):

(i) recognition of a need for change;
(ii) identification of aims;
(iii) formulation of proposals, involving alternative forms of land use, and recognition of their main requirements;
(iv) recognition and delineation of the different types of land present in the area;
(v) comparison and evaluation of each type of land for the different uses;
(vi) selection of a preferred use for each type of land;
(vii) project design, or other detailed analysis of a selected set of alternatives for distinct parts of the area;
(viii) decision to implement;
(ix) implementation;

(x) monitoring of the operation.

In this process of land use planning, land evaluation has the technical function to feed the planner and decision maker with options for alternative land use types, taking into account the physical condition of the land and the socio-economic situation. Thus land evaluation covers steps (iii), (iv), and (v) in the land use planning as outlined above.

Much attention has been focused on the evaluation of the land for crop production, rainfed as well as irrigated (Vink 1975). Soils and water are in this case the key attributes of the land to be investigated and the development of the 'framework' is largely due to this application. Fundamental in the approach, however, is that the land is evaluated for alternative types of land use. For large parts of the world grazing of the (semi-natural) vegetation, in other words use as rangeland in whatever form, provides such an alternative or rather a set of alternatives. Under such conditions vegetation must be considered the key attribute of the land. In only a few cases the 'framework-approach' been employed in areas where grazing is the dominant type of land use. Beek (1978) gives examples from Latin America where the use as rangeland is worked out as an alternative. A detailed, straightforward and successful application is the study of Baig (1977) in Baluchistan, Pakistan, which clearly illustrates the applicability of the procedure for the evaluation of arid rangelands. The present chapter aims at outlining how land use planning and the land evaluation procedure as described in the framework (FAO 1976) could be applied to the desert shrub rangelands of Iraq. The way of using the framework as described below may have a wider application than for Iraq alone and as such it could be another step towards a framework for rangeland evaluation.

Resources surveys are an integral part of land evaluation and for quantitative rangeland evaluation the measurement of certain vegetation attributes is of crucial importance. These two aspects are dealt with in XIII.C and XIII.D, respectively.

XIII.B. Application of land use planning including land evaluation to the desert shrub rangelands of Iraq

XIII.B.1. *Proposed overall land use planning*

In view of the overall deterioration and the total area to be covered in the desert rangelands of Iraq, a two-phase inventory and evaluation approach, followed by an implementation and monitoring phase, as shown in Fig. XIII-1 should be considered.

Upon recognition of the need for change, for which the chapters VII–XI of the present study could be of use, in phase I the relevant major *land utilization types* should first be worked out as possible alternatives of future use (see XIII.B.3.).

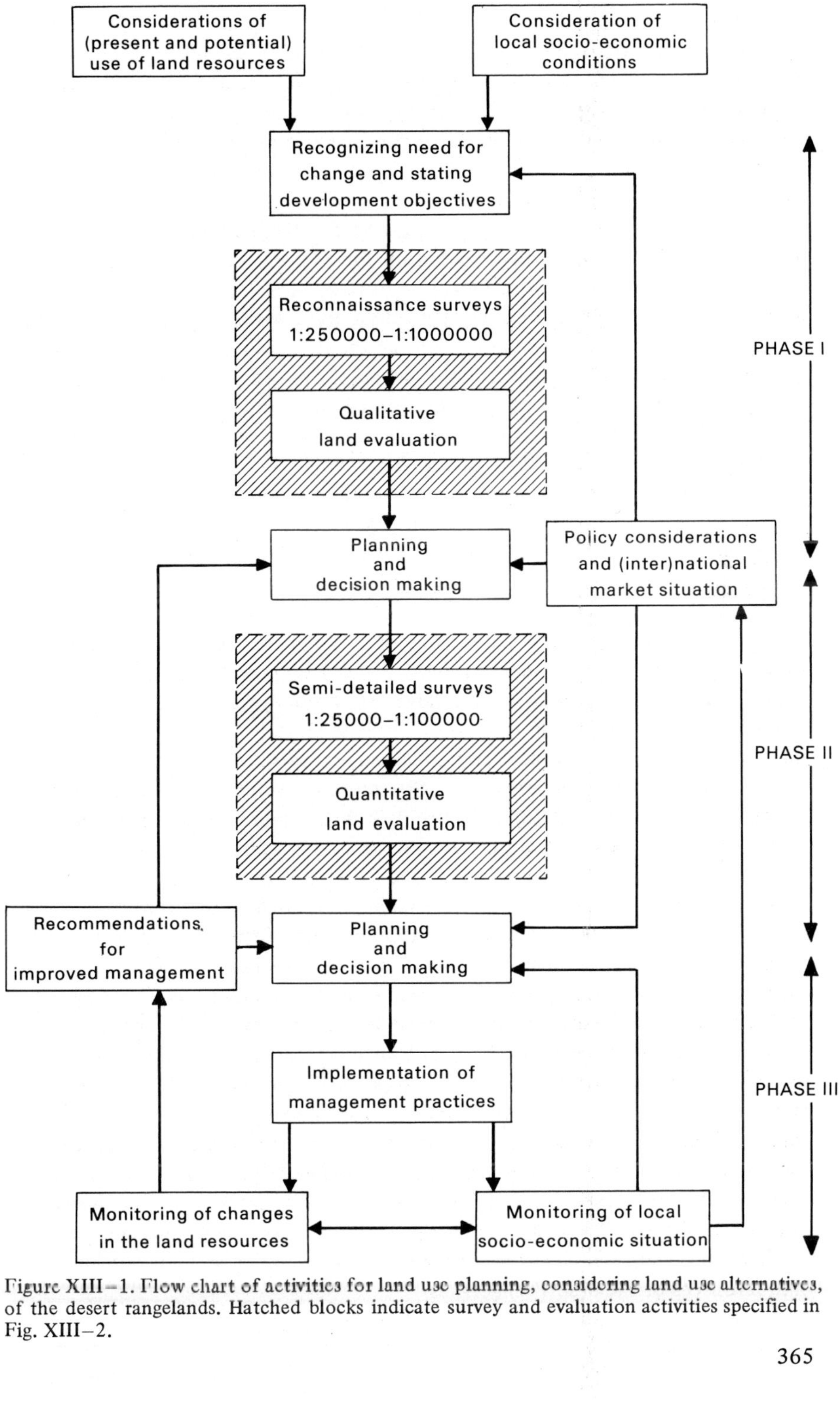

Figure XIII–1. Flow chart of activities for land use planning, considering land use alternatives, of the desert rangelands. Hatched blocks indicate survey and evaluation activities specified in Fig. XIII–2.

Keeping in mind the requirements of the utilization types, systematic basic rangeland surveys at reconnaissance level (see XIII.C.2.) must be carried out for an assessment of the land qualities (see XIII.B.4.). A survey of the socio-economic conditions could be made at the same time, or postponed to phase II when more detailed information is required for an economic evaluation of the land use alternatives. This will depend on the knowledge of socio-economic conditions in the area already available. The surveys required for the assessment of the physical attributes of the land will also depend on the existing knowledge. The geological map 1 : 1 000 000 and the soil maps of Buringh (1960) and Altaie (1968), at the same scale of the whole country, provide a good base. The outcome of phase I should be a qualitative evaluation of the land, *i.e.* an evaluation in which the relative suitability of the land for the different land utilization types is expressed in qualitative terms, without a calculation of costs and returns. Because of time and cost involved, such overall surveys are often neglected in land use planning. In these cases areas are directly selected for a detailed or semi-detailed evaluation. Local policy considerations are sometimes considered more important for the choice of the land than its actual condition and the consequence may be a waste of money and efforts after all. For the desert rangelands of Iraq rangeland survey at reconnaissance scale (*e.g.* 1 : 500 000) would require the input of a team of five specialists for about two years. This estimate is based on a similar survey at present conducted of the Kalahari Desert in Botswana (Thalen, in: DHV 1978). Based on the outcome of such a (country-wide) programme for the desert rangelands, areas should be selected for larger scale surveys as required for a specific purpose- (quantitative) land evaluation in phase II. In this phase economic criteria must be included and the cost and expected return of improvements worked out for a quantitative comparison of the current suitability and the potential suitability of the land at some future date when improvements have been completed (Fig. XIII-2). For this purpose the land utilization types must be specified in great detail. For an evaluation of the returns from rangeland, the approach presented in the previous chapter may be helpful. Such a quantitative comparison of the suitability of each part (land mapping unit) of the selected area should be followed by a phase III, introduction of management practices as dictated by the selected land utilization types. Implementation will require a strong organization with adequate field staff which can most probably be realized in Iraq by simply extending the already existing D.G. of Range Management. This, most important, organizational aspect falls outside the scope of the present study. Active management will have an impact on the land resources and the changes should be monitored and periodically assessed to find trends and to arrive at an optimal use of the rangeland resources. It could lead to recommendations for other management practices. It could even lead to new surveys and a revision of the quantitative evaluation, a sequence as shown in Fig. XIII–1. The scheme presents thereby a flexible approach for inventory leading to management and monitoring of the land resources of the desert rangelands.

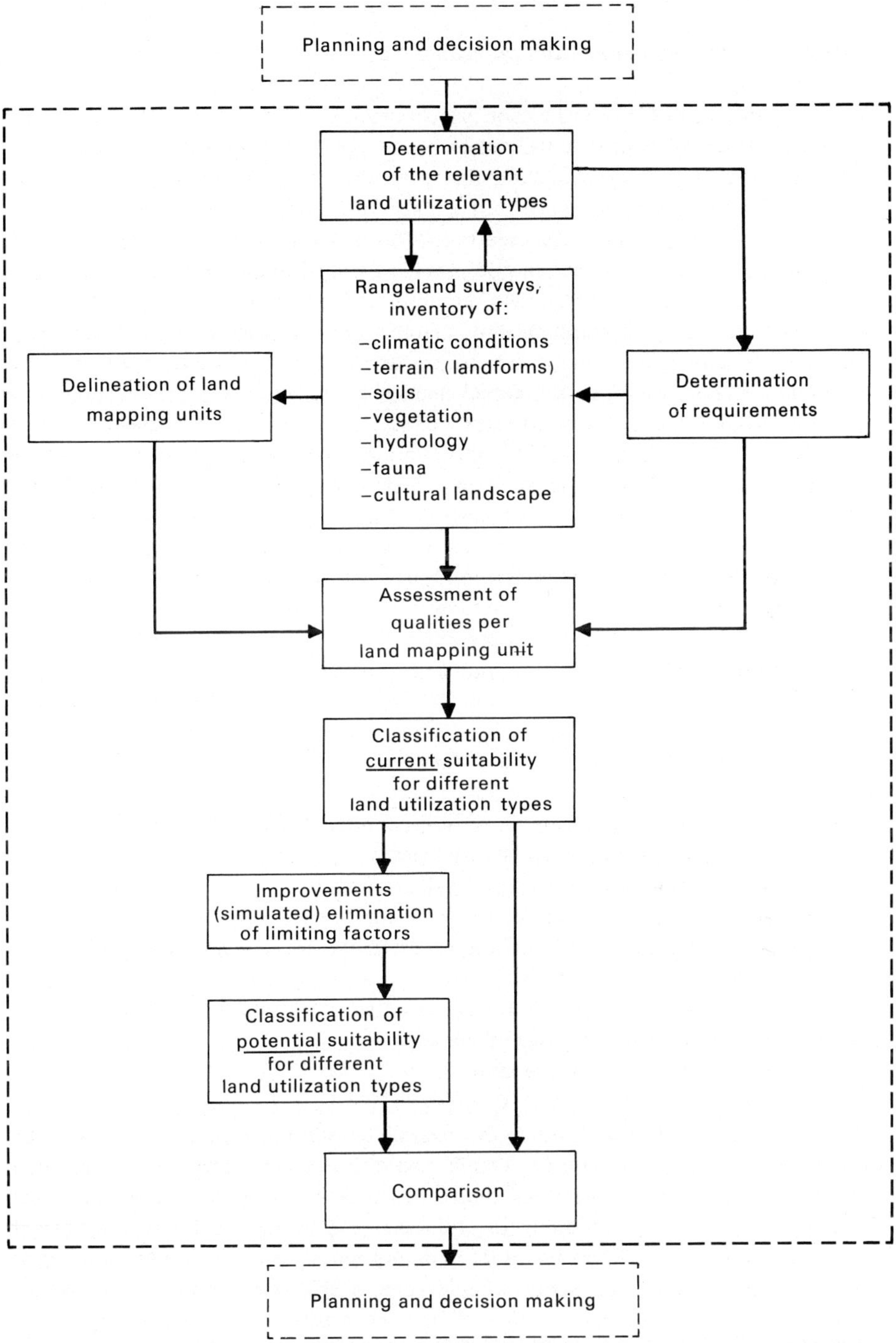

Figure XIII–2. Flow chart of activities in rangeland survey and rangeland evaluation for land use planning.

XIII.B.2. *The land evaluation procedure*

The activities in the land evaluation procedure are shown in Fig. XIII–2. Determination of the relevant land utilization types (see XIII.B.3.) is the start, together with an initiation of surveys. Preliminary results of the surveys can be used to refine the land utilization types, but usually the development objectives as laid down in the governmental plans together with a general knowledge of the area will be adequate. The surveys that should provide the basic data for the evaluation are separately discussed in XIII.C.. The requirements as dictated by the land utilization types are, per land mapping unit, matched with the survey data. For important aspects land quality maps can be prepared. Their use could go beyond the evaluation phases, well into the actual management. Maps of current forage production, erodibility, and groundwater quality, are examples. For the suitability classification the qualities have to be integrated. Overruling constraints such as endemic diseases and accessibility should first be considered and then the picture should be gradually refined. For a quantitative approach the formulas presented in XIII.B.5. are proposed. As a last activity the potential suitability should be assessed from the current suitability combined with improvements. Such improvements could include the drilling of new wells, reseeding, introduction of veterinary care, etc. The final result is presented to the planner and decision maker as a set of alternatives. The selection and implementation is a governmental matter.

XIII.B.3. *Relevant land utilization types*

For a selection and description of relevant land utilization types, Beek (1978) mentions the following points to be considered:

(i) the terms of reference of the land evaluation study
(ii) overall development objectives and land suitability criteria
(iii) the overall development situation, in particular present land use, past trends and outlook for the future
(iv) the time period during which the land evaluation results should be relevant
(v) the socio-economic attributes of the land
(vi) the physical attributes of the land.

In a sound governmental system, the overall development objectives should be found in the terms of reference. Trends and outlook must have been considered much earlier and have determined the recognition of a need for change as shown in Fig. XIII–1. For determining the relevant land utilization types, the present land use is important. When phase III is incorporated in the land use planning as shown in Fig. XIII–1, the period of relevance of the evaluation is not a consideration for the land utilization types. When the market for a certain commodity changes, a re-evaluation may be necessary, as indicated in Fig. XIII–1. These considerations leave the following determinants for the land utilization types in Iraq:

(i) terms of reference

(ii) present land use

(iii) socio-economic attributes of the land

(iv) physical attributes of the land.

Point (i) depends to a great extent on governmental policy considerations. A longterm objective to be included in the case of the Iraqi rangelands is the restoration and conservation of the production potential of the land. The present land use, point (ii), is well covered in the earlier chapters of this study, where also trends and outlook for the future can be read. The socio-economic aspects, point (iii), are closely linked to the policy considerations in point (i). In the course of land use planning and land evaluation they become towards the implementation phase increasingly important. Monitoring of the local socio-economic situation in phase III as shown in Fig. XIII–1 is, therefore, as important as monitoring the land. Information on major trends, to be taken into account, has been given in IV.H. and IV.G. This leaves as major point of consideration the physical attributes of the land: climate, geology and geomorphology as reflected in the terrain, soils, vegetation, hydrology and fauna. These attributes have, for the study area, been discussed in chapter IV. Considering the constraints imposed by the climatic conditions and the nature of the terrain and soils, any form of large scale cultivation is ruled out. For an exploitation of the primary production potential of the area the grazing animal has to be used. Extensive grazing in one or another form, closely related to the present and past land use will remain for most of the area the most relevant land utilization type. However, further specification is required for a proper land evaluation.

For a qualitative evaluation, following the reconnaissance scale surveys, the following five *major land utilization types* might theoretically be considered relevant:

- dryland farming (rainfed cereal cultivation or growing of foddercrops)
- cultivation under supplementary irrigation
- extensive grazing
- recuperation; allowing to recover from earlier damage
- conservation for (re-introduction of) wildlife.

For the sake of completeness the exploitation of the non-renewable resources such as oil, gypsum, etc. should be mentioned, but 'mining' as relevant land utilization type is here left out of the discussion. It has its own, very specific requirements and is usually of overruling importance in economic terms, whereby all other land utilization types are neglected as alternatives. Moreover, its use of land is very limited; no large areas are involved.

In the qualitative land evaluation the scale of the surveys will only allow a broad delineation of areas. Not much detail can be shown (see XIII.C.2.), and for the suitability classification some of the above mentioned land utilization types may have to be combined, thereby considering *compound land utilization types* (FAO 1976). The list could include amongst other things:

a. dryland farming (rainfed cereal cultivation or growing of foddercrops)
b. extensive grazing with local dryland farming (where possible growing of foddercrops)
c. extensive grazing with local cultivation under (supplementary) irrigation
d. extensive grazing with local recuperation
e. recuperation, later to be followed by extensive grazing (for this temporal aspect the monitoring of phase III is required)
f. conservation for (re-introduction of) wildlife.

Type a. will only apply to the northernmost part of the area in the Jezira (see IX.C.). Type b. could be considered in parts of the Jezira and locally in the Western and Southern Deserts where very favourable runoff collection areas exist. However, in these areas reseeding with a mixture of high value grazing species should also be considered. Growing of crops under irrigation is already practiced in parts of the Jezira (Al-Kasab 1966) and an extension seems possible. In this respect the suggestions for wadi Batin area in the Southern Desert by van der Sluijs (1969) and for local irrigation in the Southern and Western Deserts as suggested by INGRA (1964), mentioned before in IV.E.5. may be recalled. Type c. could be applicable to a rather large area. Considering the present status of the range as described in the previous chapters, types d. and e. will be the land utilization types most suitable for the major part of the area. Type f., the protection of certain areas, combined with re-introduction of wildlife and excluding hunting, could mean a restoration of the natural desert ecosystem. This could be for study leading to a better understanding of the structure and functioning of the system. In the future it could perhaps be for tourism and recreation, but it could also serve as a distribution centre for the now almost eradicated game (see IV.F.1.), once a valuable resource.

For the quantiative evaluation of the desert rangelands, phase II of Fig. XIII–1, the alternative land utilization types to be considered must be defined in great detail. In the types a.–f., indicated above, socio-economic aspects were not included. For a quantitative evaluation this has to be done and the various related determinants are listed and discussed in Beek (1978). In a further specification the crops to be grown under rainfed conditions and irrigation should be specified. For Iraq there is ample information. In Kaul & Thalen (1971a) a list is given of 17 fodder crops with their scope of cultivation, while more detailed information on yields etc. can be found in the reports mentioned in the review of II.C.3. The strong seasonal and annual fluctuations in the supply of forage from the natural range make an additional reserve-forage highly desired and cultivation of fodder crops could supply such a safe reserve. An integration of the cultivation and livestock sector should therefore be given due attention in the choice of the land utilization types.

Extensive grazing can be subdivided by livestock type (sheep, camel, goat) but also in sedentary (in its extreme: ranching), semi-nomadic and nomadic grazing, all three types already occurring in the area. Governmental policy considerations and the national market situation for the produce will probably be decisive and working out all possible alternatives here is of little use. In defining the land utilization types related to grazing, the old Arab 'HEMA'-system must be seriously considered. In this system certain areas ('AHMIAS') receive protection the year round or seasonal, against grazing and/or forage cutting under the responsibility of the tribal or village head (in Sprague 1971). The system does probably still exist in parts of Saudi Arabia and has recently been successfully re-introduced in Syria (Draz 1978).

XIII.B.4. *Land qualities to be considered*

A land quality is an attribute (usually compound) of land which acts in a manner distinct from the sections of other land qualities in its influence on the suitability of land for a specified kind of use, and is used to characterize and distinguish land classes of different value (FAO 1976, corrected in Zonneveld i.p.). The land qualities related to agriculture are many and by and large part of the prevailing soil and climatic conditions, *e.g.* nutrient availability, moisture availibility, and temperature regime. They can also be grouped in a few major qualities such as crop yield, the resultant of many others. The qualities related to cultivation are relatively unimportant for an evaluation of the desert shrub rangelands and are amply covered in Beek (1978). The most relevant qualities to be assessed for grazing, especially for domestic animal production, are briefly discussed below.

Accessibility. It is of overruling importance whether the area under evaluation is from the outside accessible to the grazing animal. Steep slopes or the absence of stock watering points along the tracks to the area may be limiting. If at all, this quality will form a limitation only for a minor part of the desert shrub rangelands of Iraq.

Relief (Topography). The different kinds of animals will react differently to slope conditions. Slopes can be measured on aerial photographs. The use potential of the different slope classes will have to be worked out. Studies in the USA clearly showed the importance of slope for range utilization by cattle (Mueggler 1965). In the desert rangelands of Iraq there will be little or no restriction of this quality for goat grazing, but for sheep and camels some correction factors for available forage may have to be applied.

Surface conditions. Dune bodies can be hardly accessible to sheep. Areas can be strawn with sharp, angular stones, boulders etc., making the terrain difficult to traverse for the animals.

Climatic hardships. The climatic conditions in the area as described in IV.B., indicate that particularly in summer (high temperatures) and winter (frost) animal condition can be directly affected. However, this does apply to the desert shrub rangelands as a whole and in an assessment of this quality per land mapping unit only the possibility to find shade and shelter could cause a differentiation.

Endemic diseases. This quality can be an overruling constraint and requires an assessment of the officials of the, in Iraq already existing, veterinary organization. The availability of veterinary services could cause differences between the land mapping units.

Forage production. In itself the resultant of a number of qualities, to a great extent the same as to be considered for crop yield, forage production is most easily directly assessed (see XIII.D.). The problem in assessing this quality is its direct dependence on the degree of use and the amount of rainfall, as shown in chapter VII and chapter X. For a proper assessment long-term records on production in relation to rainfall must be kept. In several countries such work is now underway as part of range monitoring programmes. Knowing the variability is as important as knowing the average. Unless adequate supplementary forage from a safe forage-reserve can be supplied, or the livestock can be easily removed from the range, the average prediction should not be used for a grazing capacity assessment. This could lead to disasters for the herds as well as the range. Some lower value has to be advised, based on variability, the risk that is accepted and the possibilities for relief-measures in emergencies. Although the conditions are not entirely comparable, the tragedy of the Sahelian countries following the drought of the early seventies, should be a lesson. Not only the annual variability is important, the discrepancies in available forage during the seasons in the same year should also be taken into consideration. For an easy assessment, forage production should be expressed in forage produced per unit area per time as required for management. This time unit could be an entire season, for practical reasons for Iraq as follows (not coinciding with the seasons of IV.B.2.):

- spring: February, March, April, first half of May
- summer: second half of May, June, July, August, September
- autumn/winter: October, November, December, January.

In the assessment of spring forage production the ephemeral component will be extremely important (see VII.E.4.).

Forage quality, palatability and nutritive value. In VII.C. the information on this aspect for the desert shrub rangelands of Iraq has been compiled and discussed. As for the previous quality, a very strong seasonal aspect must be taken into consideration. Expression of the nutritive value in one rating may not be possible without feeding experiments and a separate assessment for such aspects as protein, minerals, energy value, etc. may be required. A palatability rating, as mentioned before in VII.C., will always have only a relative value.

Proper use factor, resilience of the vegetation in relation to grazing. This quality is closely related to the previous two. It is determined by the species composition, in other words specific for a particular vegetation type. How much of the produced material can be safely taken without affecting the production potential, is a problem that can only be solved by cutting or grazing experiments, e.g. as reported in this study for *Haloxylon salicornicum* in chapter X and chapter XI. For practical purposes often a value of 50% has been used; an extensive discussion is given in Stoddart *et al.* (1975). For some of the desert shrubs which shed their leaves and branches to reduce respiration (see chapter VI), utilization of 60% to 70% of the current season's growth would probably do no harm.

Toxicity of the forage. The occurrence of poisenous plants can sometimes lead to great losses in livestock. *Halogeton*-poisening of sheep in the USA is a well known example (see Stodddart *et al.* 1975). Bedouins in Iraq told that camels eating only green shoots of *Haloxylon salicornicum* could be seriously affected by diarrhoea. The toxicity of *Peganum harmala* when green and the fruits of *Citrullus colocynthis* when dry were often mentioned. The toxicity may, therefore, be of a seasonal nature only. Usually death follows only after eating large quantities.

Occurrence of large predators. The once common large carnivores have been discussed in IV.F. Their present role is insignificant.

Availability of drinking water. This major quality has three aspects:

(i) spatial distribution of watering points

(ii) quantity of available water, and

(iii) quality of available water.

Aspects (i) and (ii) in particular have a temporal aspect when surface water is included in the assessment (see IV.E.2.). This quality usually requires separate mapping and is one of the first considered for 'improvements'.

Resistance to soil erosion. The importance of this quality, one of the 'conservation qualities' of the land or 'environmental control qualities' (Beek 1978), may be clear from the discussions in the chapters on utilization.

Besides the above qualities, directly related to the land and the animal performance, there are a number of qualities indirectly related to animal production, e.g. slaughtering facilities, distance to markets, etc., that should be given due consideration. For 'conservation' as land utilization type, the possibilities for control of protection measures in the area are important. It is a quality partly related to the terrain, partly to the human activities. When 'tourism' or 'recreation' are incorporated, the possibilities for accommodating people overnight must be assessed.

XIII.B.5. *Integration of quality ratings*

Once the qualities have been assessed, they have to be compared for each land mapping unit to arrive at a classification of suitability for the different land utilization types (Fig. XIII–2). In most cases this suitability is expressed in relative, qualitative terms, following a standard procedure for the map legend (see FAO 1976 and Baig 1977). When a quantitative assessment has to be given of the suitability for a certain type of grazing, it will be necessary to work out the grazing capacity per livestock type *e.g.* as in phase II of the suggested procedure for the desert shrub rangelands of Iraq.

Condon (1968) described a technique developed in Australia where areas are compared with a standard land class of known grazing capacity. Values for other areas are calculated by using multipliers derived from rating scales for a number of factors (in fact land qualities), e.g. fertility, moisture relationships and erodability of soils; topography; tree density, influencing pasture growth; drought forage; pasture type; range condition; rainfall; (pastorally) barren areas. In a formula this can be expressed as follows (Thalen 1978):

$$G_a = S_a \times f_1 \times f_2 \times \ldots f_n$$

$G_a =$ unknown grazing capacity for animal type (a), expressed in animal units per unit area

$S_a =$ known capacity for animal type (a) for the reference (standard) area, expressed in animal units per unit area

$f_1, f_2 \ldots f_n =$ multipliers (bonus or penalty ratings) for the factors mentioned above.

For most extensive grazing areas, including the desert shrub rangelands of Iraq, such a standard is not available. A crude estimate can, nevertheless, be obtained by relating the available forage to the animal food requirement and applying multipliers for the non-vegetation qualities, as in Condons approach. In a formula expressed this could read as follows (modified from Thalen 1978):

$$G_a = \frac{(P_h \times p_h \times n_h) + (P_b \times p_b \times n_b)}{R_a} \times f_1 \times f_2 \times \ldots f_n$$

$G_a =$ unknown grazing capacity for animal type (a) for a land mapping unit expressed in animal units per unit area

$P_h =$ production of forage in the herbaceous layer in the land mapping unit

$p_h =$ proper use factor for the herbaceous layer

$n_h =$ correction factor for nutritive value in the herbaceous layer

$P_b =$ production of forage in the form of browse

$p_b =$ proper use factor for the browse

$n_b =$ correction factor for nutritive value of the browse

$R_a =$ forage requirement of animal type (a)

$f_1, f_2, \ldots f_n =$ multipliers for relevant land qualities

It has been pointed out (Thalen 1978) that in case the land mapping unit is a complex mapping unit, wherein several land elements are combined, an assessment per element is required for an accurate estimate. This will especially apply to heterogeneous, large areas, mapped at small scale. The formula is a simplification and cannot provide more than a *crude estimate* of the grazing capacity. Tested in Pakistan by Baig (1977) the resulting grazing capacity figures were in good agreement with the outcome of an animal census in the same area. With an increasing knowledge, more and more accurate values will be found for the multipliers. Monitoring, as proposed in phase III of Fig. XIII–1 should yield such data. Only with grazing trials can the entire expression of forage supply divided by animal requirement be replaced by one single figure, as in the case of the Condon calculation.

An advantage of the formula approach is that a comparison between current and potential suitability in terms of grazing capacity is easily made. In chapter VII the potential for higher forage production in the desert shrub rangelands was shown. Such higher values would enter the grazing capacity estimates as changes in the factors P_h and P_b. Changes in water supply would alter the multiplier for that quality, etc. The very important temporal aspect of forage (and surface water) supply, can be entered by an evaluation per season as mentioned under 'forage production' in the previous section. The often used 'animal unit month' would in this case be a suitable way of expressing the grazing capacity. Another gain would be that an integration of natural forage and cultivated fodder could be easier worked out in the evaluation.

XIII.C. Survey

XIII.C.1. *Inventory and monitoring*

The database required for rangeland evaluation and rangeland management is to a great extent provided by survey. Survey data are not the only input. Grazing trials can yield valuable data too and studies of the present land use in selected areas. In an evaluation the spatial aspect is easiest expressed in maps. Only in exceptional cases can an other presentation be used *e.g.* where areas are very distinctly known already and the boundaries of the units to be evaluated so clear that there can be no misunderstanding as to which areas or part of an area is referred. For the Iraqi rangelands, however, maps will have to be produced. From Fig. XIII–1 can be seen that in all three proposed phases survey is an essential, integral part of the activities. A distinction should be made between a one time detection, characterization and mapping of the features of the land, and the detection of trends for which repetition of observation is required, indicated as *inventory* and *monitoring,* respectively (ASP 1975).

An inventory for rangeland evaluation should as far as possible cover all the qualities to be taken into consideration for a suitability classification for the different land utilization types (XIII.B.4.). For accessibility, relief and surface conditions, and also for an estimate of erosion hazard, terrain analysis and soils survey is required. For the availability of drinking water hydrological, especially hydrogeological survey is necessary. Vegetation survey must provide the data on forage production, forage quality, toxicity of forage and the proper use. To include the predators and large wild herbivores, a general inventory of the fauna would have to be carried out. The already existing patterns, resulted by man's presence in the area, the so called 'cultural landscape', should be included for several reasons, *e.g.* for an assessment of the existing land use. The survey of climatic conditions will usually mean an elaboration of the existing data and a presentation of the results in mapform. The survey of endemic diseases is somewhat similar, a specialist task for a veterinarian.

For monitoring, aimed at detecting trends in order to direct the management, in principle the same aspects as listed above for inventory are important. Some attributes related to the 'solid earth' hardly change or do not change at all. Lithology and landforms are examples. Soil conditions can change by such processes as vegetation removal, cropping, surface sealing and erosion. The prime objects for monitoring however are vegetation, animal users and (surface) water conditions.

Surveys for both inventory and monitoring can be done at almost any scale, but the different scales ask for a different approach, as briefly indicated below.

XIII.C.2. *Reconnaissance survey*

For the Iraqi desert rangelands reconnaissance scale should mean scales not larger than 1 : 250 000. A scale of 1 : 500 000 would probably be most suitable. At smaller scales not enough detail (wadis, depressions) can be shown. A larger scale, e.g. 1 : 250 000 cannot be justified in view of the amount of groundsampling that would be required, while also the actual mapping and cartography would become extremely costly. The area of 240 000 km^2 mapped at 1 : 500 000 can be represented at a map of approximately 1 m^2, still easy to handle. At 1 : 250 000 scale, some 4 m^2 map would have to be produced, meaning high cost and a map that is difficult to oversee and to handle.

For the proposed surveys the application of remote sensing techniques is essential. Panchromatic black and white aerial photographs could be used, provided there is adequate manpower for interpretation and the scale of photography is small enough (1 : 70 000 or smaller). At scale 1 : 50 000 many thousands of photographs would have to be handled. The best alternative would be to use satellite imagery supplemented with aerial photography of selected areas. There is adequate coverage of Iraq by LANDSAT – 1 and – 2 and in geological surveys in Iraq the imagery is already intensively used. For such work a Remote Sensing Centre has been established in Baghdad. With the assistance of such a centre digital processing of imagery could be included and image enhancement techniques particularly suitable for the Iraqi conditions could be applied. The possibilities to show even narrow strips of vegetation along wadis has already been demonstrated (Hielkema pers. comm.). For an application of such techniques to the entire 240 000 km^2 the cost might, for the time being, be prohibiting if the processing cannot be done with Iraqi facilities. The usefulness of satellite imagery for small scale rangeland surveys in particular has been sufficiently pointed out in the literature (*e.g.* Poulton *et al.* in ASP 1975) and has perhaps even been overemphasized. However, the requirement mentioned in other studies (van Dijk *et al.* 1977; ITC 1978) that a good deal should already be known of the area and that regular fieldchecks are desired, is already met or easy to meet in the Iraqi situation.

At the proposed scale the final reconnaissance rangeland map should be a sort of holistic landscape map with legend units described in terms of geology, landforms, soil and vegetation. Rarely will the map units be homogeneous: in many cases complexes will have to be delimited or units with a certain percentage 'noise'. In such cases the constituting elements should be given due attention for the evaluation and the contribution of the complex elements should, therefore, be indicated. The bias that could be caused by omissions on this point has been demonstrated elsewhere (Thalen 1978). The advantages of an holistic approach at the proposed scale as distinct from single attribute surveys have been discussed in detail in Zonneveld (1972). A detailed description of the methodology for a rangeland reconnaissance survey in which satellite imagery, panchromatic black and white aerial photography scale 1 : 70 000, large scale (1 : 1000) colour photography, direct observation from low flying aircraft and groundsampling are integrated, is given in DHV (1978).

XIII.C.3. *Semi-detailed and detailed survey*

Scales from 1 : 25 000 to 1 : 100 000 could be called 'semi-detailed' and scales larger than 1 : 25 000 'detailed' for the desert rangelands of Iraq. Such surveys would have to be undertaken for areas selected by the governmental authorities concerned. The nature of the area, the land utilization types to be considered in the evaluation and time and facilities at hand should determine the final map scale. Most useful would be (i) a map of terrain conditions and soils, (ii) a vegetation and land use map, and (iii) a (geo-)hydrological map. For the preparation of each of these maps a different approach is required (e.g. Goosen 1967; ASP 1975; Verstappen 1977). The vegetation map should show the vegetation structure, floristic community types and aerial biomass and/or production. The methods for such surveys where aerial photography and ground sampling are used as complementary tools in a distinct sequence have now been worked out in detail (Zonneveld *et al.* i.p.). Such maps have a lasting value, although the vegetation map in particular may require a regular updating. On the basis of these maps the land mapping units for evaluation are delineated. These land mapping units are in fact landscape units with a certain internal homogeneity. The concept is similar to that of the *range sites* used in the USA (USDA 1976) and advocated by American workers for other areas as well (see Allred 1968 for Saudi Arabia). The following remarks on using range sites to inventory rangeland in USDA (1976) demonstrate this:

'Range sites are a basic component of rangeland inventories. They are ecological subdivisions into which rangeland is divided for study, evaluation, and management. The range site map provides the basic ecological data for planning the use, development, rehabilitation, and management of rangeland.'

In semi-detailed and detailed survey aerial photography should be used for delineation of the unit boundaries and stratification of the field sampling. Panchromatic black and white photography at scales 1 : 5 000 to 1 : 25 000 would probably be most useful, depending on the nature of the land, the total area to be mapped, the manpower available and the final map scale. Large scale (appr. 1 : 500 to 1 : 1 000) full colour or infra-red false colour sample strips would provide useful additional data and be a start for a rangeland monitoring programme (see DHV 1978). The set up for such a programme can only be suggested when area and facilities are specified. Techniques for rangeland monitoring in the USA are being developed (Tueller *et al.* 1972; ASP 1975); ecological monitoring of wildlife management and conservation areas is already applied in East Africa, although more on an experimental than on an operational basis. Standard techniques have been developed for this purpose (Grimsdell 1978).

XIII.D. Measurement

XIII.D.1. *Approach and methods*

Besides the species composition, the following attributes of the desert shrub vegetation may be regarded of importance for an assessment of the vegetation with respect to inventory, monitoring and management: (i) height, (ii) crown cover, (iii) crown volume, (iv) aerial biomass, and (v) net aerial production, current season's growth or available forage.

Height can be measured directly for individual shrubs (see III.B.3.c.). Crown cover can be most easily calculated from the diameters (see III.B.3.d.) or by line intercept (see Brown 1954). Crown volume can be worked out from height and diameters (see III.B.3.e.).

Aerial biomass and net aerial production can be measured by cutting, drying and weighing. This is destructive, time-consuming and costly. Other techniques have been suggested but seem less suitable for operational application under desert conditions far away from any maintenance facilities for sophisticated equipment (e.g. Morris *et al.* 1976). Work on shrub vegetation types elsewhere (USA, USSR and North Africa) has shown the scope for using easy to measure attributes such as height and crown diameters to calculate aerial biomass and production from regression equations (Kinsinger & Strickler 1961; Mason & Hutchings 1967; Whittaker & Woodwell 1968; Kharin *et al.* 1971; Floret 1971; Uresk *et al.* 1977; Wierenga 1977). In Iraq this work was independently taken up in 1970 (Thalen 1972; Kaul & Thalen in UNESCO 1975). Results of further measurements to work out methods for estimating aerial biomass and available forage or current season's growth from height, crown diameters and combinations thereoff, are presented below. The aim was to develop *efficient and relatively simple quantitative techniques, directly applicable in the field.* Most of the measurements were done on the shrubs of the three key species treated before in chapter VI, *viz. Haloxylon salicornicum, Rhanterium epapposum* and *Artemisia herba-alba.*

Table XIII–1. Mean ($\overline{x}$), standard deviation (s) and coefficient of variation (c.v.) for some of the attributes of the samples of *Haloxylon salicornicum*, used for regression analysis.

		Height (m)	Crown cover (m²)	Crown volume (m³)	Aerial biomass (kg dry)	Current season's growth (kg dry)
May 1971 (n= 30)	$\overline{x}$	0.258	0.346	0.212	0.264	0.068
	s	0.127	0.454	0.451	0.499	0.123
	c.v.	49.1	131.1	212.7	189.1	180.2
Aug. 1971 (n= 30)	$\overline{x}$	0.280	0.414	0.264	0.180	0.099
	s	0.105	0.476	0.489	0.250	0.140
	c.v.	37.4	114.8	185.4	138.8	141.7
Oct. 1971 (n= 30)	$\overline{x}$	0.321	0.376	0.206	0.171	0.116
	s	0.094	0.328	0.264	0.171	0.111
	c.v.	29.4	87.3	128.0	99.6	96.1
Dec. 1971 (n= 30)	$\overline{x}$	0.338	0.427	0.220	0.187	0.099
	s	0.081	0.249	0.190	0.124	0.055
	c.v.	23.9	58.3	86.7	66.0	56.2
May 1972 (n= 30)	$\overline{x}$	0.331	0.405	0.226	0.155	0.068
	s	0.100	0.367	0.311	0.180	0.071
	c.v.	30.1	90.7	137.5	116.2	103.9
July 1972 (n= 30)	$\overline{x}$	0.392	0.490	0.309	0.197	0.135
	s	0.145	0.436	0.435	0.230	0.152
	c.v.	36.9	89.1	140.5	116.8	112.1
Oct. 1972 (n= 30)	$\overline{x}$	0.356	0.477	0.283	0.245	0.186
	s	0.107	0.370	0.323	0.293	0.194
	c.v.	30.2	77.4	114.0	119.7	104.2
Dec. 1972 (n= 30)	$\overline{x}$	0.394	0.620	0.442	0.221	0.156
	s	0.111	0.603	0.743	0.333	0.216
	c.v.	28.2	97.2	168.2	150.6	138.4

Table XIII–2. Mean ($\overline{x}$), standard deviation (s) and coefficient of variation (c.v.) for some of the attributes of the samples of *Rhanterium epapposum*, used for regression analysis.

		Height (m)	Crown cover (m^2)	Crown volume (m^3)	Aerial biomass (kg dry)	Current season's growth (kg dry)
March 1972 (n = 50)	$\overline{x}$	0.332	0.306	0.168	0.218	0.017
	s	0.134	0.346	0.279	0.341	0.023
	c.v.	40.2	113.2	166.7	156.5	132.7
May 1972 (= 30)	$\overline{x}$	0.389	0.431	0.273	0.248	0.125
	s	0.143	0.433	0.405	0.325	0.135
	c.v.	36.7	100.4	148.4	131.1	105.9

At different dates height and crown diameters of at least about 30 shrubs of a certain species were measured. After cutting at surface level, the total fresh weight was determined. Current season's growth was separated from the older material by hand and the fresh weight of both portions was determined in the field. In the laboratory, after drying, the ovendry weight was recorded. In the selection of the shrubs, plants were taken randomly within some distinct volume classes, to make sure that the entire range from very small to very large was represented in the sample. The descriptive statistics given in tables XIII–1, XIII–2 and XIII–3 are, therefore, only characteristics for the samples, not necessarily for the population

Table XIII–3. Mean (x), standard deviation (s) and coefficient of variation (c.v.) for some of the attributes of the samples of *Artemisia herba-alba*, used for regression analysis.

		Height (m)	Crown cover (m^2)	Crown volume (m^3)	Aerial biomass (kg dry)	Current season's growth (kg dry)
April 1972 (n= 28)	$\overline{x}$	0.416	0.291	0.134	0.154	0.015
	s	0.168	0.273	0.159	0.171	0.019
	c.v.	40.5	93.8	118.5	111.2	126.6
July 1972 (n= 30)	$\overline{x}$	0.453	, 0.263	0.114	0.332	0.100
	s	0.137	0.232	0.150	0.324	0.083
	c.v.	30.3	88.1	131.5	97.5	82.5
Sept. 1972 (n= 30)	$\overline{x}$	0.441	0.318	0.147	0.240	0.064
	s	0.137	0.262	0.183	0.242	0.059
	c.v.	31.0	82.4	124.8	101.1	92.5
Dec. 1972 (n= 30)	$\overline{x}$	0.426	0.342	0.162	0.222	0.053
	s	0.125	0.232	0.155	0.212	0.038
	c.v.	29.4	67.7	96.1	95.4	71.0

from which they were sampled. With the measured attributes height (h), largest diameter (D1), crown cover (cc, calculated from the diameters as shown in III.B.3.d.) and crown volume (cv, calculated from height and diameters as shown in III.B.3.e.) as independent variables (X) and aerial biomass dry (Bd) and net

Table XIII–4. Simple correlation coefficients, showing the closeness of fit of first order (1), second order (2) and third order (3) regression equations where aerial biomass dry (Bd) and current season's growth dry (Pd) are respectively the dependent variables and height (h), diameter (D1), crown cover (cc) and crown volume (cv), respectively, the independent variables, of *Haloxylon salicornicum* shrubs.

Dependent var.		Bd				Pd			
Independ. var.		h	D1	cc	cv	h	D1	cc	cv
May 1971 (n= 30)	1	0.671	0.861	0.977	0.991	0.657	0.856	0.965	0.983
	2	0.784	0.983	0.992	0.992	0.766	0.971	0.984	0.983
	3	0.815	0.987	0.993	0.993	0.812	0.975	0.986	0.986
Aug. 1971 (n= 30)	1	0.677	0.875	0.973	0.972	0.637	0.859	0.960	0.974
	2	0.709	0.969	0.979	0.978	0.661	0.971	0.978	0.974
	3	0.716	0.971	0.979	0.978	0.666	0.979	0.978	0.974
Oct. 1971 (n= 30)	1	0.703	0.915	0.960	0.935	0.660	0.875	0.897	0.856
	2	0.712	0.947	0.960	0.954	0.661	0.887	0.904	0.897
	3	0.713	0.947	0.960	0.960	0.663	0.889	0.904	0.911
Dec. 1971 (n= 30)	1	0.631	0.902	0.929	0.907	0.477	0.870	0.875	0.842
	2	0.644	0.919	0.930	0.930	0.478	0.872	0.882	0.874
	3	0.644	0.919	0.931	0.931	0.487	0.874	0.884	0.877
May 1972 (n= 30)	1	0.694	0.934	0.971	0.972	0.714	0.946	0.964	0.955
	2	0.769	0.954	0.972	0.977	0.790	0.967	0.966	0.960
	3	0.794	0.956	0.975	0.978	0.825	0.967	0.966	0.960
July 1972 (n= 30)	1	0.834	0.900	0.966	0.942	0.836	0.911	0.957	0.915
	2	0.865	0.953	0.966	0.978	0.857	0.950	0.960	0.975
	3	0.876	0.955	0.979	0.983	0.875	0.957	0.978	0.981
Oct. 1972 (n= 30)	1	0.747	0.875	0.928	0.954	0.765	0.904	0.950	0.964
	2	0.859	0.956	0.967	0.970	0.850	0.954	0.965	0.966
	3	0.874	0.959	0.975	0.981	0.861	0.954	0.965	0.967
Dec. 1972 (n= 30)	1	0.710	0.902	0.956	0.970	0.722	0.905	0.950	0.956
	2	0.873	0.972	0.971	0.970	0.861	0.960	0.959	0.957
	3	0.922	0.972	0.971	0.970	0.905	0.960	0.959	0.958

production dry (Pd) as dependent variables (Y), a regression analysis was carried out. The following statistical relationships were determined:

- first order (linear) regression equations : $Y = b_1X + b_0$
- second order (quadratic) regression equations : $Y = b_2X^2 + b_1X + b_0$
- third order regression equations : $Y = b_3X^3 + b_2X^2 + b_1X + b_0$

The numerical processing was done with a PDP 11/45 computer. The simple correlation coefficient (r = the signed square root of the index of determination) is used as an indicator to show how well the regression equations fit the data.

XIII.D.2. *The samples*

For *Haloxylon salicornicum* eight data-sets of 30 shrubs collected over the seasons in 1971 and 1972 were used for the analysis (table XIII–1). For *Rhanterium epapposum* data sets of 50 shrubs from March 1972 and of 30 shrubs from May

Table XIII–5. Second order regression equation parameters to estimate aerial biomass (Bd) and available aerial production or current season's growth (Pd) from the plant diameter (D1) in *Haloxylon salicornicum*. Based on data from 30 shrubs per date, measured in the Ukhaidir – Abu Dibbis area.

		b_0	b_1	b_2	r*
May 1971	Bd	0.2202	– 1.0065	1.2776	0.983
	Pd	0.0521	– 0.2294	0.3025	0.971
Aug. 1971	Bd	0.0685	– 0.3208	0.5403	0.969
	Pd	0.0545	– 0.2288	0.3285	0.971
Oct. 1971	Bd	0.0012	– 0.0796	0.4070	0.947
	Pd	– 0.0305	0.0877	0.1575	0.887
Dec. 1971	Bd	0.0002	– 0.0722	0.3785	0.919
	Pd	– 0.0349	0.1203	0.0637	0.872
May 1972	Bd	– 0.0268	0.0160	0.2846	0.954
	Pd	– 0.0010	– 0.0023	0.1188	0.967
July 1972	Bd	0.0179	– 0.2022	0.4774	0.953
	Pd	– 0.0122	– 0.0551	0.2716	0.950
Oct. 1972	Bd	0.2317	– 0.8403	0.9739	0.956
	Pd	0.0743	– 0.3047	0.5108	0.954
Dec. 1972	Bd	0.0797	– 0.3314	0.4849	0.972
	Pd	0.0193	– 0.1261	0.2783	0.960

* Simple correlation coefficient, showing the closeness of fit.

1972 were processed (table XIII–2). Data from shrubs of *Artemisia herba-alba* used in the analysis include data from 28 shrubs, collected in April, 30 shrubs collected in July, 30 shrubs in September and 30 shrubs in December (table XIII–3).

The *Haloxylon*-samples were collected in the Ukhaidir – Abu Dibbis area, the *Rhanterium*-samples near Samah, and the *Artemisia*-samples in the Rutba fenced area. All plants used were *ungrazed* in the 'current' growing season.

XIII.D.3. *Results*

The correlation coefficients found for the *Haloxylon*-samples are shown in table XIII–4. The unduly high values are caused by the purposive sampling. As mentioned before, both, very small and very large shrubs were purposely included in the sample, a correct procedure for the aim for which the sampling was undertaken (Schumacher & Chapman 1954). Height is a relatively poor indicator for aerial biomass and production. Diameter, cover and volume of the crown are all three highly correlated with the weight values. For all three the fit of a second order equation is considerably better than that of a first order. The gain by using a third order equation instead of a second order one is negligible. For crown cover and crown volume two, respectively three, measurements per shrub have to be made, for the diameter only one. In view of the time involved and the only slight gain in accuracy of the third order equations, a second order regression equation with the largest diameter (D1) as independent variables (X) seems the most efficient method to estimate aerial biomass (Bd) or production (Pd) as the dependent variable (Y). The parameters for the regression equations of this type are given in table XIII–5.

Table XIII–6. Simple correlation coefficients, showing the closeness of fit of first order (1), second order (2) and third order (3) regression equations where aerial biomass dry (Bd) and current season's growth dry (Pd) are respectively the dependent variables and height (h), diameter (D1), crown cover (cc) and crown volume (cv), respectively, the independent variables, of *Rhanterium epapposum* shrubs.

Dependent var.		Bd				Pd			
Independ. var.		h	D1	cc	cv	h	D1	cc	cv
March 1972 (n= 50)	1	0.741	0.848	0.924	0.927	0.776	0.873	0.938	0.937
	2	0.859	0.935	0.936	0.928	0.882	0.940	0.947	0.938
	3	0.864	0.940	0.936	0.929	0.894	0.947	0.953	0.940
May 1972 (n= 30)	1	0.868	0.904	0.961	0.956	0.872	0.904	0.901	0.850
	2	0.952	0.970	0.966	0.959	0.892	0.913	0.909	0.883
	3	0.953	0.970	0.967	0.973	0.892	0.914	0.923	0.948

Table XIII–7. Second order regression equation parameters to estimate aerial biomass (Bd) and available aerial production or current season's growth (Pd) from the plant diameter (D1) in *Rhanterium epapposum*. Based on data from 50 shrubs in March and 30 shrubs in May, measured near Samah.

		b_0	b_1	b_2	r*
March 1972	Bd	0.0942	– 0.5772	1.0102	0.935
	Pd	0.0050	– 0.0258	0.0595	0.940
May 1972	Bd	0.0533	– 0.3801	0.7632	0.970
	Pd	– 0.0500	0.1552	0.1141	0.913

* Simple correlation coefficient, showing the closeness of fit.

The data for *Rhanterium epapposum* show about the same picture as given for *Haloxylon salicornicum* (table XIII–6). The parameters for the relevant regression equations are presented in table XIII–7.

The results for the *Artemisia herba-alba* samples as shown in table XIII–8 show that the correlations for aerial biomass are higher than for the other two species, while the correlations for the production are low compared to the values found for *Haloxylon salicornicum* and *Rhanterium epapposum*. A relatively good linear correlation with the diameter can be seen. This might be expected while *Artemisia herba-alba* plants first form a few shoots growing to a certain height (see VI.G.3.).

Table XIII–8. Simple correlation coefficients, showing the closeness of fit of first order (1), second order (2) and third order (3) regression equations where aerial biomass dry (Bd) and current season's growth dry (Pd) are respectively the dependent variables and height (h), diameter (D1), crown cover (cc) and crown volume (cv), respectively, the independent variables, of *Artemisia herba-alba* shrubs.

Dependent var.		Bd				Pd			
Independ. var.		h	D1	cc	cv	h	D1	cc	cv
April 1972 (n= 28)	1	0.816	0.943	0.956	0.939	0.740	0.819	0.816	0.787
	2	0.835	0.974	0.959	0.942	0.780	0.847	0.816	0.801
	3	0.838	0.975	0.960	0.952	0.780	0.847	0.817	0.807
July 1972 (n= 30)	1	0.795	0.907	0.951	0.928	0.728	0.836	0.876	0.858
	2	0.834	0.914	0.951	0.953	0.734	0.844	0.876	0.867
	3	0.834	0.919	0.953	0.957	0.747	0.844	0.880	0.868
Sept. 1972 (n= 30)	1	0.811	0.965	0.985	0.970	0.687	0.837	0.830	0.795
	2	0.824	0.985	0.985	0.983	0.690	0.838	0.842	0.834
	3	0.835	0.985	0.985	0.983	0.714	0.840	0.842	0.852
Dec. 1972 (n= 30	1	0.790	0.920	0.940	0.949	0.780	0.843	0.845	0.829
	2	0.833	0.950	0.957	0.950	0.782	0.843	0.845	0.833
	3	0.835	0.951	0.957	0.950	0.789	0.849	0.853	0.844

Table XIII–9. First order (1) and second order (2) regression equation parameters to estimate aerial biomass (Bd) and available aerial production or current season's growth (Pd) from the plant diameter (D1) in *Artemisia herba-alba*. Based on data from 30 shrubs per date (for April 28 shrubs), measured in the Rutba exclosure.

			b_0	b_1	b_2	r*
April 1972	Bd	(1)	– 0.1180	0.4459		0.943
	Bd	(2)	0.0035	– 0.0602	0.3759	0.974
	Pd	(1)	– 0.0114	0.0438		0.819
	Pd	(2)	0.0008	– 0.0071	0.0378	0.847
July 1972	Bd	(1)	– 0.3216	1.0918		0.907
	Bd	(2)	– 0.1604	0.5570	0.3713	0.914
	Pd	(1)	– 0.0534	0.2567		0.836
	Pd	(2)	– 0.0084	0.1071	0.1039	0.844
Sept. 1972	Bd	(1)	– 0.3238	0.8508		0.965
	Bd	(2)	– 0.0356	– 0.0164	0.5593	0.985
	Pd	(1)	– 0.0538	0.1801		0.837
	Pd	(2)	– 0.0457	0.1510	0.0188	0.838
Dec. 1972	Bd	(1)	– 0.3263	0.8124		0.920
	Bd	(2)	0.0424	– 0.2931	0.7383	0.950
	Pd	(1)	– 0.0362	0.1322		0.843
	Pd	(2)	– 0.0329	0.1222	0.0066	0.843

* Simple correlation coefficient, showing the closeness of fit.

As the plant grows older there is relatively little increase in height but new sprouts develop from the base and enlarge the crown cover. In this species a simple diameter measurement entered into a linear regression equation should give an accurate estimate of aerial biomass and a somewhat less accurate estimate of the current season's green material. Parameters for the linear and quadratic regression equations for four months as found in 1972 in the Rutba exclosure are given in table XIII–9.

XIII.D.4. *Concluding remarks*

The results presented above clearly show the scope for a development of techniques for an accurate estimate of aerial biomass and available forage from other, relatively easy to measure, plant attributes. Shrub diameter seems the single best attribute for the three species tested, but for other species height may be equally important.

Production, and thereby biomass, varies with the rainfall; it may also vary according to the soil type, although in the present study no data on this point could be collected. For accurate and efficient forage estimates, it would be useful to work out standard tables per species with rainfall, soil type and plant size (diameters, if necessary also height) as variables, for calculating the forage production. Until such tables are available, regressions will have to be worked out in the field for each situation separately. In view of the difference in time required to measure a few diameters compared to determining the dry weight of the net aerial production, the method of regression is of great utility.

The use of very large scale aerial photography (e.g. scale 1 : 500) as a sampling tool to estimate crown volume classes or actually measure crown diameters of larger shrubs, analogous to techniques already well developed in forestry (ASP 1975) may prove to be of great value. Such special purpose aerial photography could be taken as part of a rangeland monitoring programme (DHV 1978; Thalen i.p.). Besides the small scale satellite imagery, the large scale aerial photography should therefore be incorporated in the management planning of the desert shrub rangelands of Iraq.

CHAPTER XIV

PAST, PRESENT AND FUTURE

XIV. PAST, PRESENT AND FUTURE

Before man permanently inhabited the present desert shrub rangelands of Iraq, probably large herds of gazelles roamed the area and ostrich was common; droves of onagers crossed the Jezira and groups of oryx ranged the Southern and Western Deserts. They were preyed upon by lion, cheetah, hyena and other large carnivores (see IV.F.). Surface water was only available for part of the year. Some species could do almost without water, but others had to migrate in a continuous search for food and water. The picture of the desert vegetation at that time is a speculative one. Considering some older records and remnants found today, there must have been stands of trees in depressions and the wadis may have been lined with a tall woody vegetation, as still found in the nineteenth century along wadi Tharthar in the Jezira (Ainsworth 1841). Perennial grasses must have covered extensive areas, at least in the Jezira and in other areas the shrub vegetation was much denser than today. There must have been a dynamic equilibrium between animal use of the land on the one hand and food, water, and shelter supply on the other. When the latter diminished because of a too high pressure resulting from increasing animal numbers, migration of at least part of the animals was required for mere survival and the range could regenerate.

When man appeared on the scene, he was at first as much part of this ecosystem as any of the other large animal species, and like these species he had to cope with the limitations set by the environment. First as a hunter and gatherer he is known from the area, but some 10 000 years ago he lived already in small settlements at the fringe of the desert. His presence in the desert rangelands of Iraq must have been incidental. It could, with exception of places with permanent surface water, never last long and the impact on the land was slight, often only temporal, a minor local change of the resources supporting him. Little is known for sure about man's way of life in the area. Artifacts accidentally found and some systematic excavations have proved his presence, but for his way of life we have few indications and therefore extrapolate the knowledge from groups living under more or less similar conditions elsewhere, such as the bushmen of the Kalahari.

From a gatherer, man became a producer, a process documented especially for this country. He established a new relationship with certain animals: providing protection against predators, perhaps providing food and water when needed, and receiving an assured supply of meat, milk, material for clothing, later for tents and a means of transport. Pastoralism was a new way to use the land and probably not much later certain measures for a sustained forage production of the land were applied. That was range management in its earliest form, although the term was coined only thousands of years later. What would have been called 'management' now, was at the time probably the most natural thing to do. As pointed out for the early pastoral cultures in Stoddart *et al.* (1975):

'Although different patterns emerged from place to place, there was a common element best described as that of man in nature in which man strives to live in

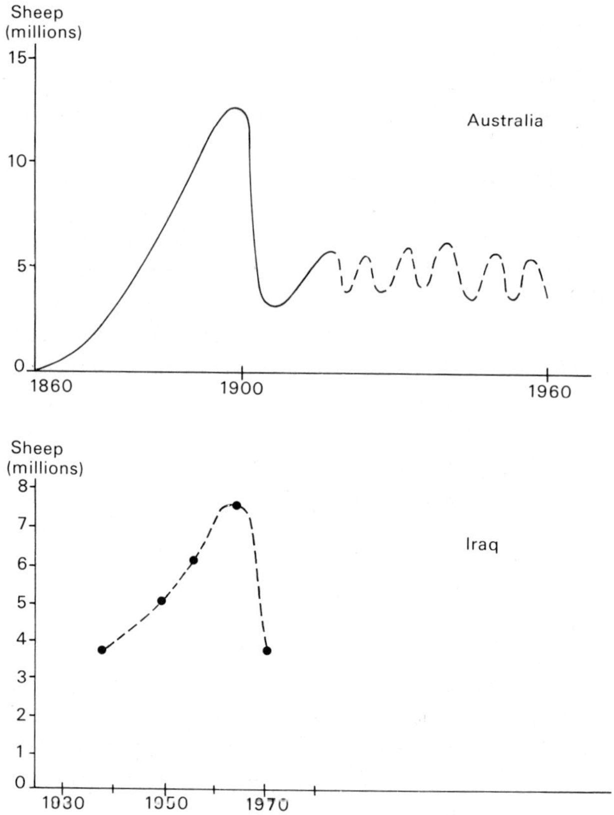

Figure XIV–1. Trend of sheep numbers in rangeland areas of Australia (after Perry, as shown in Stoddart *et al.* 1975) in comparison to the trend in Iraq.

harmony with his surroundings not for the sake of conservation, but for his very survival. A balance is struck among man, livestock, and the environment'.

What this pattern was in the present desert shrub rangelands of Iraq and how and where it affected the natural rangelands ecosystem is not known and may never be known. The appreciation of the early Bedouins for their land and its resources, the plants and wildlife, but especially for their domestic animals is for pre-islamic times recorded from the old poems (Jacob 1967) where for instance 'RIMTH' (= *Haloxylon salicornicum*) and many large tree and other shrub species are already mentioned. The movements of people north and northeastward from the Arabian peninsula have been discussed in IV.G.3. and IV.G.4. A particular vivid description of this process is given by T.E. Lawrence in his 'Seven pillars of wisdom' (Lawrence 1926):

> 'Finally, under a last impulse from the straining population behind them, the border people (now almost wholly pastoral), were flung out of the furthest crazy oasis into the untrodden wilderniss as nomads.
>
> . . .
>
> . . . people were pushed gradually north, up between Medina of the Hejaz and Kasim of Nejd, till they found themselves at the fork of two roads. They could strike eastward, by Wadi Rumh or Jebel Shammar, to follow eventually the Batn to Shamiya, where they would become riverine Arabs of the Lower Euphrates;
>
> . . .
>
> So we see clans, born in the highlands of Yemen, thrust by stronger clans into the desert, where, unwillingly, they became nomad to keep themselves alive. We see them wandering, every year moving a little further north or a little further east as chance has sent them down one or other of the wellroads of the wilderness, till finally this pressure drives them from the desert again into the sown, with the like unwillingness of their first shrinking experiment in nomad life. This was the circulation which kept vigour in the Semitic body. There were few, if indeed there was a single northern Semite, whose ancestors had not at some dark age passed through the desert. The mark of nomadism, that most deep and biting social discipline, was on each of them in his degree.'

To trace and evaluate the rights and restrictions for the use of the desert rangelands, established and respected over the centuries by the Bedouins, requires a study in itself. The already mentioned 'hema'-system is an example. Watering rights were a form of control and only a few decades ago the powerful and respected Sheikh Mahruth al Hadhal divided the families of his tribe over the watering points in the Western and Southern Deserts. For centuries such measures must have worked. The factors controlling the number of livestock have been given in IV.H., where also was discussed how, with the political changes and the introduction of new technology the entire picture drastically changed.

The drilling of new wells combined with motorised transport opened up new areas and under the safe, controlled conditions livestock numbers could build up after World War I (table IV–14). Not only by grazing (chapter VII) but also by an increase in ploughing for hazardous rainfed cereal cultivation (chapter IX) and fuel cutting (chapter VIII), particularly in combination, the vegetation resource and with it the vulnerable soils were depleted.

The present status of the range, the livestock it feeds in comparison to twenty years ago, is the result. Such developments on overstocked rangelands are not confined to the desert shrub rangelands of Iraq. Similar situations have been described for the USA (Stoddart *et al.* 1975) and for Australia (Perry in Stoddart *et al.* 1975, see Fig. XIV–1) where new land was occupied and cattle and sheep numbers built up rapidly, exceeding the grazing capacity and resulting in large die-offs. There are indications that in some Sahelian countries similar processes are going on (Warren & Maizels 1977). In the USA the disasters resulted at about the turn of the century in the establishment of range science and the range management profession.

Realising the present condition of the range and its potential productivity, Iraq is now initiating programmes aimed at an increased production and better use of the desert areas. To work out the proper stocking rates and overall grazing capaci ties for the various parts of the desert rangelands, a systematic land evaluation as part of an overall land use planning as outlined in chapter XIII should be undertaken, starting with those areas where presently new villages are being established and roads constructed. Chapter XI and chapter XII provide strong enough arguments and the methods are known (*e.g.* Zonneveld *et al.* i.p.; ITC 1978; D.H.V. 1978).

Carrying out a survey and evaluation is not a prerequisite for an introduction of all management measures. Given the existing governmental bodies in charge can handle the task, a number of such measures could be implemented instantly. In chapter VIII the need to stop the cutting of shrubs for fuel has been shown. In an oil and gas producing country the provision of substitutes should be relatively cheap. With the improved system of roads and tracks the distribution of fuel to the villages should not be difficult. Chapter IX has indicated the necessity for a strict enforcement of the law on the southern boundary of rainfed agriculture. General management measures have been proposed by the foreign rangeland specialists working in Iraq some 20–25 years ago. In Kaul & Thalen (1971a) range improvement measures have been outlined. In UNESCO (1975) the same authors have given a number of recommendations for surveys and investigations that should lead to an improved management and listed about one hundred areas found most suitable for an implementation of direct management and improvement measures. The basic information is already available. Moreover, many measures already successfully applied or developed in the neighbouring countries can be directly copied in Iraq. An active exchange of information on desert shrub rangeland management, also outside the conference room, could be well rewarding for the countries of South West Asia.

The management of a few hundred thousands of square kilometres desert rangelands is not an easy task. The present stage of degradation is mainly due to, from an ecological point of view rather thoughtless, application of facilities provided by new technology. That same technology could now be used for restoration and improved management, from the satellite for the survey to the lorry carrying the fence posts. In the first sentences of this study was stated that action is required to save what is left and possibly to restore what was once there and that for such action knowledge, organization and coordination is required. It may be repeated here, adding that the present study cannot do more than, in a sort of synthesis, contribute to the knowledge. Unless far reaching land use planning and sound management practices are introduced at a larger scale than at present, the desert shrub rangelands of Iraq will further deteriorate. Oil revenues place the country in a privilaged position compared to most other countries with arid lands. The means and methods are available. For the future of the desert shrub rangelands of Iraq, a country known as the 'cradle of civilization', with a rich history including systems of sophisticated land management, not more can be said at present than what Hitti (1968) stated as follows for the Arab world as a whole: 'The achievement of the past is the promise of the present for the future'.

SUMMARY

SUMMARY

'Desertification' has been recognized as a worldwide problem for the arid lands, particularly since the subject was dealt with at the United Nations conference in 1977 in Nairobi. The situation in South West Asia provides a clear example. In the present century, mainly in the last decades, several changes in land use have taken place in this area. Such changes often cause degradation and deterioration of natural resources, especially of fauna, vegetation and soils. The value of the land decreases and ultimately the land may even be of no further use to man.

When grazing is the accepted land use, vegetation is the key resource. The present study deals with the desert shrub rangelands of Iraq, which contain the major characteristics of such an area, having been under grazing for many centuries. Emphasis is given to the ecology and utilization of the vegetation and to man as an ecological factor. The aim of this study is to contribute to a well planned land use of these areas under proper management.

The field data were collected from April 1970 to June 1973 when the author worked for UNESCO in the UNDP/SF/UNESCO supported Institute for Applied Research on Natural Resources near Baghdad. The study area comprises the Southern Desert, Western Desert and Lower Jezira (Fig. I–3), in total some 240 000 km^2. The vegetation mainly consists of low shrubs, growing in open stands. In between the shrubs an ephemeral vegetation develops after the winter and spring rains. Shrub vegetation types cover about 10–15% of the earth's land area. They are an important natural resource, especially for livestock grazing. The area investigated covers some 55% of the total land area of Iraq, a country well known for its ancient civilizations which existed in and near the Mesopotamian Plain. At present oil is the main source of income. With a few exceptions, the soils, vegetation and fauna of the desert areas of Iraq have hardly been investigated due to the limitations of the area for an intensive land use. The floristic and phyto-geographical aspects have been relatively closely studied by plant collectors.

Mean annual rainfall in the study area varies from about 50 to 350 mm. There can be great differences from year to year. Most of the rain falls in a few heavy showers. In winter frost is recorded while in summer the temperatures can rise to 50°C. Annual evaporation ranges from two to three and a half metres and relative air humidity can in summer be less than ten per cent. Seven physiographic units have been distinguished, based on relief, occurrence of wadis, sandy or stony nature, soil type and elevation (about 0–915 metres a.s.l.). These units and the soil types are described. Until early this century water for man and livestock came from the following sources: a few deep wells, some shallow wells in wadibeds, a few springs, and rainwater which had temporarily accumulated at the surface. Starting in the Thirties a large number of wells has been drilled so far, in total a few hundred. The wells are often very deep and equipped with pumps and storage tanks. This improvement in water supply and the introduction of terrain cars have made the entire desert area accessible to man en his livestock. After the introduction of modern firearms and with the new fast means of transport, the larger wild animals were exterminated. Oryx, ostrich, lion and cheetah have disappeared. Gazelles still occur, but only in very small numbers.

Artifacts prove that man has been living in these areas for a long time but at first probably only temporarily and in small numbers. In the second or first millennium B.C. when nomadism came into existence and after the domestication of the dromedary or one-humped camel, in places there may have been a pressure on the land. For many centuries nomadic grazing was the only land use in the desert rangelands. The Bedouin tribes had their own strict rules for the right to use certain areas, in which drinking water supply played a major role. Lack of watering points and raiding were two factors which controlled the total number of livestock. A profound change in this pattern took place after World War I. The boundaries of the present Iraq were fixed, police control became effective and fast transport and improved water supply 'opened up' the desert rangelands. The number of camels and sheep increased; sheep from 5 525 000 in 1938 to 11 040 000 in 1965 for Iraq as a whole.

The impact of this process on the vegetation has been investigated in this study. Based on a 4000 kilometres' reconnaissance survey for which a description of terrain, soils, vegetation and land use was made, a detailed vegetation sampling was completed at 54 subjectively selected locations. In 50 quadrats, of 4 m^2 per location, density, height, crown cover and (total) aerial biomass of the perennials was measured. For each location soil samples were taken from two sites, from 0–30 cm and 30–60 cm depth. The precision of the sampling and the data processing for the classification are discussed. To quantify the relative contribution of the species to the stand, RIMP-values (Relative IMportance Percentages), mean value of relative density, relative frequency and relative crown cover, were calculated. In total 15 community types and 14 socio-ecological species groups were distinguished. A climatic gradient, differences in soil texture and land use (former ploughing) partly explain the types ecologically. Per cent cover ranges from 1% to 20% with a few higher values. The total number of (perennial) species is low.

Aspects of the autecology of three key shrub species, *viz. Haloxylon salicornicum, Rhanterium epapposum* and *Artemisia herba-alba* were studied, in particular seed production, seed dispersal, germination and seedling development. The first two species have their northernmost distribution in Iraq. Germination and seedling development demonstrate the adaptive characteristics of the species to the harsh environment. The germination of *Rhanterium epapposum*, a member of the Compositae family growing on sandy soils, is described for the first time. In the flowerheads, after falling, several seeds germinate simultaneously while remaining together in the flowerhead. The rootlets penetrate through the receptacle into the moist sand. More than one seedling will usually develop simultaneously to form a new shrub. A comparison of the adaptive characteristics shows how each species is best adapted to a particular set of environmental conditions.

The utilization of the desert shrub rangelands by man and his livestock is treated in a separate part in this study.

Grazing by sheep, at present numbering a few millions, and to a less extent by camels, goats and donkeys, has taken place since olden times. Due to an increased grazing pressure the vegetation has in recent years been affected by overgrazing and grazing too early. The existing data on palatability and selectivity have been gathered and are presented together with newly collected information. Chemical analyses of plant samples show that the parts of the shrubs selected by the animals are high in crude protein and also in minerals, even to the extent that the content is higher than in cultivated fodder crops. Moisture content varies strongly according to the seasons. In spring it can be so high for the range vegetation as a whole, including the annuals, that no drinking water is required for the livestock. Seasonal forage production of selected species has been measured and for *Haloxylon salicornicum* a comparison of the production in 1971 and 1972 has been made. An analysis of the 4000 km reconnaissance survey transects in combination with the forage production measurements allows a rough estimate of the total forage production in the 240 000 km^2 study area. For 1971 this amounts to 2.91 x 10^9 kg dry. A comparison with the forage demand as estimated from the 1971 livestock census figures (2.52 – 3.31 x 10^9 kg dry) shows that for 1971 the figures are of the same magnitude. The animal-census figures of 1965 indicate that in that year the livestock numbers were much higher and it is, therefore, likely that at that time and probably also in the earlier years forage production was much higher. Field observations providing further evidence for a decreased production are discussed. Moreover, the production per millimetre rainfall is only about one kilogram per hectare, while figures of three to four kilograms are given in the literature for similar areas in North Africa.

Cutting of shrubs for fuel is another vegetation utilization process that is doing damage to the desert shrub rangelands. Around villages, wells, etc. a continuous cutting takes place and often the young shoots of the regrowth are being grazed. This results in bare areas around such settlements. Experiments to study the impact of cutting on *Haloxylon salicornicum* and *Rhanterium epapposum* were carried out. *Haloxylon* shows a relatively high resistance to cutting but the frequency of this treatment as well as the uprooting of the plants by the fuel gatherers also has the inevitable effect on this species, resulting in the death of the plants. *Rhanterium epapposum* shows very little resistance to cutting.

Areas receiving runoff water are locally ploughed for cereal cultivation. In such areas the most productive natural vegetation is found. This is destroyed by the ploughing; regeneration is poor. Crop failures are common because of the unreliable rainfall, but the introduction of heavy machinery has made the ploughing and harvesting relatively easy. In particular in the northern Jezira, the part of the study area with the highest annual rainfall, almost the entire natural vegetation has been destroyed by this process and been replaced by a relatively useless vegetation in which *Peganum harmala* and *Artemisia scoparia* are the most common dominants. In the Western and Southern Deserts degradation of the vegetation in wadibeds and depressions is taking place by the same process.

Observations inside exclosures in comparison to the outside situation show the increase in cover and aerial biomass of the vegetation under protection from (over)utilization. Succession was found to be a very slow process for the desert shrub vegetation. Aerial biomass and crown cover inside exclosures was more than three times higher than outside. *Haloxylon salicornicum* showed a decreased production under full protection. Light to moderate grazing seems to stimulate the growth of this species.

Overgrazing, fuel cutting and ploughing, in various combinations, are causing an extensive degradation of the vegetation of the desert shrub rangelands and an increasing soil erosion. Regeneration of the vegetation in many areas will already pose problems or may even be impossible. The once productive desert rangelands of Iraq are increasingly converted to barren, lifeless desert. Desertification is here not a 'desert creep' but rather a process occurring from the inside whereby the influence of degraded land grows until the entire area is affected.

The last part of the study describes an approach that could lead to improved management. An economic justification for management of the desert shrub rangelands, especially in relation to livestock production, is presented. For the period 1953–1961, for which the required data were available, the desert rangelands contributed about 10–15% in the total contribution of the agricultural sector to the national income. The increasing demand for livestock products, caused by a growing population with an increasing per capita income, is reflected in the continuously increasing meatprices.

An approach in three phases to achieve a proper management and use of the study area is presented. In the first phase reconnaissance surveys of terrain, soil and vegetation should produce small scale maps and a qualitative land suitability classification for different land utilization types. The FAO land evaluation procedure provides a useful methodology. The relevant land utilization types and the land qualities that should be assessed for the rangeland evaluation are discussed. The government should then select priority areas and desired land utilization types. In the next phase the selected areas should be mapped at a scale ten times larger than that of the first phase and a quantitative land evaluation should follow, including the socio-economic factors and a cost-benefit prediction. This should lead to an implementation of land management practices in the third phase. This would require a strong governmental organization dealing with these areas. Some aspects of the required mapping and the measurement of production in the shrub vegetation are discussed separately. With modern survey techniques the surveys could be carried out in a short period of time. Aerial biomass and production of the vegetation can be calculated from the height and diameters of individual shrubs, attributes which are easy to measure.

The developments in the area are briefly summarized. The same technology causing the extensive degradation of the desert shrub rangelands could be used for a proper management. The methods to be used are by and large known and with the oil revenues Iraq is in a better position to deal with the problem of desertification than most other countries with arid areas.

SAMENVATTING

SAMENVATTING

Vooral na de conferentie van de Verenigde Naties in 1977 in Nairobi is het duidelijk geworden dat 'woestijnvorming' een probleem is dat voor alle aride gebieden over de gehele wereld geldt. De situatie in grote delen van Zuidwest Azië, waar in deze eeuw, met name in de laatste decennia, veranderingen zijn opgetreden in de wijze en intensiteit van het landgebruik, is een duidelijk voorbeeld. Veel van de veranderingen leiden op grote schaal tot een ernstige aantasting en uitputting van natuurlijke hulpbronnen, met name van fauna, vegetatie en bodem. Hierdoor wordt in toenemende mate land minder bruikbaar of zelfs onbruikbaar.

Als begrazing het algemeen aanvaarde landgebruik is, neemt de vegetatie als natuurlijke hulpbron daarbij een sleutelpositie in. Tegen deze achtergrond behandelt deze studie de al eeuwenlang begraasde woestijngebieden van Irak, waarbij de nadruk ligt op de oecologie en het gebruik van de vegetatie en op de mens als oecologische faktor. Het doel van de studie is het leveren van een bijdrage aan een verantwoord gebruik onder een gericht beheer van deze gebieden.

Het veldwerk is uitgevoerd van april 1970 tot juni 1973 vanuit het Institute for Applied Research on Natural Resources, nabij Bagdad. Het onderzoeksgebied wordt gevormd door de Southern Desert, Western Desert en Lower Jezira (Fig. I–3), tezamen ongeveer 240 000 km^2 groot. Het plantendek bestaat er voor het overgrote deel uit lage, open struikvegetaties waartussen, na de winter- en voorjaarsregens, annuellen en bolgewassen opbloeien. Min of meer overeenkomstige vegetatietypen bedekken ongeveer 10–15% van het landoppervlak van de aarde; zij vormen een natuurlijke hulpbron van grote betekenis, vooral in verband met extensieve veeteelt. De onderzochte woestijngebieden vormen ongeveer 55% van het oppervlak van Irak, een land dat vooral bekend is door de vroegere culturen in en langs de Mesopotamische vlakte. Olie vormt er nu de belangrijke bron van inkomsten. Bodem, vegetatie en fauna van de woestijngebieden van Irak zijn, door de marginaal beschouwde gebruiksmogelijkheden, over het algemeen nog in geringe mate of in het geheel niet onderzocht. Enkele kleine gebiedjes vormen hierop een uitzondering. Aan de floristische en plantengeografische aspecten van de vegetatie is relatief veel aandacht besteed, vooral door verzamelaars tijdens studiereizen.

Het in deze studie onderzochte gebied heeft een gemiddelde jaarlijkse regenval variërend van ongeveer 50 mm tot 350 mm, met enorme verschillen tussen de jaren. Het grootste deel van de regen valt in enkele hevige buien. De temperaturen dalen 's winters soms tot beneden het vriespunt en lopen 's zomers op tot ongeveer 50°C. De jaarlijkse verdamping bedraagt ongeveer twee tot drie en een halve meter en de relatieve luchtvochtigheid kan in de zomer overdag dalen tot beneden tien procent. Het gebied kan worden verdeeld in zeven fysiografische eenheden die zich vooral van elkaar onderscheiden door reliëf, het meer of minder doorsneden zijn met wadis, zandig of stenig karakter en, daarmee samenhangend, bodemtype en hoogteligging (ongeveer 0–915 meter boven gemiddeld zeeniveau). Deze eenheden en de voorkomende bodemtypen worden kort besproken. Van groot belang voor de exploitatie van de vegetatie is de watervoorziening. Oorspronkelijk was men daarvoor aangewezen op enkele diepe gegraven putten, ondiepe putten veelal in of nabij wadis, enkele natuurlijke bronnen en (slechts van korte duur) verzameld regenwater dat aan het oppervlak bleef staan. Vanaf de jaren dertig zijn er putten geslagen. In de loop der jaren is het aantal daarvan uitgegroeid tot enkele honderden. Deze zijn soms tot honderden meters diep geboord en voorzien van pompen en opslagtanks. Deze verbetering in de watervoorziening en het gebruik van vrachtwagens en ander gemotoriseerd vervoer hebben vrijwel de gehele woestijn binnen het bereik van mens en vee gebracht. Na het in gebruik raken van vuurwapens en vooral met de invoering van snelle transportmogelijkheden werd er meedogenloos gejaagd, vooral op de grotere diersoorten. Oryx, struisvogel en andere diersoorten die hier eens in grote groepen voorkwamen zijn thans uitgeroeid evenals de grote roofdieren zoals de leeuw en het jachtluipaard. Nog slechts gazellen komen sporadisch voor.

De mens moet al zeer lang geleden in dit gebied aanwezig zijn geweest, getuige de gevonden artefacten. Van een duidelijke invloed op de vegetatie kan echter pas sprake geworden zijn na het ontstaan van het nomadisme, toen Bedoeïenen vanuit het Arabisch schiereiland naar het Noorden trokken in het eerste en tweede millennium v. Chr., na het domesticeren van de dromedaris, de éénbultige kameel. Eeuwenlang is daarna de nomadische begrazing het landgebruik geweest voor de woestijngebieden. De Bedoeïenenstammen hadden hun eigen strikte regels voor het gebruiksrecht van bepaalde streken. Vooral de beperkingen in de drinkwatervoorziening en de vijandelijkheden tussen de stammen hielden de totale begrazingsdruk laag. Na de eerste wereldoorlog is dat beeld snel veranderd. De grenzen van het huidige Irak werden vastgelegd, er kwam leger- en politietoezicht, en met snelle transportmogelijkheden en zich steeds uitbreidende drinkwater-voorzieningen werd de woestijn 'opengelegd'. De kuddes kamelen en vooral schapen groeiden; voor geheel Irak nam het aantal schapen toe van 5 525 000 in 1938 tot 11 040 000 in 1965. Dit proces en de gevolgen ervan voor de vegetatie zijn nagegaan.

Gebaseerd op een verkenning van het gebied, waarvoor van ongeveer 4000 km een globale beschrijving van terrein, bodem, vegetatie en landgebruik werd gemaakt, is, na een subjectieve selektie, op 54 plaatsen een gedetailleerde vegetatie bemonstering uitgevoerd. In 50 kwadraten van 4 m^2 per plaats werden dichtheid, hoogte, bedekking en (als totaalgetal) biomassa van de voorkomende overblijvende planten bepaald. Tevens werd de bodem bemonsterd op twee punten per plaats op de dieptes 0–30 cm en 30–60 cm. De betrouwbaarheid en de bewerking van deze gegevens voor een classificatie worden besproken. Bij de verwerking van de gegevens is een nieuwe kwantitatieve maat gebruikt. Daarbij is in een percentage (het gemiddelde van relatieve dichtheid, relatieve frequentie en relatieve bedekking) de bijdrage van de soort in de totaalopname uitgedrukt. Uiteindelijk werden 15 vegetatietypen en 14 socio-oecologische groepen onderscheiden. De oecologische basis voor deze typen is met name gevonden in een klimaatsgradient, een verschil in bodemtextuur, en het landgebruik, waarbij vooral het ploegen van de bodem in het verleden meespeelt. De bedekkingspercentages liggen in de grootte orde van 1–20% met slechts enkele hogere waarden. Het aantal soorten is laag.

De autoecologie van de drie belangrijkste soorten, *Haloxylon salicornicum*, *Rhanterium epapposum* en *Artemisia herba-alba*, wordt behandeld, waarbij de nadruk ligt op de oecologie van de voortplanting. De eerste twee soorten vinden hun noordelijkste verspreidingsgrens in Irak. Het kiemingsproces en de ontwikkeling van de kiemplanten van de soorten illustreren hun aanpassing aan de extreme omstandigheden. Voor *Rhanterium epapposum*, een composiet van zandige bodems, wordt voor het eerst het kiemingsproces beschreven. In de gevallen bloemhoofdjes kiemen na de regens meerdere zaden terwijl ze in het hoofdje blijven zitten. De worteltjes dringen door de gemeenschappelijke bloembodem in het vochtige zand. Er is geen reden om aan te nemen dat door concurrentie over het algemeen slechts één kiemplant verder uit zou groeien. Uit een vergelijking van de strategie van de voortplanting wordt duidelijk hoe elk van de soorten aan een bepaald milieu is aangepast. In een apart deel worden het gebruik van de woestijngebieden door de mens en vee, in het bijzonder de vegetatie, en de gevolgen van dit gebruik, uitgewerkt.

Begrazing door kamelen en schapen, thans vooral door schapen (enkele miljoenen), en in mindere mate geiten en ezels, heeft van oudsher plaatsgevonden, maar door de toegenomen begrazingsdruk is de vegetatie recentelijk sterk aangetast. De nieuw verzamelde en de reeds bekende gegevens over de voorkeur van de veesoorten voor de meest algemene plantensoorten zijn samengevat. Chemische analyse van plantenmateriaal toonde aan dat de bij voorkeur gegeten delen van de struiken een hoog ruw eiwit gehalte hebben en vaak rijk zijn aan mineralen; de gehaltes liggen zelfs hoger dan bij gekweekte voedergewassen. Het vochtgehalte varieert sterk over de seizoenen. In het voorjaar kan de vegetatie (inclusief de annuellen) zoveel vocht bevatten dat extra drinkwater voor het vee overbodig wordt. Van geselecteerde soorten is de produktie over de seizoenen en, in één geval, voor twee jaren vergeleken. Door de ongeveer 4000 km verkenningstochten als transecten te analyseren en door gebruik te maken van de beschikbare produktiegetallen, is een globale schatting gemaakt van de totale produktie in de 240 000 km^2 in 1971. Dit cijfer ($2{,}91 \times 10^9$ kg drooggewicht) is vergeleken met de behoefte aan veevoer, afgeleid uit de 1971 – census gegevens ($2{,}52 - 3{,}31 \times 10^9$ kg drooggewicht). Deze cijfers zijn voor het begin van de jaren zeventig goed met elkaar in overeenstemming. De produktie in de midden zestiger jaren, en waarschijnlijk ook daarvoor, moet echter veel hoger zijn geweest gezien de aantallen stuks vee, en de daarmee samenhangende behoefte aan voer, vermeld voor de 1965 census. Verdere veldgegevens, die een aanwijzing zijn voor de afgenomen produktie, worden besproken. Tevens blijkt de produktie per millimeter regen voor de overblijvende vegetatie slechts ongeveer één kilogram per hectare te bedragen terwijl in de literatuur voor vergelijkbare gebieden in Noord Afrika drie tot vier kilogram worden vermeld.

Het kappen van struiken voor brandstof (koken, verwarming) is een gebruik dat ondanks zijn omvang op zichzelf geen ernstige gevolgen hoeft te hebben. Dat dit wel het geval is komt doordat het voortdurend plaatsvindt in dezelfde beperkte gebieden, rondom dorpen, putten, bronnen, etc., en bovendien omdat de eventuele hergroei weer aan begrazing blootstaat. Experimenten, waarin struiken van *Haloxylon salicornicum* en *Rhanterium epapposum* (herhaaldelijk) werden geknipt, tonen enerzijds, zolang het knippen niet te frequent plaatsvindt, de relatief grote weerstand van de eerste soort, maar tonen anderzijds de desastreuze gevolgen van het gebruik zoals dat door de bevolking, in combinatie met het vee, plaatsvindt.

De meest direkte vernietiging van de natuurlijke vegetatie heeft plaatsgevonden doordat lager gelegen gebieden, die aflopend regenwater ontvangen, mechanisch worden geploegd in (onverantwoorde) pogingen een graangewas te verbouwen. De meest produktieve vegetaties vallen hier onder en regeneratie vindt niet of nauwelijks plaats. Dit proces wordt besproken voor de noordelijke Jezira, het gebied met de hoogste regenval-verwachting, waar vrijwel geen natuurlijke vegetatie is overgebleven, en tevens voor de wadis en later gelegen terreinen in de woestijn in het zuiden. Na ploegen blijkt in veel gevallen een vegetatie op te komen waarin *Peganum harmala* en *Artemisia scoparia* domineren, twee soorten die voor begrazing of ander gebruik vrijwel geen waarde hebben.

Waarnemingen aan gebieden die een aantal jaren strikt beschermd zijn geweest of slechts licht begraasd tonen aan hoe hoog de bedekking, biomassa en produktie van bepaalde vegetaties kunnen worden onder dergelijke omstandigheden. Suksessie treedt wel op, maar blijkt voor de struikvegetaties een uiterst langzaam proces. Vergelijkingen van gebieden binnen en buiten afrasteringen geven een driemaal meer of nog hogere biomassa en bedekking voor de gebieden binnen de afrasteringen. Onder strikte bescherming blijkt de produktie van *Haloxylon salicornicum* terug te lopen. Een niet te intensieve begrazing stimuleert bij deze soort de groei.

De gecombineerde en elkaar versterkende gevolgen van overbegrazing, kappen voor brandstof en ploegen, in diverse combinaties, tonen duidelijk aan hoe ver de aantasting van de vegetatie, en ook van de bodem door enorme erosie, al is voortgeschreden. Voor veel gebieden zal regeneratie moeilijk zijn of zelfs onmogelijk blijken. De verwachting is daarom dat, als de aantasting kan doorgaan in het huidige tempo, de eens zo produktieve woestijn-begrazingsgebieden van Irak over enkele decennia zullen zijn omgezet in een kale, levenloze woestijn. Deze 'woestijnvorming' is geen proces dat zich over een zeker front in een bepaald tempo afspeelt. Het is veeleer een aantasting van binnen uit, waarbij de invloedssferen van de aangetaste gebieden elkaar tenslotte gaan raken, en waardoor geleidelijk de natuurlijke hulpbronnen van het gebied in zijn geheel worden aangetast en vernietigd.

In het laatste deel van de studie wordt een aanpak uitgewerkt die tot een beter beheer kan leiden. Een argument hiervoor kan ook gevonden worden in de economische betekenis van het gebied, in velerlei opzicht, maar vooral voor de veeteelt. Deze betekenis is afzonderlijk uitgewerkt. Voor de periode 1953–1961 waarvan voldoende gegevens beschikbaar waren, blijkt dat de woestijngebieden alleen al door de veeteelt 10–15% bijdroegen tot het totale aandeel van de landbouwsector aan het nationaal inkomen. Tevens blijkt uit de snel stijgende vleesprijzen, door de groeiende bevolking en het toenemend inkomen, de toenemende vraag naar veeteeltprodukten.

Om tot een verantwoord beheer te komen wordt een aanpak in drie fasen uitgewerkt waarin de landevaluatie-procedure, zoals die onlangs door de FAO is voorgesteld, een belangrijke rol speelt. In de eerste fase wordt op basis van inventarisatie en kleinschalige kaarten van de belangrijkste onderdelen van het land, zoals terreingesteldheid, bodem, vegetatie, etc. voor het gehele gebied een globale geschiktheidsclassificatie voor verschillend landgebruik uitgewerkt. Deze landgebruikstypen en de daarvoor in aanmerking te nemen faktoren worden afzonderlijk besproken. De overheid zou hieruit een keuze moeten maken en prioriteitsgebieden aangeven. In de volgende fase worden geselecteerde gebieden op een tienmaal grotere schaal gekarteerd. Er wordt dan ook een evaluatie met een kosten/baten verwachting gemaakt, waarbij rekening wordt gehouden met de sociaal-economische faktoren. Op basis hiervan moeten in de derde fase concrete beheersmaatregelen worden doorgevoerd waarvoor een sterke organisatie in het leven geroepen zou moeten worden. De ontwikkeling in de sociaal-economische sector zowel als veranderingen in de vegetatie en bodemgesteldheid dienen dan nauwkeurig gevolgd te worden waarbij een voortdurend overleg tussen de waarnemers, overheid, beheerders en gebruikers tenslotte een optimaal landgebruik zou moeten garanderen. Enkele aspecten van de kaarteringen, en het zo belangrijke meten van de produktie van de vegetatie, worden afzonderlijk behandeld. Onder toepassing van moderne survey-technieken zouden de kaarteringen relatief snel en efficiënt kunnen worden uitgevoerd. De bovengrondse biomassa en produktie van de struikvegetaties blijken met vrij grote betrouwbaarheid af te leiden uit de gemakkelijk te meten parameters diameters en hoogte van de individuele struiken.

Tenslotte worden de ontwikkelingen kort samengevat. Dezelfde technische ontwikkelingen die geleid hebben tot de enorme aantasting en degradatie van de woestijnbegrazingsgebieden zouden kunnen worden aangewend om tot een verantwoord beheer van de nog overgebleven natuurlijke hulpbronnen te komen. De methoden daarvoor zijn in grote lijnen voorhanden en de inkomsten uit olie-export plaatsen Irak in een relatief gunstige positie.

REFERENCES

REFERENCES

Aart, R. van (1974), Aspects of land use planning in the Lower Mesopotamian Plain - a problem analysis. Neth. J. Agric. Sci. 22 : 54 - 65.

Abul-Fatih, Hussain A. (1970), The vegetation of saline swampy areas of Bahr-al-Najaf, Iraq. M.Sc.Thesis Univ. Baghdad, in: Bull.Coll.Sci. Baghdad 17 (1976): 49 - 52.

Abul-Fatih, H. (1975), Vegetation and soil of saline depressions near Najaf, Central Iraq. Vegetatio 30 : 107 - 115.

Agnew, A.D.Q. (1960), The protected range areas at Khidr el Mai and Shubaichi. Report to Dir. Gen. Research and Projects, Min. Agric., Iraq, 7 pp. (typescript).

Agnew, A.D.Q. (1961), Studies on the Plant Ecology of the Jezira of Central Iraq II.Hummock formation by *Haloxylon salicornicum* (Moq.) Bge. Bull.Coll.Sci. Baghdad 6 : 42 - 59.

Agnew, A.D Q. (1962a), Flora of the Baghdad District, Part I: Monocotyledons, with a general account of the vegetation of the area. Suppl. to Bull. Coll. Sci. Baghdad 6 : 1 - 170.

Agnew, A.D.Q. (1962b), A study of the oak forests of Gara Mountain, Mosul Liwa. Proc. Iraq Sci. Soc. 5 : 31 - 43.

Agnew, A.D.Q. & John E.C. Flux (1970), Plant dispersal by hares (*Lepus capensis* L.) in Kenya. Ecology 51 : 735 - 737.

Agnew, A.D.Q. & R.W. Haines (1960), Studies on the Plant Ecology of the Jezira of Central Iraq I. The influence of perennials on annuals in the Jezira. Bull. Coll. Sci. Baghdad 5 : 41 - 60.

Ahmed, Adnan M. & Walter Kraft (1972), A contribution to the hydrogeology of the Western Desert of Iraq. J. Geol. Soc. Iraq V : 135 - 148.

Ahmed, Mohammed M.A. (1968), Mutton and beef price analysis, central region, Iraq. Iraqi J. Agric.Sci. III : 49 - 61.

Ahmed, Mohammed M.A. & Hikmat A.Al-Roumi (1967), Demand and price analysis of meat in Iraq. Iraqi J. Agric.Sci. II : 61 - 69.

Ainsworth, W. esq. (1841), Notes of an excursion to Kal'ah Sherkát, the U'r of the Persians, and to the ruins of Al Hadr, the Hutra of the Chaldees, and Hatra of the Romans. J. Roy. Geog. Soc. XI : 1 - 20.

Al-Amari, Ahmed Shehab, Jamal Nuri Al-Barazanchi, N.C. Keulemans, W.S. Post, Khairi Habib Tammo & Jabbar Kadhem Tukhaikh (1968), Investigations into the production of forage under irrigation for animal feed at Abu Ghraib, Iraq. Tech. Rep. no. 19 Animal Husbandry Research and Training Project. UNDP/FAO - D.G. of Animal Resources and Veterinary Services, Abu Ghraib, Iraq. 35 pp.

Al-Ani, Hikmat A. & Sharif Y. Haddad (1973), Seasonal changes in the plant behaviour of Falluja and Iskanderiya gypsum desert flora. Iraqi J. Agric.Sci. VII : 1 - 11.

Al-Ani, Tariq A. & Maida S. Jawad (1973), Range Resources of Iraq VIII. Height-nutrient composition relationship of five native range species. Tech. Rep. no. 47 Inst. for Appl. Research on Natural Resources, Abu Ghraib, Iraq, 20 pp. (mimeo).

Al-Ani, Tariq A.& Maida S. Jawad (1974), Range Resources of Iraq X. Seasonal variation in nutrient content of *Haloxylon salicornicum* and *Artemisia herba-alba*. Tech. Rep. no. 56 Inst. for Appl. Research on Natural Resources, Abu Ghraib, Iraq 12 pp. (mimeo).

Al-Ani, T.A., I.M. Habib, A.I. Abdulaziz & N.A. Ouda (1970a), Plant indicators in Iraq Part II - Mineral composition of native plants in relation to soils and selective absorption. Techn. Rep. no. 15 Inst. for Appl. Research on Natural Resources, Abu Ghraib, Iraq, 8 pp. (mimeo).

Al-Ani, Tariq A., M.M. Al-Mufti, N.A. Ouda, R.N. Kaul & D.C.P. Thalen (1970b), Range Resources of Iraq I. Range cover types of Western and Southern Desert. Tech. Rep. no. 16 Inst. for Appl. Research on Natural Resources, Abu Ghraib, Iraq, 8 pp. (mimeo).

Al-Ani, T.A., A.I. Abdulaziz, M.M. Al-Mufti, F.M.R. Al-Charchafchi, R.N. Kaul & D.C.P. Thalen (1971a), Range Resources of Iraq II. Studies on palatability and nutritive value of some range species. Tech. Rep. no. 19 Inst. for Appl. Research on Natural Resources, Abu Ghraib, Iraq, 10 pp. (mimeo).

Al-Ani, Tariq A., A.I. Abdulaziz, M.M. Al-Mufti, F.M.R. Al-Charchafchi, R.N. Kaul & D.C.P. Thalen (1971b), Studies on nutrient composition of some range species. Iraqi J. Agric.Sci 6 : 3 - 19.

Al-Ani, Tariq A., M.M. Al-Mufti & F.M.R. Al-Charchafchy (1971c), Range Resources of Iraq IV. Germinability of some native range species of Iraq. Tech. Rep. no. 30 Inst. for Appl. Research on Natural Resources, Abu Ghraib, Iraq, 14 pp. (mimeo).

Al-Ani, Tariq A., M.M. Al-Mufti, N.A. Ouda, R.N. Kaul & D.C.P. Thalen (1974), A reconnaissance survey of range cover types in the Western and Southern Deserts of Iraq. Proc. XII Int. Grassl. Congr. Moscow 1974, Vol. I Part II, pp. 583-594.

Al-Hashimi, M.H. (1972), An evaluation of animal husbandry in Iraq. The Baghdad Observer V no. 1231, 14 Febr. 1972.

Al-Hilli, Majeed R. (1970), Ecological study in Baiji sand dunes. M. Sc. thesis Univ. Baghdad, in : Bull.Coll.Sci. Baghdad 17 (1976) pp. 20 - 25.

Ali, K.T. (1955), Agricultural potentialities of Iraq. Section II : Animal production pp. 7 - 11. Paper no. 13 - 12 FAO Centre on Land Problems in the Near East, Salahuddin, Iraq.

Al-Kasab, Nafi Nasser (1966), Die Namdenansiedlung in der Irakischer Jezira. Tübinger Geographische Studien Heft 20, Geogr. Institut der Universität Tübingen, 148 pp.

Allen, Shirley Walter (1959), Conserving Natural Resources. Principles and Practice in a democracy (2nd ed.) Mc. Graw-Hill, New York, 370 pp.

Allouse, B.E. (1954-1968), A bibliography on the vertebrate fauna of Iraq and neighbouring countries. Iraq Nat. Hist. Mus. Publ.; no. 4 Mammals pp. 1-31 (1954); no. 5 Birds pp. 1-45 (1954); no. 6 Reptiles & Amphibians pp. 1 - 23 (1955); no. 7 Fishes pp. 1 - 31 (1955); no. 25 Mammals (sec. comp.) pp. 1- 36 (1968).

Allred, B.W. (1968), Range management training handbook for Saudi Arabia. Report FAO, Rome, 216 pp.

Al-Mufti, M.M., Tariq A. Al-Ani & R.N. Kaul (1970), Preliminary results from trials of some exotic grass species. Tech. Rep. no. 4 Inst. for Appl. Research on Natural Resources, Abu Ghraib, Iraq, 5 pp. (mimeo).

Al-Mufti, M.M., T.A. Al-Ani, D.C.P. Thalen & R.N. Kaul (1971a), Range Resources of Iraq V. Studies on adaptability of some exotic range species in the environs of Fudhaliya. Tech. Rep. no. 29 Inst. for Appl. Research in Natural Resources, Abu Ghraib, Iraq, 24 pp. (mimeo).

Al-Mufti, M.M., T.A. Al-Ani, D.C.P. Thalen & R.N. Kaul (1971b), Range Resources of Iraq III. The herbarium of the Institute for Applied Research on Natural Resources, Iraq - Collection, Identification and Preservation of range plants and seeds. Tech. Rep. no. 21 Inst. for Appl. Research on Natural Resources, Abu Ghraib, Iraq, 21 pp. (mimeo).

Al-Rawi, Ali & H.L. Chakravarty (1964), Medicinal Plants of Iraq. Tech. Bull. no. 15 Min. Agric., Dir. Gen. Research and Projects, Iraq.

Altaie, F.H. (1968), The Soils of Iraq. Ph. D. thesis, Ghent, Belgium.

Altaie, Flayeh, H. C. Sys & G. Stoops (1969), Soil Groups of Iraq. Their Classification and Characterization. Pedologie XIX : 65 - 148.

Anderson, C.A. (1958), Suggested procedure for range activities in Iraq. Report ICA / Iraq, 20 pp. (mimeo).

Ashraf, A. (1976), Land evaluation for rural planning in the Sousse area. M.Sc. thesis ITC, Enschede, The Netherlands, 169 pp.

ASP (1975), Manual of Remote Sensing (2 vols.) Robert G. Reeves (ed. in chief) American Society of Photogrammetry, 2144 pp.

Baig, M. Shabbir (1977), Inventory and evaluation of rangeland. An example for rangelands in arid zones worked out for Quetta-Pishin area, Baluchistan, Pakistan. M.Sc. thesis ITC, Enschede, The Netherlands, 132 pp. + app.

Bannink, J.F., H.N. Leys & I.S. Zonneveld (1974), Akkeronkruidvegetatie als indicator van het milieu, in het bijzonder de bodemgesteldheid. (Weeds as environmental indicators, especially for soil conditions). Meded. Stichting Bodemkartering, Bodemk. Studies nr. 11. Wageningen, The Netherlands.

Barnes, R.F. (1973), Laboratory Methods of Evaluating Feeding Value of Herbage. (chapter 32, pp. 179-214) in: Chemistry and Biochemistry of Herbage. G.W. Butler & R.W. Bailey (eds.), Academic Press, London and New York.

Batanouny, K.H. (1973), Soil properties as affected by topography in desert wadis. Acta Bot. Acad. Sc. Hung. 19 : 13 - 21.

Batanouny, K.H. (1974), The desert vegetation in Egypt. Joint Symposium on Arid Lands Ecosystems (Egyptian and American Scientist) 12 - 18 May 1974 Cairo Univ., Inst. of African Research and Studies, 52 pp. (mimeo).

Batanouny, K.H. & S. Abu El-Souod (1972), Ecological and phytosociological study of a sector in the Lybian Desert. Vegetatio 25 : 335 - 356.

Batanouny, K.H. & M.R. Hilli (1973), Phytosociological study of Ghurfa desert, central Iraq. Phytocoenologia 1 : 223 - 249.

Batanouny, K.H. & M.Y. Sheikh (1972). Ecological Observations along Baghdad - Huseiba Road, Western Desert, Iraq. Feddes Repert. 83 : 245 - 263.

Batanouny, K.H. & M.A.F. Zaki (1969), Root development of two common species in different habitats in the Mediterranean subregion in Egypt. Acta Bot. Acad. Sc. Hung. 15 : 217 - 226.

Bazilevich, N.I., L.Y. Rodin & N.N. Rozov (1970), Geographical aspects of biological productivity, in: Sov. Geogr. Rev. Transl. 12 : 293 - 317 (1971), transl. from Mat. V. Syezda Geogr. Obshch. USSR, Leningrad.

Beard, John S. (1973), The Physiognomic Approach, pp. 355 - 386 in: Ordination and Classification of Communties, Handbook of vegetation science Vol. 5, R.H. Whittaker (ed.) Dr. W. Junk b.v. Publ. The Hague, The Netherlands.

Beaudelaire, J.P. (1972), Water for livestock in semi arid zones. World Animal Rev. 3 : 1 - 9.

Beaumont, Peter, Gerald H. Blake & J. Malcolm Wagstaff (1976), The Middle East: A Geographical Study. John Wiley & Sons, London, 572 pp.

Beek, K.J. (1978), Land evaluation for agricultural development. Publ. 23 International Institute for Land Reclamation and Improvement, Wageningen, The Netherlands, 333 pp.

Beek, K.J. & J. Bennema (1972), Land evaluation for agricultural land use planning. An ecological methodology. Dept. Soil. Sci. and Geol., Agric. University, Wageningen, The Netherlands, 70 pp.

Beek, M.A. (1960), Atlas van het tweestromenland. Elsevier, Amsterdam, 164 pp.

Blair Rains, A. (1970), Chemical analysis of fodder samples and the interpretation of results. Appendix 3, pp. 127-133 in: An Ecological Survey of Western Province, Zambia, with Special Reference to the Fodder Resources. vol. 2. Land Resource Study no. 8. D.O.S. Tolworth, England.

Bodenheimer, F.S. (1943), Studies on the Ecology and Control of the Moroccon Locust *(Docoistaurus marcoccanus)* in Iraq. Dep. Agric. Iraq. Bull. 29, 121 pp.

Boissier, P.E. (1867–1888), Flora Orientalis, 5 vols. & Suppl., Basel - Geneva - Lyon.

Bor, N.L. (1968), Gramineae, Volume nine, Flora of Iraq, C.C. Townsend, Evan Guest & Ali Al-Rawi (eds.) Min. Agric., Iraq, 588 pp.

Bormann, F.H. (1953), The statistical efficiency of sample plot size and shape in forest ecology. Ecology 34 : 474 - 487.

Braidwood, R.J. (1952), From Cave to Village, birth of agriculture and animal husbandry. Scientific American 187 : 62 - 66.

Braidwood, R.J. (1954), The Iraq - Jarmo Project. Sumer X : 120 - 138.

Braun - Blanquet, J. (1928), Pflanzensoziologie, Grundzüge der Vegetationskunde. Biologische Studienbücher 7. 1 Ed. Berlin, 330 pp.

Braun - Blanquet, J. (1964), Pflanzensoziologie, Grundzüge der Vegetationskunde, 3 Ed. Springer, Wien - New York, 865 pp.

Brice, W.C. (1966), South-West Asia. A systematic regional geography. Univ. of London Press, London.

Brichambaut, G. Perrin de (1956a), Note regarding the main perennial grasses observed in the low and middle mountains of Kurdistan, 2 pp. (mimeo).

Brichambaut, G. Perrin de (1956b), Report to the Government of Iraq on the improvement of fodder production. FAO/56/11/8595 Report no. 562, Rome, 49 pp. (mimeo).

Brichambaut, G. Perrin de & C.C. Wallén (1963), A study of agroclimatology in semi-arid and arid zones of the Near East Tech. Note no. 56, WMO, Geneva, 64 pp.

Brinkman, R. & A.J. Smyth (eds) (1973), Land evaluation for rural purposes. Summary of an expert consultation, Wageningen, The Netherlands. Publ. 17 International Institute for Land Reclamation and Improvement, Wageningen, The Netherlands, 116 pp.

Brown, Dorothy (1954), Methods of Surveying and Measuring Vegetation. Comm. Bureau of Pastures and Field Crops, Bull. 42, Farnham Royal, Bucks, England, 223 pp.

Bryan, H.M. & H.W. Springfield (1955), Range Management in Iraq - Findings, Plan and Accomplishment. J. Range Manage. 8 : 249 - 256.

Burdon, David J. (1971), Exploitation of Groundwater for Agricultural Production in Arid Zones. pp. 290 - 300 in: Food, Fiber and the Arid Lands, William G. Mc Ginnies, Bram J. Goldman and Patricia Paylore (eds.) The University of Arizona Press, Tucson, Arizona.

Buringh, P. (1960), Soils and soil Conditions in Iraq. Min. Agric. Dir. Gen. Research and Projects, Iraq. 322 pp.

Buringh, P. (1968), Introduction to the study of soils in tropical and subtropical regions. PUDOC, Wageningen, The Netherlands, 118 pp.

Butzer, K.W. (1958), The Near East during the last glaciation : a palaeogeographical sketch. Geogr. J. 124 : 367 - 369.

Butzer, Karl W. (1964), Environment and Archaeology, An introduction to Pleistocene Geography. Aldine Publ. Comp. Chicago, 524 pp.

Cain, S.A. & G.M. de O. Castro (1959), Manual of vegetation analysis. Harper, New York.

Central Statistical Organization (1970), Annual Abstract of Statistics 1970, Min. Planning, Iraq 575 pp.

Central Statistical Organization (1973), Results of the 1971 census of agriculture, Vols. I and II, Min. Planning, Baghdad, 278 + 127 pp.

Chapline, W.R. & R.O. Whyte (1947), The use and misuse of shrubs and trees as fodder. Imp. Agri. Bur. Joint Publ. 10. Imp. Bur. of Pastures and Field Crops, England.

Chapman, G.W. (1957) Forests and Forestry in Iraq. Report D.G. of Forests, Min. Agric. Iraq (mimeo).

Chapman, V.J. (1960), Iraq and the Middle East, pp. 222-226 in: Salt Marshes and Salt Deserts of the World, Leonard Hill, London and New York.

Chaudri, I.I., H.A. Kareem, A. Al-Zubaidi & A.Y. Hanna (1971), Plant indicators of alluvial soils of central Iraq. Vegetatio 23 : 315 - 322.

Cheesman, R.E. (1920), Report on the Mammals of Mesopotamia collected by Members of the Mesopotamian Expeditionary Force, 1915 to 1919. J. Bombay Nat. Hist. Soc. 27 : 323 - 346.

Clawson, M., H.H. Landsberg & L.T. Alexander (1971), The agricultural potential of the Middle East. Am. Elsevier Publ. Comp. Inc., New York, 312 pp.

Clifford, H.T. (1976), Dendrograms and their Interpretation. pp. 96 - 101 in: Pattern analysis in agricultural science, W.T. Williams (ed.) CSIRO Melbourne, Elsevier Sci. Publ. Comp., Amsterdam, Oxford, New York.

Clor, M.A., T.A. Al-Ani & F. Charchafchi (1973), Range Resources of Iraq IX, Seed viability and seedling vigour of *Haloxylon salicornicum* as affected by time lapse, storage conditions and seed size. Tech. Rep. no. 52 Inst. for Appl. Research on Natural Resources, Abu Ghraib, Iraq, 7 pp. (mimeo).

Clor, M.A., T.A. Al-Ani & F. Charchafchi (1974a), Range Resources of Iraq XIII, Germination and seedling development of *Haloxylon salicornicum*. Tech. Rep. no. 57 Inst. for Appl. Research on Natural Resources, Abu Ghraib, Iraq, 33 pp. (mimeo).

Clor, M.A., T.A. Al-Ani & F. Charchafchi (1974b), Range Resources of Iraq XV. Germination, storage conditions and after-ripening of the seeds of *Artemisia herba-alba*. Tech. Rep. no. 76 Inst. for Appl. Research on Natural Resources, Abu Ghraib, Iraq, 17 pp. (mimeo).

Clor, M.A., T.A. Al-Ani & F. Charchafchi (1976), Germinability and seedling vigor of *Haloxylon salicornicum* as affected by storage and seed size. J. Range Manage. 29 : 60 - 62.

Condon, R.W. (1968), Estimation of grazing capacity on arid grazing lands. pp. 112 - 124 in: Land evaluation, G.A. Stewart (ed.) Mac Millan, Melbourne, Australia.

Cook, C. Wayne (1972), Comparative nutritive values of forbs, grasses and shrubs, pp. 303 - 310 in: Wildland Shrubs - Their Biology and Utilization. An Int. Symp., Logan, Utah, USDA Forest Service General Techn. Rep. INT - I, 1972.

Cooke, Ronald U. & Andrew Warren (1973), Geomorphology in Deserts, B.T. Batsford Ltd. London, 374 pp.

Crampton, E.W. & L.E. Harris (1969), Applied Animal Nutrition (2nd ed.). W.H. Freeman and Comp. San Francisco, 753 pp.

Dakshini, K.M.M. (1972), Indian subcontinent, pp. 3-15 in: Wildland Shrubs - Their Biology and Utilization, An Int. Symp., Logan, Utah, USDA Forest Service General Tech. Rep. INT - I, 1972.

Dapper, O. (1680), Nauwkeurige beschrijvingen van Azië, behelsende de gewesten van Mesopotamië, Assyrië, Anatolia, Klein Azië, Arabië. Amsterdam (in Kon. Bibl., The Hague).

Davis, P.H. (1975), Flora of Turkey and the East Aegean Islands. Vol. I (1965) - Vol. V (1975). Edinburgh University Press.

Dennis, P.E. (1953), Report to the Government of Iraq on the Investigations and Development of Groundwater Resources. Report no. 189, FAO, Rome.

D.H.V. (1978), Methodology Report Botswana Countrywide Animal and Range Assessment Project. Report by D.H.V. Consulting Engineers Amersfoort, The Netherlands, to Dept. of Wildlife, National Parks and Tourism, Republic of Botswana, Gaborone, 73 pp.

Dickson, V. (1955), The wild flowers of Kuwait and Bahrain, Allen and Unwin, London, 144 pp.

Dietz, Donald R. (1972), Nutritive value of shrubs, pp. 289 - 302 in: Wildland Shrubs - Their Biology and Utilization,A Int. Symp., Logan, Utah, USDA Forest Service General Tech. Rep. INT - I, 1972.

Doughty, Charles M. (1908), Wanderings in Arabia, an abridgment of 'Travels in Arabia Deserta' made with the author's sanction by Edward Garnett, 2 vols., Duckworth, London.

Dougrameji, J. & R.N. Kaul (1971), Report on the possibilities of reclamation of sand dunes and proposed future research. Tech. Rep. no. 37 Inst. for Appl. Research on Natural Resources, Abu Ghraib, Iraq, 10 pp. (mimeo).

Draz, Omar (1978), Revival of the 'Hema' system of range reserves as a basis for Syrian Range Development Programme. Paper presented at 1st International Rangeland Congress, Denver, Colorado, August 1978.

Dregne, Harold E. (1967), Inventory of Research on Surface Materials of Desert Environments. pp. 287 - 377 in Deserts of the World. William G. McGinnies, Bram J. Goldman & Patricia Paylore (eds.) University of Arizona Press, Tucson.

Dudley Stamp, L. (ed.) (1966), A glossary of geographical terms (2nd ed.) Longman Group Ltd., London.

Dijk, A. van, D.C.P. Thalen & I.S. Zonneveld (1977), The application of satellite multi spectral scanning in land survey - preliminary experiences with optical and digital image (pre-) processing of an area near Kahama, Tanzania. Printed Report, D.H.V. Consulting Engineers, Amersfoort, ITC, Enschede, The Netherlands, 76 pp.

Dijksterhuis, E.J. (1949), Condition and management of rangeland based on quantitative ecology. J. Range Manage. 2 : 104 - 115.

Dzerdzeevskii, B.L. (1958), On some climatological problems and microclimatological studies of arid and semi-arid regions in U.S.S.R., pp. 315 - 325 in: Climatology and microclimatology, Proc. Canberra Symposium. Arid Zone Research XI, UNESCO, Paris.

Eig, A. (1935). Ecologie du Criquet Maroccain en Iraq. Bull. Ent. Res. 26 : 293 - 309.

Emberger, L. (1955), Afrique du Nord-Ouest. pp. 219 - 294 in: Plant Ecology Reviews of Research, Arid Zone Research VI, UNESCO, Paris.

Emberger, L. (1958), The study of the vegetation and its bearing on the national economy, with special reference to the Mediterranean region and Iraq. Prosp. Iraq Biol. 1 : 4 - 14.

Engel, C.G. and R.P. Sharp (1958), Chemical data on desert varnish, Geol. Soc. Am. Bull. 69 : 487 - 518.

Europa Publications Ltd. (1975), The Middle East and North Africa 1975 - 1976. (Iraq pp. 362 - 400).

Everist, Selwyn L. (1972), Australia, pp. 16 - 25 in: Wildland Shrubs - Their Biology and Utilization, An Int. Symp. Logan, Utah, USDA Forest Service General Tech. Rep. INT - I, 1972.

FAO (1970), Report in Group Fellowship Study Tour on Settlement in Agriculture of Nomadic, Semi-Nomadic and other pastoral people, U.S.S.R. Report no TA 2810, FAO, Rome, 46 pp.

FAO (1971), Integration of the grain and livestock sectors in the Near East. FAO Monthly Bull. of Agric. Econ. and Stat. 20 (2) : 1 - 8.
FAO (1972), Report on the FAO experts consultation on the settlement of nomads in Africa and the Near East. Cairo, 4 - 12 Dec. 1971. FAO Report no. RP 20, Rome, 35 pp.
FAO (1974), Approaches to land classification. Soils Bulletin no. 22, FAO, Rome, 120 pp.
FAO (1976), A framework for land evaluation. Soils Bulletin no. 32, FAO, Rome, 72 pp.
FAO/UNESCO (1977), Soil map of the world, sheet VII-1, South Asia, UNESCO, Paris.
Fine, N.C., Khalid T. Ali, Mohammed Faraj & Omar Amin Ali (undated), Industrial Dates and Ground Date Pits for Fattening Cattle. Min. Agric. Iraq, Div. Animal Husb., 6 pp. (mimeo).
Firmin, R. & M.D. Kernick (1964), Notes on a new *Acacia* species in Kuwait. Paper presented to the 10th Int. Bot. Congress, Edinburgh, 5 pp. (mimeo).
Fisher, W.B. (1950), The Middle East, a Physical, Social and Regional Geography. Methuen, London (3rd ed. 1956).
Floret, Christian (1971), Recherches phyto-écologiques entreprises par le C.N.R.S. sur le biome 'Zone Aride' en Tunisie. Doc. no. 57 C.N.R.S., CEPE, Montpellier, 26 pp.
Floret, C. & R. Pontanier (1974), Etude de trois formations végétales naturelles du Sud tunesien. Production bilan hydrique des sols, Inst. Mat. Rechn. Agrond., Tunis, 44 pp. (mimeo).
FMC. Corporation (1964), Preliminary study and report on Land Reclamation and Irrigated Agriculture in the Western Desert of Iraq. Vol. 1 San Jose, California, 25 pp. + 20 pp. app.
Fosberg, F.R. (1967), A classification of vegetation for general purposes. pp. 73 - 120 in: Guide to the check sheet for I.B.P. Areas. I.B.P. Handbook no. 4, Blackwell Sc. Publ. Oxford and Edinburgh.
Freitag, H. (1971a), Die natürliche Vegetation Afghanistans, beiträge zur Flora und Vegetation. Vegetatio XXII: 285 - 344.
Freitag, H. (1971b), Studies in the natural vegetation of Afghanistan. pp. 89 - 106 in: Plant life of South-West Asia. Proc. Symp. Bot. Soc. of Edinburgh. P.H. Davis, P.C. Harper & I.C. Hedge (eds).

Gehrke, Ulrich & Gustav Kuhn (1963), Die Grenzen des Irak. Historische und rechtliche Aspekte des irakischen Anspruchs auf Kuwait und des irakisch-persischen Streites um den Schatt al-Arab. W. Kohlhammer GMBH, Stuttgart.
Gillett, J.B. (1948a), The Desert Vegetation of Iraq. 5 pp. (mimeo).
Gillett, J.B. (1948b), The Steppe Vegetation of Iraq. 7 pp. (mimeo).
Gillett, J.B. (1948c), The Pastures and Wild Fodder Plants of Iraq. 10 pp. (mimeo).
Gimingham, C.H. & G.R. Miller (1968), Measurement of the Primary Production of Dwarf Shrub Heaths. (Part II, pp. 43-51) in: Methods for the Measurement of Primary Production of Grassland. IBP Handbook no. 6. Blackwell Sci. Publ. Oxford and Edinburgh.
Goodall. D.W. (1970), Statistical Plant Ecology. Ann. Rev. Ecol. and Syst. 1: 99 - 124.
Goodall, D.W. (1973a), Sample Similarity and Species Correlation. pp. 107 - 156 in: Ordination and Classification of Communities, Handbook of vegetation science Vol. 5, R.H. Whittaker (ed.) Dr. W. Junk b.v. Publ., The Hague, The Netherlands.
Goodall, D.W. (1973b), Numerical Classification. pp. 575 - 615 in: Ordination and Classification of Communities, Handbook of vegetation science 5, R.H. Whittaker (ed.) Dr. W. Junk b.v. Publ., The Hague, The Netherlands.
Goodin, Joe R. and Syrus M. McKell (1971), Shrub productivity - a reappraisal of arid lands. pp. 235 - 246 in: Food, Fibre and the Arid Lands, William G. McGinnies, Bram J. Goldman and Patricia Paylore (ed.) The University of Arizona Press, Tucson, Arizona.
Goosen, D.J. (1967), Aerial photo-interpretation in soil survey. Soils Bulletin no. 6, FAO, Rome.
Gounot, M. (1969), Methodes d'étude quantitative de la végétation. Masson et Cie., Editeurs, Paris. 314 pp.
Greig-Smith, P. (1964), Quantitative plant ecology (2nd ed.) Butterworths, London, 256 pp.
Grimsdell, J.J.R. (1978), Ecological monitoring. Handbook no. 4 African Wildlife Leadership Foundation, Nairobi, Kenya, 73 pp.
Gronovius, J.F. (1755), Flora Orientalis, Leiden.
Guest, E.R. (1932), Notes on Trees and Shrubs for Lower Iraq. Dep. Agric. Iraq. Bull. 26: 1 - 18.

Guest, E.R. (1933), Notes on Plants and Plant Products, with their Colloquial Names in Iraq. Dep. Agric. Iraq Bull. 27 : 1 - 111.
Guest, E.R. (1953a), The Rustam Herbarium Part VI. General and Ecological account. Kew Bull. 8 : 383 - 403.
Guest, E.R. (1953b), The Rustam Herbarium Part VII. Economics. Kew Bull. 8 : 535 - 542.
Guest, E.R. (ed.) (1966), Flora of Iraq volume one. Introduction to the Flora, an account of the geology, soils, climate and ecology of Iraq with gazetteer, glossary and bibliography. Min. Agric., Iraq, 313 pp.
Guest, E.R. & R.A. Blakelock (1954), The Rustam Herbarium Part VIII. Bibliography of Iraq. Kew Bull. 9 : 243 - 249.

Habib, I.M., T.A. Al-Ani, M.M. Al-Mufti, B.H. Al-Tawil & B.A. Takessian (1970), Plant indicators in Iraq I. Native vegetation as indicator for soil salinity and waterlogging. Tech. Rep. no. 14 Inst. for Appl. Research on Natural Resources, Abu Ghraib, Iraq, 10 pp. (mimeo).
Hadac, Emil (1966), Bibliographia Phytosociologica: Iraq Pars II. Excerpta Botanica 7: 102 - 104.
Hadac, E. & A.D.Q. Agnew (1962), Report on the 1960 University of Bahgdad Scientific Excursion to Helgurd Mountain, Erbil Liwa, Iraq. I. A floristic and plant ecological survey of the Helgurd Region. Bull. Iraq Nat. Hist. Inst. 2 : 5 - 10.
Hadac, E. & A.D.Q. Agnew (1963), Report on the 1960 University of Baghdad Scientific Excursion to Helgurd Mountain, Erbil Liwa, Iraq. II. Plant communities of the Helgurd Region. Bull. Iraq Nat. Hist. Inst. 2 (6) : 1 - 18.
Haines, R. Wheeler (1951), *Haloxylon schweinfurthii* Asch. (Chenopodiaceae) near Cairo. Bull. de l'Inst. Fouad 1er du Desert (1951) : 77 - 80.
Haines, R. Wheeler (1957), Flowers of the Falluja Desert. Prosp. Iraq Biol. 1 : 30 - 40.
Haines, R. Wheeler (1958), Autumn flowering on the gypsum deserts of Baghdad. Iraq Nat. Hist. Mus. Publ. 13 : 1 - 12.
Hammouda, M.A. & Z.Y. Bakr (1969), Some aspects of germination of desert seeds. Phyton 13 : 183 - 201.
Handel-Mazzetti, H.F. von (1912, 1913, 1914a), Wissenschaftliche Ergebnisse der Expedition nach Mesopotamien, 1910: Pteridophyta und Anthophyta aus Mesopotamien und Kurdistan sowie Syrien und Prinkipio. Ann. Naturh. Mus. Wien, 26 : 120 - 154 (1912), 27 : 41 - 92 and 391 - 459 (1913), 28 : 14 - 39 (1914).
Handel-Mazzetti, H.F. von (1914b), Die Vegetationsverhältnisse von Mesopotamien und Kurdistan. Ann. Naturh. Mus. Wien 28 : 48 - 111.
Harris, S.A. (1960), The distribution of certain plant species in similar desert and steppe soils in central and northern Iraq. J. Ecol. 48 : 97 - 105.
Harris, Grant A. (1977), Changing philosophies of rangeland management in the United States. J. Range Manage. 30 : 75 - 78.
Harrison, D.L. (1956), Gerbils from Iraq, with description of a new Gerbil. J. Mamm. 37 : 417 - 422.
Harrison, D.L. (1964, 1968, 1972), The mammals of Arabia. Vol. I (1964) Introduction, Insectivora, Chiroptera, Primates. Vol. II (1968) Carnivora, Hyracoida, Artiodactyla. Vol. III (1972) Lagomorpha Rodentia. Ernest Benn Ltd. London, 670 pp.
Hasan, M.S. (1958), Growth and structure of Iraq's population 1867 - 1947. Bull. Oxford University 20 : 339 - 350.
Haseeb, K. (164), The national income of Iraq 1953 - 1961. Oxford University Press, London, 1964 New York and Toronto 184 pp.
Hassan, A. Hassan & Mustafa Mashkour (1972), Mean annual precipitation in Iraq. J. Geol. Soc. Iraq V : 119 - 127.
Hassawy, G.S., S.A. Tammimi & H. Al-Izzi (1968), Weeds in Iraq. Techn. Bull. no. 167 Bot. Div., Min. Agric. Iraq, 256 pp.
Hatt, Robert T. (1959), The mammals of Iraq. Misc. Publ. Mus. Zool. Univ. Michigan no. 106 113 pp.
Heady, Harold F. (1972), Ecological consequences of Bedouin settlement in Saudi Arabia. pp. 683 - 693 in: The Careless Technology, Ecology and International Development. M. Taghi

Farvar & John P. Milton (eds.), The Center for the Biology of Natural Systems, Washington University.
Heady, Harold F. (1975), Rangeland Management. McGraw-Hill Book Company, 460 pp.
Hills, E.S. (ed.) (1966), Arid Lands, a geographical appraisal. Methuen & Co. Ltd. London, 461 pp.
Hitti, Philip K. (1968) The Arabs, a short history. MacMillan, London, Melbourne, Toronto, 212 pp.
Holt, A.L. (1923), The future of the North Arabian Desert. Geogr. J. (1923) : 259 - 271.
Hooker, Joseph Dalton (1881), Hooker's Icones Plantarum or Figures, with descriptive characters and remarks, of new and rare plants, selected from the Kew Herbarium. Vol. XIV part III (1351 - 1375), October 1881. Williams and Norgate, London and Edinburgh.
Hooker, J.D. (1885), The Flora of British India, Vol. V Chenopodiaceae to Orchideae, L. Reeve and Co. Ltd., Brook, Ashford, England.
Horwitz, W. (1960), Offical methods of analysis of the Association of Agricultural Chemists. 9th ed., Ass. Off. Agric. Chem. Washington D.C., USA.
Houérou, H.N. le (1959), Recherches écologiques et Flonstiques sur la végétation de la Tunisie méridionale. Inst. Rech. Sahar. Univ. Alger, Mém. hs., 510 pp.
Houérou, H.N. le (1959), Recherches écologiques et floristiques sur la végétation de la Tunisie la production fourragère et pastorale. Rapport PNUD / AT 2386, FAO, Rome, 14 pp.
Houérou, H.N. le (1969), North Africa: Past, present and future. pp. 227 - 278 in: Arid lands in transition, A.A.A.S., Washington D.C., USA.
Houérou, H.N. le (1972a), Africa - the Mediterranean region, pp. 26 - 36 in: Wildland Shrubs - Their Biology and Utilization, An Int. Symp., Logan, Utah, USDA Forest Service General Tech. Rep. INT - I, 1972.
Houérou, H.N. le (1972b), An assessment of the primary and secondary production of the arid grazing lands ecosystems of North Africa. pp. 168 - 172 in: Ecophysiological foundation of ecosystems productivity in arid zone. L.E. Rodin (ed.), Nauka, Leningrad.
Houérou, H.N. le (1975), The rangelands of North Africa: Typology, yield, productivity and development. Proc. Seminar on Evaluation and Mapping of tropical African Rangelands, pp. 41 - 55. I.L.C.A., Addis Abeba.
Houérou, H.N. le (in press), Africa north of the Sahara. in: Rangelands in Arid Zones, R.N. Kaul & D.C.P. Thalen (eds.) Dr. W. Junk b.v. Publishers, The Haque, The Netherlands.
Houérou, H.N. le & C.H. Hoste (1977), Rangeland Production and Annual Rainfall Relations in the Mediterranean Basin and in the African Sahelo - Sudanian Zone. J. Range Manage. 30 : 181 - 189.
Huzayyin, S. (1956), Expansion of the Arabs: its relation to climatic changes and other factors. Bull. de l'Inst. d'Egypte XXXVIII : 37 - 52.
Hyder, D.N. (ed.) (1973), Arid Shrublands - proceedings of the third workshop of the United States / Australian rangelands panel, Tucson, Arizona, March 26 - April 5, 1973, 148 pp. Society for Range Management, Denver, Colorado.

Ibrahim, K.M. (1967), The pasture, range and fodder crop situation in the Near East. Summary report and bibliography. Report no. PL: PFC/1, FAO, Rome, 159 pp.
INGRA (1960), Explanatory Report, One hundred well programme in Southern and Northern Desert. INGRA-Consulting Dept. Report, Zaghreb, Yugoslavia, 20 pp.
INGRA (1964), Statistical Report Northern and Southern Deserts, One hundred ten wells programme. INGRA-Consulting Dept. Report, Zaghreb, Yugoslavia, 54 pp.
INGRA (1967), Mosul Liwa 100 + 14 wells programme. Final Report, vol. I and II. INGRA-Consulting Dept. Report, Zaghreb, Yugoslavia.
International Bank (1952), The economic development of Iraq. Publ. for International Bank for Reconstruction and Development by the John Hopkins Press, London, 463 pp.
ITC (1978), Projet de développement rural integré de la region du Kaarta, Republique du Mali, Rapport Final, Phase de reconnaissance du volet VIII: cartographie. ITC, Enschede, Pays-Bas (The Netherlands), 329 pp.

Jaccard, P. (1902), Gesetze der Pflanzenverteilung der Alpinen Regio. Flora 90 : 349 - 377.
Jackson, M.L. (1958), Soil chemical analysis. Prentice Hall Inc., Englewood Cliffs, USA.

Jacob, Georg (1897), Altarabisches Bedouinenleben. Georg Olms Verlagbuchhandlung, Hildesheim, W-Germany, (Reprografischer Nachdruck) 278 pp.

Jamali, M.F.J. (1934), The new Iraq: Its problem of Bedouin education. Bureau of Publications, Teachers College, Columbia University, New York City, 160 pp.

Johnson, Douglas L. (1969), The nature of nomadism. Univ. of Chicago, Dept. of Geogr. Research paper no. 118, 185 pp.

Johnson, C.M. & A. Ulrich (1959), Analytical methods for use in plant analysis. Calif. Agric. Exp. Sta. Bull. 766 : 25 - 78.

Jones, L. (1969), Rapid population growth in Baghdad and Amman. Middle East Journal 23 : 209 - 215.

Jones, Quentin & A.S. Barclay (1972), Industrial raw materials from shrubs, pp. 101 - 108 in: Wildland Shrubs - Their Biology and Utilization, An Int. Symp., Logan, Utah, USDA Forest Service General Tech. Rep. INT - I, 1972.

Jones, Richard G., Ismail M. Al-Sabi, Klaas Koster, Rajaa Y. Toma & Khalid S. Al-Oubaidi (1967), Survey of sheep and goat slaughter receipts in Baghdad. Tech. Rep. no. 1 Animal Husbandry Research and Training Project UNDP/FAO - D.G. of Animal Resources and Veterinary Services, Abu Ghraib, 7 pp.

Kappen, L., O.L. Lange, E.D. Schulze, M. Evenari & U. Buschbom (1972), Extreme water stress and photosynthetic activity of the desert plant *Artemisia herba-alba* Asso. Oecologia 10 : 177 - 182.

Kassas, M. & W.A. Girgis (1970), Habitat and plant communities in the Egyptian Desert. VII. Geographical facies of plant communities. J. Ecol. 58 : 335 - 350.

Kassir, S.M., N. Bashir, J. Alikhan, F. Al-Jaff, F. Al-Dessouby & D.G. McFetridge (1967), Survey of cattle, buffalo and camel slaughter receipts in Baghdad, Iraq, 1966. Tech. Rep. no. 8 Animal Husbandry Research and Training Project. UNDP/FAO - D.G. of Animal Resources and Veterinary Services, Abu Ghraib, 33 pp.

Kaul, R.N. (1969a), Iraq - Range Ecology (Institute of Applied Research on Natural Resources). UNESCO/FR/SF/Report Serial no. 1127/BMS.RD/AVS. Paris, 12 pp. + App.

Kaul, R.N. (1969b), Institute for Applied Research on Natural Resources, Abu Ghraib, Iraq. Progress Report in Plant Ecology. Report to UNESCO (Paris), 7 pp. + Annex I - III and App. I - IV. (typescript).

Kaul, R.N. (ed.) (1970), Afforestation in Arid Zones. Dr. W. Junk N.V. Publ. The Hague, The Netherlands, 435 pp.

Kaul, R.N. (1973), An integrated natural resources survey in northern Iraq. Nature and Resources IX. 13 - 18.

Kaul, R.N. & M.M. Al-Mufti (1974), Range Resources of Iraq XI. A preliminary ecological appraisal of *Artemisia herba-alba* vegetation. Tech. Rep. no. 55 Inst. for Appl. Research on Natural Resources, Abu Ghraib, Iraq, 29 pp. (mimeo).

Kaul, R.N. & D.C.P. Thalen (1971a), Range Resources of Iraq. A Problem Analysis. Tech. Rep. no. 20 Inst. for Appl. Research on Natural Resources, Abu Ghraib, Iraq, 28 pp. (mimeo).

Kaul, R.N. & D.C.P. Thalen (1971b), Range ecology at the Institute for Applied Research on Natural Resources, Abu Ghraib, Iraq. Nature and Resources VII : 10 - 15.

Kellersohn, Heinrich (1952), Die Landwirtschaft im Irak. Erdkunde VII : 276 - 288.

Kernick, M.D. (1966), Plant Resources, range ecology and fodder plant introduction. Report to the Government of Kuwait. Rep. no. TA 2181, FAO, Rome, 95 pp.

Kershaw, Kenneth A. (1964), Quantitative and dynamic ecology. Edward Arnold Publ. Ltd. London, 183 pp.

Kevork, Ohannes K. & Hilmi Sabir Al-Uthman (1972), Ecological observations on the Egyptian Spiny-tailed Lizard *Uromastix aegyptius* : 1. On food and feeding habits. Bull. Iraq Nat. Hist. Mus. V. : 26 - 44.

Kharin, N.G., A. Babayev & I.A. Grigorshenko (1971), Interpretation methods for the determination of the volume of desert shrubs. Problems of Desert Development 1971 (1): 44 - 52 (Russian).

Khawaja, A.K., S.A. Matti, R.F. Asadi, K.M. Mokhtar & S.H. Aboona (undated), The compo-

sition and nutritive value of Iraqi feedstuffs. Report D.G. of Animal Resources. Min. Agric. Iraq, 34 pp.
Kinsinger, Floyd (1977), Sampling precision and probability. Resource Inventory Notes, May 1977, pp. 1 - 5. USDI, Bureau of Land Management, Denver Service Centre, Colorado.
Kinsinger, Floyd E. & Gerald S. Strickler (1961), Correlation of production with growth and ground cover of Whitesage. J. Range Manage. 14 : 274 - 278.
Kittani, H. (1956), Grazing resources on forest lands (progress since the Baghdad Meeting 1954). Paper 1st meeting FAO-working party on development of grazing and fodder resources of the Near East, Cairo, 2 pp. (mimeo).
Knetsch, G. & P. Gounot (1963), Arid zone research in Iraq. Arid zone (Newsletter) 22 : 9 - 18, UNESCO, Paris.
Köppen, W. (1931), Grundriss der Klimakunde (2nd ed.) Walter de Gruyter Co. Berlin, 388 pp.
Kozlowski, T.T. (ed.) (1968), Water Deficits and Plant Growth. Vol. I. Development, Control and Measurement. Academic Press, New York.
Krader, Lawrence (1959), The ecology of nomadic pastoralism. Int. Soc. Sci. J. 11 : 499 - 510.
Kreeb, Karlheinz (1961), Bibliographia phytosociologica: Iraq. Excerpta Botanica 3 : 78.
Krochmal, Arnold (1972), Medicinal values. pp. 98 - 100 in: Wildland Shrubs - Their Biology and Utilization. An Int. Symp., Logan, Utah, USDA Forest Service General Tech. Rep. INT - I, 1972.

Labib, A.I. & Taha, S. Matti, R. Asadi, K.M. Salek & S.H. Yossef (1968), Iraqian Feedstuffs Composition Tables and nutrition figures for animals and poultry. Rep. no. 184 Animal Husb. Div. Min. Agric. J. Iraq, 11 pp.
Lawless, R.I. (1972), Iraq: changing population patterns Ch. 4. in: Populations of the Middle East and North Africa, a geographical approach, J.I. Clarke and W.B. Fisher (eds.) University of London Press Ltd.
Lawrence, T.E. (1926), Seven pillars of wisdom. Reprinted edition (1967) published by Dell Publishing Co., Inc. New York.
Lebon, J.H.G. (1953), Population Distribution and the Agricultural Regions of Iraq. Geogr. Rev. XLIII : 223 - 228.
Lebon, J.H.G. (1964), Recent research on the landpotential of Iraq. Geogr. Rev. LIV : 104 - 109.
Lees, G.M. & N.L. Falcon (1952), The geographical history of the Mesopotamian Plains. The Geogr. J. CXVIII (I) : 24 - 39.
Lewis, James K. (1969), Range management viewed in the ecosystem framework. pp. 97 - 187 in: The ecosystem concept in natural resource management. George M. Van Dyne (ed.). Academic Press, New York and London.
Lieth, Helmut & Robert H. Whittaker (eds.) (1975), Primary productivity of the biosphere. Ecological Studies 14, Springer-Verlag. Berlin, Heidelberg, New York, 339 pp.
Long, G.A. (1955), First preliminary report to the Government of Iraq on ecology and grazing problems. FAO - Report to Min. Agric. Iraq, 5 pp. (mimeo).
Long, G.A. (1956), Second preliminary report to the Government of Iraq. Ecology and grazing problems. FAO - Report to Min. Agric. Iraq, 9 pp. (mimeo).
Long, G.A. (1957), Rapport aux Gouvernements de l'Egypte, de l'Irak, de l'Iran, de la Jordanie et du Liban sur 'Les études phyto - écologiques de base et leurs relations avec les problèmes pastoraux'. FAO - Report no. 627, 36 pp.
Longrigg, S.H. (1925), Four centuries of modern Iraq. Clarendon Press, Oxford, 378 pp.
Longrigg, S.H. (1953), Iraq 1900 to 1955, a political, social and economic history. Oxford University Press. London, New York, Toronto, 436 pp.

Mac Fadyen, W.A. (1938), Water Supplies in Iraq. Water supply report no. 7. Southern Desert. Iraq. Geol. Dep. Publ. no. 1, Min. Econ. and communities, Gvt. of Iraq, 156 pp.
McGinnies, William G. (1972), North America. pp. 55 - 66 in: Wildland Shrubs - Their Biology and Utilization. An int. Symp., Logan, Utah, USDA Forest Service General Tech. Rep. INT - I, 1972.

McGinnies, William G., Bram J. Goldman & Patricia Paylore (eds.) (1967), Deserts of the World, an inventory of geographical research on desert environments. 788 pp. The University of Arizona Press, Tucson, Arizona.

McGinnies, William G., Bram J. Goldman & Patricia Paylore (eds.) (1969), Arid lands in Perspective. The University of Arizona Press. Tucson, Arizona.

McGinnies, William G., Bram J. Goldman & Patricia Paylore (eds.) (1971), Food, Fiber and the Arid Lands. The University of Arizona Press. Tucson, Arizona, 437 pp.

McKell, Cyrus M., James P. Blaisdell & Joe R. Goodin (eds.) (1972), Wildland Shrubs - Their Biology and Utilization. An Int. Symp., Logan, Utah, USDA Forest Service General Tech. Rep. INT - I, 1972, 494 pp.

Mahdi, Nuri & P.V. Georg (1969), A Systematic List of the Vertebrates of Iraq. Iraq Nat. Hist. Mus. Publ. no. 26, 104 pp.

Mahmood, Ghazi (1967), Determination of the carrying capacity in the Western Desert (Rutba) 1967. Report to Dir. Gen. Research and Projects. Range Management Division Min. Agric. Iraq. 47 pp. (typescript).

Martonne, E. de (1926), Aréisme et indices d'aridité. Académie des Sciences, Paris, Comptes Rendus 182 : 1395 - 1398.

Martonne, E. de & L. Larfrère (1927), Map of interior basin drainage. Geogr. Rev. XVII (3), pl. 2.

Mason, Lamar R. & Selar S. Hutchings (1967), Estimating foliage yields on Utah juniper from measurements of crown diameter. J. Range Manage. 20 : 161 - 166.

Meigs, P. (1953), World distribution of arid and semi-arid homoclimates. pp. 203 - 210 in: Hydrology, Reviews of Research on Arid Zone. Arid Zone Research no. I UNESCO, Paris.

Meigs, P. (1964), Classification and occurrence of Medeterranean type dry climates. pp. 17 - 21 in: Land use in semi-arid Mediterranean climates. Arid Zone Research XXVI. UNESCO, Paris.

Meteorological Department Iraq (1962), Climatological Atlas for Iraq. Publ. no. 13. Iraq Met. Service and Dept. Baghdad.

Milner, C. & R. Elfyn Hughes (1968), Methods for the Measurement of the Primary Production of Grassland. IBP Handbook no. 6. Blackwell Sc. Publ. Oxford and Edinburgh, 70 pp.

Mitchell, Raoul C. (1956), Aspects géologiques du désert occidental de l'Irak. Bull. Soc. Géol. Fr. (6) VI: 391-406.

Moen, Aaron N. (1973), Wildlife Ecology, W.H. Freeman and Comp., San Francisco, 458 pp.

Mobayen, S. & V. Tregubov (1970), Carte de la végétation naturelle de l'Iran, Proj. UNDP/ FAO IRA 7 Bull. 14, 18 pp.

Monakhov, M.A. (1972), On the associations of plant communities with definite ground conditions in the Western Desert of Iraq. Bull. Moscow Ass. for Use of Nature, section Biology, Vol. 2 : 90 - 95 (Russian).

Mooney, H.F. (1956), Southwestern Asia : Iraq. pp. 428 - 433 in: A world geography of forest resources. S. Haden-Guest, J.K. Wright & Eileen M. Teclaff (eds.). American Geogr. Soc. publ., New York.

Moore, R.M. & R.A. Perry (1970), Vegetation in Australian grasslands, R.M. Moore (ed.) Australian National University Press, Canberra.

Moore, R.M., R.W. Condon & J.H. Leigh (1970), Semiarid woodlands in: Australian Grasslands. R.M. Moore (ed.) Australian National University Press, Canberra.

Morris, Meredith J., Kendall L. Johnson & Donald L. Neal (1976), Sampling shrub ranges with an electronic capacitance instrument. J. Range Manage. 29 : 78 - 81.

Mountfort, Guy (1965), Portait of a Desert. The Story of an Expedition to Jordan. Collins, London, 192 pp.

Mueggler, Walter F. (1965), Cattle distribution on steep slopes. J. Range Manage. 18 : 255 - 257.

Mueller-Dombois, Dieter & Heinz Ellenberg (1974), Aims and Methods of Vegetation Ecology. John Wiley & Sons, New York and London, 547 pp.

Musil, Alois (1928), The manners and customs of the Rwala bedouins. America Geogr. Soc. New York.

Nabelek, Fr. (1923), Le désert et le steppe en Mésopotamie. Acta Bot. Bohemica 2 : 1-2.

Nasserula, Sabah M. (1971), Ecological studies on the vegetation around Ukhaidir, Iraq. M.Sc. thesis Univ. Baghdad, in: Bull. Coll. Sci. Baghdad (1967) pp. 58 - 60.
Naval Intelligence Division (1944), Iraq and the Persian Gulf. Geographical Handbook Series B.R. 524, 682 pp.
Neale, P.E. (1948), Preliminary report of livestock work in Egypt, Iraq, Syria and Lebanon. FAO regional conf. Near East, Cairo, 15 pp. (mimeo).
Noble, A.H. (1926), The subsurface water resources of Iraq. Publ. of the Dept. of Agric. Iraq, 30 pp.
Noy-Meir, Imanuel (1973), Desert Ecosystems: Environment and Producers. Ann. Rev. Ecol. and Syst. 4 : 25 - 51.
Novikoff, G., F.H. Wagner & M. Skouri (1975), Tunisian presaharan project. System analysis of the presaharan ecosystem of southern Tunisia. US/IBP Desert Biome. Progress Report, 179 pp. (mimeo).

Oedekoven, K.H. (1970), Iraq. pp. 98 - 109 in: Afforestation in arid zones. R.N. Kaul (ed.) Dr. W. Junk N.V. Publishers, The Hague, The Netherlands.
'Oraim, Abdul Jabbar (1965), Sedentarization of Bedouins in Iraq. Meeting of experts of the Arab League on the Settlement of Nomads (Arabic, quoted in Yacoub 1971).
Orshan, G. (1972), Morphological and physiological plasticity in relation to drought. pp. 245 - 254 in: Wildland Shrubs - Their Biology and Utilization. An Int. Symp. Logan, Utah, USDA Forest Service General Tech. Rep. INT - I, 1972.
Orshan, G. & S. Dishin (1968), Seasonal changes in productivity under desert conditions. pp. 191 - 201 in : Functioning of terrestrial ecosystems at the primary production level. Proc. Copenhagen Symposium. UNESCO Natural Resources Research Series no. 5. UNESCO, Paris.
Orshan, G. & G. Zand (1962), Seasonal body reduction of certain desert halfshrubs. Bull. Res. Counc. Israel 11 D. : 35 - 42.

Pabot, H. (1962), Méthodes d'étude de la couverture végétale applicables dans les pays du Proche et du Moyen-Orient, 21 pp. (typescript).
Parsons, The Ralph M. Company (1955), Groundwater Resources of Iraq. (9 vols.)
1955a. Jezira Desert (Vol. 5)
1955b. Southern Desert (Vol. 7)
1955c. Northern Desert (Vol. 8)
1955d. Hydrologic aspects of selected areas (Vol. 9).
Pearse, C. Kenneth (1970), Range deterioration in the Middle East. Proc. XI Int. Grassl. Congr. Surfers Paradise, pp. 26 - 30.
Pechanec, J.F. (1967), The range society is at the crossroads. J. Range Manage. 20 : 125 - 128.
Peppelenbosch, P.G.N. (1968), Nomadism on the Arabian Peninsula, a general appraisal. Tijdschr. Econ. Soc. Geogr. LIX : 335 - 346.
Perry, R.A. (1962), Measurement of vegetation. Tech. Mem. 62/1 C.S.I.R.O. Div. of Land Research and Regional Survey, Australia, 59 pp.
Petrov, M. (1970), Central Asia. pp. 234 - 267 in: Afforestation in Arid Zones. R.N. Kaul (ed.) Dr. W. Junk N.V. Publishers, The Hague, The Netherlands.
Petrov, M.P. (1972), Asia. in: Wildland Shrubs - Their Biology and Utilization, An Int. Symp. Logan, Utah, USDA Forest Service General Tech. Rep. INT - I, 1972.
Pitman, C.R.S. (1922), Notes on Mesopotamian mammals. J. Bombay Nat. Hist. Soc. 28 : 474 - 480.
Poore, M.E.D. & V.C. Robertson (1963), An approach to the rapid description and mapping of biological habitats. Publ. subcommission on conservation of terrestrial biological communities of the I.B.P., 68 pp.
Post, G.E. (1932), Flora of Syria, Palestine and Sinai. 2nd ed. J.E. Dinsmore (ed.) Vol. I and II.
Post, W.S., R.T. Malcolm & K.H. Tammo (1968), Some economic aspects of mechanised fodder production on irrigated lands. Tech. Rep. no. 20 Animal Husbandry Research and

Training Project. UNDP/FAO - D.G. of Animal Resources and Veterinary Services, Abu Ghraib, Iraq, 32 pp.
Preston, R.L. (1972), Protein requirements for growing and lactating ruminants. Paper at mini symp. on 'Protein requirements for growing and lactating ruminants', Agric. University Wageningen, The Netherlands, Jan. 1972, 26 pp. (mimeo).

Quézel, Pierre (1965), La végétation du Sahara, du Tchad à la Mauritanie. Gustav Fischer Verlag. Stuttgart, 333 pp.
Quézel, P. & S. Santa (1962), Nouvelle flore de l'Algérie et des régions désertiques méridionales. C.N.R.S. Paris I, 565 pp.

Raswan, Carl R. (1930), Tribal areas and migration lines of the North Arabian Bedouins. Geogr. Rev. XX: 494 - 502.
Raswan, Carl R. (1934), Im Land der schwarzen Zelte. Mein Leben unter den Bedouinen. Ullstein Verlag, Berlin.
Rechinger, K.H. (1963 - 1970), Flora Iranica (in parts by families) no. 1 - 72. Akad. Druck und Verlagsanstalt, Graz, Austria.
Rechinger, K.H. (1964), Flora of Lowland Iraq. Cramer Verlag, Weinheim, 746 pp.
Regel, C. von (1953), La végétation en Iraq (Mesopotamie) et ses rélations avec la végétation de la Région Mediterranée. Euclides 13 : 431 - 438.
Regel, C. von (1954), Phytogeographie de l'Iraq. pp. 51 - 52 in: Proc. Eight Int. Bot. Congress, Paris, Sections 7 - 8.
Reitan, Clayton H. & Christine R. Green (1967), Weather and climate. part II of 'Deserts of the World, an inventory of geographical research on desert environments', William G. Mc Ginnies, Bram J. Goldman & Patricia Paylore (eds.) The University of Arizona Press Tucson, Arizona, 72 pp.
Richards, L.A. (ed.) (1954), Diagnosis and Improvement of Saline and Alkali Soils. Agric. Handbook no. 60 USDA, Washington D.C., 160 pp.
Risser, P.G. & E.L. Rice (1971), Phytosociological analysis of Oklahoma upland forest species. Ecology 52 : 940 - 945.
Robinette, W. Leslie (1972), Browse and cover for wildlife. pp. 69 - 76 in: Wildland Shrubs - Their Biology and Utilization, An Int. Symp., Logan, Utah, USDA Forest Service General Tech. Rep. INT - I, 1972.
Rodin, L.E. (ed.) (1972), Eco-physiological foundation of ecosystems productivity in arid zone. Int. Symp. USSR, June 7 - 19, 1972, Nauka, Leningrad, 232 pp.
Rodin, L.E., N.I. Bazilevich & Yu. M. Miroshnichenko (1972), Productivity and biogeochemistry of *Artemisietea* in the Mediterranean area. pp. 193 - 198 in: Ecophysiological foundation of ecosystems productivity in arid zone. L.E. Robin (ed.) Nauka, Leningrad.
Roux, Georges (1964), Ancient Iraq. George Allen & Unwin Ltd. Published in Pelican Books A 828 (1966), 480 pp.

Sankary, Mohamed Nazir, & Michael G. Barbour (1972a), Autecology of *Atriplex polycarpa* from California. Ecology 53 : 1155 - 1162.
Sankary, Mohamed Nazir, & Michael G. Barbour (1972b), Autecology of *Salsola vermiculata* var. *villosa* from Syria. Flora 161 - 421 - 439.
Sankary, Mohamed Nazir, & Michael G. Barbour (1972c), Autecology of *Haloxylon articulatum* in Syria. J. Ecol. 60 : 697 - 711.
Sarsam, N.H., K.T. Ali & G.B. McLeroy (1956), Preliminary Report on the use of date byproducts in the ration of fat-tailed sheep. Report Dept. Animal Husbandry, Abu Ghraib Agric. Exp. Sta. 7 pp. (mimeo).
Schimper, A.F.W. (1898), Pflanzengeographie auf physiologischer Grundlage. Fischer Verlag, Jena.
Schmidt - Nielsen, K. (1956), Animals and arid conditions : Physiological aspects of productivity and management. pp. 368 - 382 in: The future of arid lands. G.F. White (ed.). Publ. 43 A.A.A.S., Washington D.C..

Schumacher, F.X. & R.A. Chapman (1954), Sampling methods in forestry and range management. Duke University School of Forestry, Bull. 7 revised. Durham, North Carolina, 222 pp.

Schwan, H.E. (1955), A management program for the livestock ranges of Iraq. U.S.O.M./IRAQ Range Management Project Report, 14 pp. (mimeo).

Schwan, H.E. (1956), A report on the livestock range and related land use problems in northern Iraq. U.S.O.M./Iraq Range Management Project Report, 6 pp. (mimeo).

Schwan, H.E. (1957), Some outstanding problems of Iraq's ranges. Paper presented to U.S.O.M. meeting 16 Febr. 1957, 4 pp. (mimeo).

Serkahia, S.Y. (1956), Botanical and ecological analysis of the Western Desert. Paper presented to 1st meeting FAO working party on the development of the grazing and fodder resources in the Near East. 14 - 21 March 1956, 7 pp. (mimeo).

Serkahia, S.Y. (1957), Work of range management in Iraq. Range Management Section, Animal Husbandry Division, Min. Agric. Iraq, 3 pp. (mimeo).

Serkahia, S.Y. (1972), Textbook on range management in Iraq (in Arabic), Min. Agric. Iraq.

Shantz, H.L. (1956), History and problems of arid lands development. pp. 3 - 25 in: The future of arid lands. G.F. White (ed.), publ. no. 43, A.A.A.S., Washington D.C.

Sharif, C.M. & Neil E. West (1968), Forage moisture variations on mountain summer range. J. Range Manage. 21 : 228 - 235.

Shimwell, David W. (1971), The description and classification of vegetation. Sidgwick & Jackson, London, 322 pp.

Short, Henry L., Robert M. Blair & Leonard Burkart (1972), Factors affecting nutritive values. pp. 311 - 318 in: Wildland Shurbs - Their Biology and Untilization. An Int. Symp., Logan, Utah, USDA Forest Service General Report INT - I, 1972.

Shreve, F. (1951), Vegetation of the Sonoran Desert. in: Vegetation and flora of the Sonoran Deserts. F. Schreve & I.L. Wiggins (eds.) vol. I. Carnegie Institution of Washington, publ. no. 591, 2 Vols.

Slatyer, R.O. (1968), Measurement of the primary production of arid zone plant communities. (Part III, pp. 53 - 61) in: Methods for the measurement of the primary production of grassland, I.B.P. Handbook no. 6, Blackwell Sci. Publ. Oxford and Edinburgh.

Sluijs, A.A. van der (1969), Groundwater, a natural resource of Iraq. Final Report to UNESCO, 56 pp. (typescript).

Sobolev, L.N. & V.D. Utekhin (1973), Russian (Ramensky) approaches to community systematization. pp. 75 - 103 in: Ordination and classification of communities, Handbook of vegetation science Vol. 5, R.H. Whittaker (ed.), Dr. W. Junk b.v. Publ. The Hague, The Netherlands.

Soest, P.J. van (1975), 'Vezels', Text of three papers read at Agric. University Wageningen, The Netherlands, Febr. 1975 (in Dutch).

Sokal, R.R. & C.D. Michener (1958), A statistical method for evaluating systematic relationships. Kans. Univ. Sci. Bull. 38 : 1409 - 1437.

Solecki, Ralph S. (1954), Tar Kshaife, a possible prehistoric station near Ukhaider, Iraq. Sumer X : 62 - 68.

Solecki, Ralph S. (1955), Shanidar cave, a palaeolithic site in Northern Iraq, its relationship to the stone age sequence of Iraq. Sumer XI : 14 - 38.

Sørensen, T.A. (1948), A method of establishing groups of equal amplitude in plant sociology based on similarity of species content. K. Danske Vidensk Selsk. Biol. Skr. 5 : 1 - 34.

Soriano, Alberto (1972), South America. pp. 51 - 54 in: Wildlands Shrubs - Their Biology and Utilization, An Int. Symp., Logan, Utah, USDA Forest Service General Tech. Rep. INT - 1, 1972.

Sprague, Howard B. (1971), Improvement of livestock production by Bedouin nomads on semidesert rangelands of Saudi Arabia. Report of the Stanford Research Institute (S.R.I. Project ECH - 8680) prepared for the Ministry of Agriculture and Water, Kingdom of Saudi Arabia, 48 pp. + app.

Springfield, H.W. (1954), Natural Vegetation in Iraq. Tech. Bull. Min. Agric. Iraq, 23 pp.

Springfield, H.W. (1957), Forage problems and resources of Iraq. I.C.A. Report in coop. with USDA, Washington D.C., 37 pp.

Srivastava, A. (1969), A guide to literature on Iraq's natural resources, containing studies and reports from 1833 to 1968. UNESCO/SF/Iraq 12, Institute for Applied Research on Natural

Resources, 133 pp. (mimeo).
Stoddart, Laurence A., Arthur D. Smith & Thadis W. Box (1975), Range Management (3rd ed.) McGraw Hill, New York, 532 pp.
Sweet, Louise E. (1965), Camel raiding of North Arabian Bedouin: a mechanism of ecological adaptation. American Anthropologist 67 : 1132 - 1150.
Sykes, J.G., H. Al-Nouri, J. Suliman, A. Al-Hadad & K. Al-Morany (1969), Consumer survey of meat, dairy and poultry products in Iraq. Tech. Rep. no. 31 Animal Husbandry Research and Training Project. UNDP/FAO - D.G. of Animal Resources and Veterinary Services, Abu Ghraib, Iraq 8 pp.

Täckholm, Vivi (1956), Students' Flora of Egypt (in coll. with M. Drar & A.A. Abdel Fadeel) Cairo, 649 pp.
Täckholm, Vivi (1974), Students' Flora of Egypt (2nd ed.) Cairo University Publ., 888 pp.
Tadmor, N.H., G. Orshan & E. Rawitz (1962), Habitat analysis in the Negev desert of Israël, Bull. Res. Counc. Israël 11 D 3 : 148 - 173.
Thalen, D.C.P. (1972), A preliminary evaluation of biomass and production in the desert rangelands of Iraq. pp. 184 - 198 in: Proc. First Sci. Conf., Sci Res. Found. (25 - 30 March 1972), Min. of Higher Educ. and Sci. Research, Iraq.
Thalen, D.C.P. (1973), Final Report to UNESCO over the period March 1970 - June 1973 of the associate expert in plant ecology at the UNESCO/UNDP/SF Project Institute for Applied Research on Natural Resources, Abu Ghraib, Iraq. UNESCO, Paris 20 pp. (typescript).
Thalen, D.C.P. (1974a), Range Resources of Iraq X. An appraisal of *Rhanterium epapposum* (Oliv.) vegetation. Tech. Rep. no. 54 Inst. for Appl. Research on Natural Resources, Abu Ghraib, Iraq, 46 pp. (mimeo).
Thalen, D.C.P. (1974b), Range Resources of Iraq. Preliminary account of *Haloxylon salicornicum* (Moq.) Bge. as a forage producer. Tech. Rep. no. 60 Inst. for Appl. Research on Natural Resources, Abu Ghraib, Iraq, 38 pp. (mimeo).
Thalen, D.C.P. (1975), The Caracal Lynx (*Caracal Caracal Schmitzi*) in Iraq. Earlier and new records, habitat and distribution. Bull. Nat. Hist. Res. Centre. Univ. Baghdad IV : 1 - 23.
Thalen, D.C.P. (1978), Complex mapping units, geosyntaxa and the evaluation of grazing areas. pp. 491 - 514 in: Assoziationskomplexe (Sigmeten) und ihre praktische Anwendung (Int. Symp. Rinteln April 1977), Reinhold Tüxen (ed.), J. Cramer Verlag, Vaduz.
Thalen, D.C.P. (i.p.) On photographic techniques in permanent plot studies. Vegetatio (in press).
Thalen, D.C.P. & M.M. Al-Mufti (1973), Range Resources of Iraq VII. A record of existent data on small exclosures with particular reference to changes in the vegetation near Shabicha, Southern Desert. Tech. Rep. no. 44 Inst. Appl. Research on Natural Resources, Abu Ghraib, Iraq, 21 pp. (mimeo).
Thalen, D.C.P. & I.S. Zonneveld (1974), The vegetation component in landscape survey for land evaluation in arid areas. in: Landschaftsgliederung met Hilfe der Vegetation (Int. Symp. Rinteln April 1974), in press.
Thornthwaite, C.W. (1931), The climates of North America according to a new classification. Geogr. Rev. 21 : 633 - 655.
Thornthwaite, C.W. (1948), An approach toward a rational classification of climate. Geogr. Rev. 38 : 55 - 94.
Tigner, James R. & Gary E. Larson (1977), Sheep losses on selected ranches in Southern Wyoming, J. Range Manage. 30 : 244 - 252.
Townsend, C.C. & Evan Guest (eds.) (1966), Flora of Iraq Vol. 2, including Pteridophyta, Gymnospermae and Angiospermae (Rosaceae). Min. Agric., Iraq, 184 pp.
Townsend, C.C., Evan Guest & Ali Al-Rawi (1968), Flora of Iraq Vol. 9 Gramineae by N.L. Bor. Min. Agric. Iraq, 588 pp.
Tueller, P.T., G. Lorain, K. Kipling & C. Wilkie (172), Methods for measuring vegetation changes on Nevada rangelands. Report no. T. 16, Agric. Exp. Sta. Univ. of Nevada, Reno, Nevada, 55 pp.

UNESCO, (1955), Plant ecology. Reviews of Research. Arid Zone Research VI. UNESCO, Paris.

UNESCO (1960a), Plant- water relationships in arid and semi-arid conditions. Reviews of Research. Arid Zone Research XV. UNESCO, Paris.

UNESCO (1960b), Medicinal plants of the arid zones. Arid Zone Research XIII., UNESCO, Paris, 96 pp.

UNESCO (1969), Vegetation map of the Mediterranean region. Arid Zone Research XXX. UNESCO, Paris, 90 pp.

UNESCO (1973), International classification and mapping of vegetation. UNESCO, Paris, 96 pp.

UNESCO (1975), Iraq, contributions on natural resources research. Report no. FMR/SC/OPS/76/218 (UNDP), UNESCO, Paris, 286 pp.

United Nations (1970), Nomadic populations in selected countries in the Middle East and related issues of sedentarization and settlement. pp. 105 - 117 in: Studies on selected development problems in various countries in the Middle East, 1970. U.N. Econ. and Soc. Office, Beirut.

Uresk, D.W., R.O. Gilbert & W.H. Richard (1977), Sampling big sagebrush for phytomass. J. Range Manage. 30 : 311 - 314.

USDA (1975), Soil Taxonomy, a basic system of soil classification for making and interpreting soil surveys. USDA Soil Conservation Service Agric. Handbook no. 436, 754 pp.

USDA (1976), National Range Handbook. Soil Conservation Service, U.S. Dept. Agric., Washington D.C.

US Dept. of the Interior (1966), Salt desert shrub symposium, Cedar City, Utah, August 1 - 4, 1966, USDA, Bureau of Land Management, 292 pp.

Uvarov, B.P. (1933), Ecology of the Moroccan Locust in Iraq and Syria and the prevention of its outbreaks. Bull. Entom. Res. 24 : 407 - 418.

Vallentine, John F. (1971), Range development and improvement. Brigham Young University Press, Provo, Utah, 516 pp.

Van Dersal, William R. (1972), Soil cover and stabilization. pp. 82 - 87 in: Wildland Shrubs - Their Biology and Utilization, An Int. Symp., Utah, USDA Forest Service General Tech. Rep. INT - I, 1972.

Veen, J.P.H. van der (1957), Forage research in Iraq. FAO report, fodder Crops Section Field Crops Division, Min. Agric., Iraq, 6 pp. (mimeo).

Veen, J.P.H. van der (1959), Forage crops and cultivated pastures in Iraq. Bull. 4 Dir. Gen. Research and Projects, Min. Agric., Iraq, 40 pp.

Veen, J.P.H. van der (1960a), International development of grazing and fodder resources. J. Brit. Grassl. Soc. 15 : 137 - 144.

Veen, J.P.H. van der (1960b), Report to the Government of Iraq on pasture and fodder development. FAO / ETAP Report no. 1292, FAO, Rome, 51 pp. (mimeo).

Veen, J.P.H. van der (1962), Some aspects of land use for livestock production in southwest Asia with particular reference to Iraq. Land Economics 38 : 356 - 369.

Veen, J.P.H. van der (1964), Some aspects of plant succession in the wadi Al-Aazib range station and surrounding areas. Proc. 5th week of sciences, pp. 1 - 10. Dec. 1964, High Council of Sciences. SAR Damascus.

Veen, J.P.H. van der (1965), Research and development activities on range management and fodder production in the Syrian Arab Republic 1962 - 65. 31 pp. (typescript).

Veen, J.P.H. van der (1967), Preliminary results of a grazing trial in the Syrian Steppe. Neth. J. Agric. Sci. 15 : 198 - 206.

Verstappen, H.Th. (1977), Remote sensing in geomorphology. Elsevier Sci. Publ. Comp. Amsterdam, 210 pp.

Vesey-Fitzgerald, Desmond Foster (1955), Vegetation of the Red Sea coast south of Jedda, Saudi Arabia. J. Ecol. 43 : 477 - 489.

Vesey-Fitzgerald, Desmond Foster (1957a), The vegetation of the Red Sea coast north of Jedda, Saudi Arabia. J. Ecol. 45 : 547 - 562.

Vesey-Fitzgerald, Desmond Foster (1957b), The vegetation of Central and Eastern Arabia. J. Ecol. 45 : 779 - 798.

Vink, A.P.A. (1975), Land use in advancing agriculture. Springer, Berlin, Heidelberg, New York, 392 pp.

Voûte, Caesar Czn. (1957), A prehistoric find near Razzaza (Karbala Liwa). Sumer XIII : 135 - 156.

Wallen, C.C. (1969), Arid Zone Meteorology. Ch. III in: Arid Lands, a geographical appraisal. Methuen & Co. Ltd. London, 461 pp.

Walter, H. & H. Lieth (1960), Klimadiagramme Weltatlas. Gustav Fischer Verlag. Jena.

Walton, K. (1969), The arid zones. Hutchinson University Library, London, 175 pp.

Warren, Andrew & Judith K. Maizels (1977), Ecological change and desertification, pp. 169 - 260 in: Desertification: Its causes and concequences. Pergamon Press, Oxford.

Weinert, Erich & Majeed R. Al-Hilli (1975), The vegetation of Jabal Sanam, South Iraq. Bull. Coll. Sci. Baghdad 16 : 3 - 27.

Werger, M.J.A. (1973), Phytosociology of the Upper Orange River Valley, South Africa. A syntaxonomical and synecological study. Thesis, University Nijmegen, The Netherlands, 222 pp.

Westhoff, Victor & Eddy van der Maarel (1973), The Braun-Blanquet approach. pp. 617 - 726 in: Ordination and classification of communities, Handbook of vegetation science Vol. 5, R.H. Whittaker (ed.), Dr. W. Junk b.v. Publ. The Hague, The Netherlands.

Whittaker, Robert H. (ed.) (1973), Ordination and classification of communities, Handbook of vegetation science Vol. 5, Dr. W. Junk b.v. Publ. The Hague, The Netherlands, 737 pp.

Whittaker, R.H. & G.M. Woodwell (1968), Dimension and production relations of trees and shrubs in the Brookhaven Forest, New York. J. Ecol. 56 : 1 - 25.

White, Gilbert F. (ed.) (1956), The future of arid lands. Publ. no. 43 A.A.A.S., Washington D.C., 445 pp.

Whyte, R.O. (1961), Evolution of land use in south-western Asia. pp. 57 - 118 in: A history of land use in arid regions, L. Dudley Stamp (ed.), Arid Zone Research XVII, UNESCO, Paris.

Wierenga, T. (1977), Phytomass and species richness in a *Hammadetum* and a *Zygophylletum* on the loessial plain of Sde Zin in a dry year. Report to Lab. Plant Ecology, State University, Groningen, 21 pp. (typescript).

Williams, W.F. (1971), Principles of clustering, Ann. Rev. Eco. and Syst. 2 : 303 - 326.

Williams, W.T. (ed.) (1976), Pattern analysis in agricultural science. CSIRO Melbourne, Elsevier Sci. Publ. Comp., Amsterdam, Oxford, New York, 331 pp.

Williams, Robert E., B.W. Allred, Reginald M. Denio & Harold A Paulsen Jr. (1968), Conservation, development and use of the world's rangelands. J. Range Manage. 21 : 355 - 360.

Winkworth, R.E. (1969), Plant ecology, Institute for Applied Research on Natural Resources, Abu Ghraib, Final Report Expert to UNESCO, Paris, 33 pp. (typescript).

Winkworth, R.E. (1971), The nature of rangelands, pp. 109 - 111 in: Proc. Seminar on integrated surveys, range ecology and management. Jodhpur, India, 9 - 27 Nov. 1970. Publ. by UNESCO Field Science Office for South Asia, New Delhi.

Wirth, Eugen (1962), Agrargeographie des Irak. Hamburger Geographische Studien. Inst. für Geographie und Wirtschafts geographie der Universität Hamburg. 193 pp.

Yacoub, Salah M. (1971), Sedentarization and settlement of the nomadic populations in selected Arab countries. Pamphlet no. A.E.S. 8, Fac. Agric. Sci., American Univ. Beirut, 44 pp. (mimeo).

Yahia, H.M. (1971), Soils and soil conditions in sediments of the Ramadi Province (Iraq), their genesis, salinity, improvement and use-potential. Publ. no. 18 Fysisch-Geografisch en Bodemkundig Laboratorium van de Universiteit van Amsterdam, 227 pp.

Zohary, M. (1940), Geobotanical analysis of the Syrian Desert. Palest. J. Bot. 2 : 46 - 96.

Zohary, M. (1941), The flora of the desert to the south and west of Basra, Mesopotamia, Proc. Linn. Soc. Part I, 153 : 99 - 108.

Zohary, M. (1950), The flora of Iraq and its phytogeographical subdivision. Dep. Agric. Iraq Bull. 31 : 1 - 201.

Zohary, M. (1962), The plant life of Palestine (Israël and Jordan), Chronica bot., N.S.Pl. Sci. Books 33 : VI, The Ronald Press Co. New York, 262 pp.
Zohary, M. (1973), Geobotanical foundations of the Middle East, Vol. I and II. Gustav Fischer Verlag, Stuttgart, Swets & Zeitlinger, Amsterdam, 739 pp.
Zonneveld, I.S. (1972), Land evaluation and land(scape) science. ITC Textbook of photo-interpretation VII-4, ITC, Enschede, The Netherlands, 106 pp.
Zonneveld, I.S. (1974), On abstract and concrete boundaries, arranging and classification. pp. 17 - 42 in: Tatsachen und Probleme der Grenzen in der Vegetation. Int. Symp. Rinteln April 1968. W.H. Sommer & R. Tüxen (eds.) J. Cramer Verlag, Lehre.
Zonneveld, I.S. (i.p.), Land evaluation terminology and classification theory. A proposal for sound language use in land attribute classification and land evaluation. ITC-Journal (in press).
Zonneveld, I.S., H.A.M.J. van Gils & D.C.P. Thalen (i.p.), Aspects of the ITC-approach to vegetation survey. Festschrift Prof. R. Tüxen (in press).

INDEXES

AUTHOR INDEX

GENERAL INDEX

Figures in *italics* refer to information in tables or figures.